Routledge Revivals

An Introduction to Modern Genetics

First published in 1939 (second impression in 1950), this book provides an account of the changes in, and main principles of, genetics at
that time. These are illustrated by references to the most authoritative
and then recent investigations. Special attention is paid to the way in
which genetics overlaps with other fields of inquiry, since it is often in
these border-line subjects that the most important advances are to be
expected. The book is particularly arranged to suit the convenience of
students whose previous knowledge of genetics is small, and contains
annotated bibliographies of suggestions for further reading.

An Introduction to Modern Genetics

C. H. Waddington

Routledge
Taylor & Francis Group

First published in 1939
Second impression published in 1950
by George Allen & Unwin

This edition first published in 2016 by Routledge
2 Park Square, Milton Park, Abingdon, Oxon, OX14 4RN

and by Routledge
711 Third Avenue, New York, NY 10017

Routledge is an imprint of the Taylor & Francis Group, an informa business

© 1950 George Allen & Unwin

Publisher's Note
The publisher has gone to great lengths to ensure the quality of this reprint but points out that some imperfections in the original copies may be apparent.

Disclaimer
The publisher has made every effort to trace copyright holders and welcomes correspondence from those they have been unable to contact.

A Library of Congress record exists under LC control number: 39014464

ISBN 13: 978-1-138-95696-4 (hbk)
ISBN 13: 978-1-315-66541-2 (ebk)
ISBN 13: 978-1-138-95697-1 (pbk)

AN INTRODUCTION TO
MODERN GENETICS

BY

C. H. Waddington, Sc.D.

Fellow of Christ's College, Cambridge

WITH DIAGRAMS
AND PLATES

LONDON
George Allen & Unwin Ltd
MUSEUM STREET

FIRST PUBLISHED IN 1939
SECOND IMPRESSION 1950

PRINTED IN GREAT BRITAIN
All rights reserved
CLEMENTS, NEWLING & CO. LTD.
ALPERTON

PREFACE

FEW biologists will doubt that heredity, the subject matter of genetics, is one of the important problems of biology. Indeed, when Mendel's results were rediscovered in 1900 and the new science of genetics got under way, it appeared to some authors that a new era was dawning in biological thought. But in the years that followed it often seemed that the high hopes which had been entertained were becoming dissipated in a morass of numerical elaborations on the hackneyed theme of Mendelian segregation. Genetics was in danger of being considered, by other biologists, as a world of its own, devoted to following the comings and goings of genes whose relevance to other biological phenomena, though incontrovertible in general theory, could rarely be stated in detail and in particular.

I believe such an impression to be quite false. Experimental breeding, and the determination of the ratios between classes of offspring, is no more than one of the main techniques of genetics; and an over-emphasis on technique nearly always obscures the real interest of a science, which lies in the concepts and theories to which the experimental methods open the door. There was inevitably a period when geneticists had to concentrate their efforts on putting the basic ideas of their science on a firm foundation. In recent years genetics has been able to apply itself to wider problems, and it has produced results which cannot be overlooked by any student of evolution, development or cytology.

In this book I have tried to give an account of these recent developments. The literature of genetics has become extremely large, probably larger than that of any other branch of experimental biology except medical physiology, and it is impossible to do more than provide some help to the student who wishes to plunge into this sea of material. The bibliography is therefore rather extensive although, of course, not exhaustive. The aim has been to cite summaries, reviews and articles in which general questions are discussed and further references given, although I am conscious that this may sometimes appear to do injustice to the older authors on whose work the present-day edifice is erected.

I should add a word of explanation, and perhaps of apology, for having written a text-book of genetics although the field of my own research is usually classified as the separate science of experimental embryology. Perhaps the object of the book, as I have explained it

A*

above, may serve as some excuse; the need has arisen primarily because the different kinds of biological cobblers have in the past stuck too closely to their lasts. I want to urge that the connection between genetics and the other branches of biology, such as cytology, embryology, the study of evolution and of the biochemical nature of cell constituents, is much closer than is often admitted, and that the boundaries between these subjects deserve less attention than is usually paid to them.

It is probably inevitable that many errors, both of omission and commission, will be found in a book which, like this one, attempts to survey the present state of affairs in a large and rapidly expanding branch of science. I have, however, been extremely fortunate in receiving most generous help from many friends and colleagues, who have, I know, saved me from many errors and very greatly contributed to any merits this book may possess. I should like in particular to express my gratitude to the following, who have read and criticized the whole or large parts of the work: C. B. Bridges, C. D. Darlington, Th. Dobzhansky, C. Stern, M. Whittinghill.

C. H. W.

CAMBRIDGE
1938

CONTENTS

PART ONE

FORMAL GENETICS

PART TWO

GENETICS AND DEVELOPMENT

PART THREE

GENETICS AND EVOLUTION

PART FIVE

THE NATURE OF THE GENE

XVI. The Nature of the Gene 361

Appendix

LIST OF ILLUSTRATIONS

INTRODUCTION

In the prehistoric beginnings of intelligent life, man must have been confronted with three related problems: to win shelter, warmth, and tools from his physical environment; to control the health and sickness of his own body; and to produce food from the animals and plants which he soon domesticated for his use. His mastery of these problems progressed very unequally. Inorganic objects are in general simpler than biological organisms, and when the young sciences became something more than a mere codification of practical recipes, their theoretical structure was based on ideas derived from inanimate nature; the earth, air, fire, and water of the Greeks, and the Cartesian mechanics of the seventeenth century. The physico-chemico-mechanical system of thought can be applied naturally enough to the physiological problems of health and sickness; if we consider man's body as a going concern, its workings can be investigated like those of a machine. The biological sciences, at least those which attempted to provide causal explanations of phenomena, at first concentrated their energies on this aspect of life, and tended to neglect the more difficult problems of the reproduction and development of organisms, for which inorganic parallels did not lie so near to hand. During the nineteenth century the dominant and most progressive biological outlook was concerned with organisms as something essentially static, as machines for which the passage of time meant no more than a few extra revolutions of the wheels.

By one of those paradoxes which are so frequent in history, it was a purely mechanistic theory which directed attention to the non-mechanistic aspects of living organisms and changed the centre of gravity of biological thought. Darwin's theory of natural selection was so matter of fact, and invoked only processes which were so acceptable to a mechanistic attitude, that he was able to convince the world at large that organisms do actually evolve. As soon as this was accepted, it gradually dawned on biologists that living things cannot be considered apart from time; their essential nature changes as time passes. Interest flowed back towards the study of developmental changes and the problems of reproduction. Science once more took up the investigation of breeding and hybridization. The same current of thought spread into physics, and the indestructible timeless atoms were analysed into waves, in whose existence time is an essential element; and a temporal dimension became incorporated with the old rigid three dimensional framework of space.

Genetics as a separate science is a part of this study of the long-range temporal changes of organisms. The story of its birth is one of the most peculiar in the history of science. There had persisted, throughout the nineteenth century, a certain interest in the problems of breeding, which was connected particularly with the industry of seed production. The fundamental discovery of the existence of discrete hereditary factors was made by Mendel, a monk living in what is now Czechoslovakia. It was published in 1865–66, in a somewhat obscure journal,[1] but copies were sent to several of the important biologists of the day. An extraordinary fate overtook it; it was totally neglected. The reasons for this are still obscure. It is hard to believe that its importance could have been overlooked if the problem of heredity had been in the centre of biological interest. Perhaps those biologists who did interest themselves in the subject reacted too strongly against the prevalent mechanistic views and found Mendel's atomistic "factors" not sufficiently "biological" for their taste.

Whatever the reason, Mendel's papers were almost completely forgotten until 1900, when they were rediscovered almost simultaneously by de Vries, Correns, and Tschermak. Their significance was immediately appreciated, and confirmatory results were quickly obtained. The new science which grew up round them was, one may say, formally inaugurated by Bateson at a conference on plant breeding in 1906, when he coined the word "genetics." By this word he understood the science whose "labours are devoted to the elucidation of the phenomena of heredity and variation; in other words to the physiology of Descent."

It is important to notice that Bateson included in his definition not only heredity in the narrow sense of the transmission of characters from parent to offspring, but the whole problem of descent or reproduction. In the early years of genetics, interest was mainly concentrated on investigating the laws of inheritance. But genetics is really a much wider science than this; it cannot be separated from the study of embryology and evolution. The three aspects of the temporal changes of organisms, heredity, development and evolution, are as intimately related as, for instance, physiology and biochemistry among the sciences which deal with organisms as going concerns. Moreover, we must

---

[1] Translated in Bateson 1930.

hope eventually to be able to make a synthesis between the sciences dealing with temporal change and those which treat the day-to-day functioning of organisms. Only when these two fundamental aspects of living things are considered together can we claim to view biology as a whole.

PART ONE

Formal Genetics

The first task of genetics was to discover the rules of inheritance. The young science, beginning its life at the time Mendel's papers were re-discovered [1900], accomplished this task extremely quickly; within twenty years, the chromosome theory had been placed on a firm basis and the main outlines of the science of heredity were clear. This development is dealt with in Part I. The first chapter is devoted to the basic facts on which the chromosome theory rests, the next two to the consequences of chromosome heredity at different stages of the chromosome cycle and in abnormal chromosome cycles. The fourth chapter describes peculiarities in the behaviour of whole chromosomes, while the fifth deals with the parts of chromosomes. The study of the mechanism of crossing-over, considered in Chapter 6, is probably the most rapidly advancing part of formal genetics at the present time; it is one of the most important lines of attack in the investigation of the nature of the cell and its parts, and it links up with the study of the physico-chemical nature of the gene to which we return in Part V.

CHAPTER I

The Fundamentals of Mendelism

1. *Biological inheritance; genotype and phenotype*

It has of course been clear from very early times that there is some biological inheritance. The character of an organism depends, to some extent at least, on that of its parents. The problem of how the parental influence is exerted was for long debated in purely philosophical terms, but with the invention of adequate microscopes, and the discovery of spermatozoa and of the universal occurrence of eggs, hypotheses of a verifiable nature could be put forward. The first theory to gain general acceptance had a deceptive air of simplicity; it was supposed that the sperm (or, for the feminists, the egg) contained the complete organism in miniature, which merely had to grow to become the new adult. Elaborate theories were evolved as to how these homunculi in their turn mated and reproduced; but still more efficient microscopes soon made it clear that the eggs and sperm do not in fact contain miniature animals, and the whole elegant edifice of theory had to be abandoned.[1]

Since we can easily find examples of an animal inheriting, say, the colour of its eyes from its father; and since there are no eyes in the sperm, which is the only connection between the two individuals, it is clear that the eye colour must be represented in the sperm by something else which is responsible for passing on the father's characteristic to his son. We must therefore draw a distinction between the characters of an adult individual and the representatives of those characters which are present in the germ-cells and pass on into the next generation. The former are known collectively as the phenotype, the latter as the genotype; but these two terms will need some further discussion, since they were invented after the basic theory of genetics had been developed, and are to some extent coloured by its conceptions.

The fundamental step in the understanding of heredity depended on a bold piece of abstract thinking.[2] To the unanalytical eye of common sense, biological inheritance is singularly capricious; in some cases an animal may be more like its father, in other cases more like its mother, while in some respects it may not be like either of them. In general, one

---

[1] Cf. Punnett 1931. For a general history of embryology, cf. Needham 1934; for a chronology of genetics 1800–1934, Cook 1937.

[2] For a logistic analysis of genetical theory, see Woodger 1938.

cannot easily pretend that organisms are simply intermediate between their two parents. It seems that one must conclude that an organism does not inherit the whole of its parents' genotype, but only part of it; in any given case some of the characters which might have been inherited from the parent actually fail to appear. But theory went no farther than this till Mendel was bold enough to leave out of consideration the greater part of the characteristics of the organisms with which he was working and to concentrate entirely on one or two sharply marked features. In this way he proceeded to consider the phenotype (the appearance of the adult animal) as a set of elementary characters; and although it is obvious that an organism is not a mere assemblage of isolated anatomical structures, this analysis revealed the fundamental fact that the genotype consists of hereditary units which are very largely independent. The relation between the genotype and the phenotype, that is to say, between the hereditary constitution and the appearance of an organism, can therefore only be understood after we have discussed the facts which have been revealed by Mendelian analysis.

2. Mendel's First Law: Factors and their Segregation

Mendel's fundamental discoveries were made on the garden pea, *Pisum sativum*. The typical, oft-quoted experiment was as follows: tall-growing pea plants were crossed with short plants (P1, the first parental generation); their offspring (F1, first filial generation) were all tall, and when self-fertilized, gave a second generation of hybrids (F2) consisting of tall and short plants in the ratio of three to one. The short plants from the F2 bred true for shortness when selfed, and a third of the talls bred true for tallness, while the other two-thirds of talls again gave talls and shorts in the three to one ratio which had been found in the F2. Mendel's hypothesis was this: tallness and shortness are dependent on a pair of alternative factors, which we may call T tall and t short. Each fertilized zygote, and each cell of the organism into which it develops, contains two of these factors, and may thus be TT, or Tt or tt; but the gametes each contain only one factor selected out of the two which are contained in the germ-mother cell out of which the gamete is formed. The first cross was between TT and tt, and gave a F1 of Tt; the fact that this F1 shows as tall plants must mean that during development the T factor "dominates" over the t, which is said to be "recessive." When the F1 is selfed, each Tt plant forms equal numbers of gametes with T and with t, and if these unite at random they will give TT, Tt and tt plants in the ratio 1 : 2 : 1. Thus there will be three talls (of which one will be pure breeding TT and two Tt like the F1) to one pure breeding tt short. (Fig. 1.)

Factors of the kind postulated above are called Mendelian factors or genes. A collection of genes alternative to one another, so that normally a gamete contains only one of the set, is spoken of as a series of allelomorphic factors (the position they occupy is their "locus"); there may be more than two members of such a series, for instance, there might have been another alternative "dwarf" for Mendel's peas. Zygotes which contain two similar allelomorphs are said to be homozygous for the gene in question (e.g. the pure breeding TT or tt), while if the two

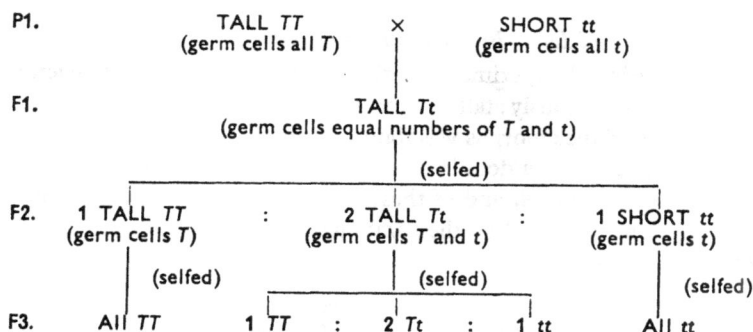

P1.   TALL *TT*   ×   SHORT *tt*
(germ cells all *T*)   (germ cells all *t*)

F1.   TALL *Tt*
(germ cells equal numbers of *T* and *t*)

(selfed)

F2.   1 TALL *TT* : 2 TALL *Tt* : 1 SHORT *tt*
(germ cells *T*)   (germ cells *T* and *t*)   (germ cells *t*)

(selfed)   (selfed)   (selfed)

F3.   All *TT*   1 *TT* : 2 *Tt* : 1 *tt*   All *tt*

Fig. 1. **Mendel's experiment of crossing tall and short peas.**

allelomorphs are dissimilar, as in the F1 plants, the organism is heterozygous. If, in a heterozygote, only one of the two allelomorphs has an effect on the character of the organism, that allelomorph is dominant over the other, which is recessive; dominance and recessiveness may be partial, when the heterozygote will show some effect of both genes, neither of which completely suppresses the other.

The fundamental points of the hypothesis developed above, which is known as Mendel's first law, are (1) that there are factors which affect development and that these factors or genes retain their individuality from generation to generation and do not become contaminated when they are mixed in a hybrid, and (2) that they become sorted out from one another when the gametes are formed.

A Note on Symbols.—There are several different systems in use for symbolizing genes. The simplest refers to a locus by a certain letter, and indicates the dominant by the capital, the recessive by the lower case letter (e.g. Aa for the heterozygote). If there are more than two allelomorphs, they can be indicated by small superscript letters (e.g. c^a, c^r, c^k, etc., for the albino series in rodents).

The convention of indicating dominance by capital letters breaks down here since there is more than one dominance relation to be considered, and anyway the dominance within such series is usually incomplete. Sometimes the most common allelomorph in the series and those dominant to it are given capital letters, those recessive to it small letters. Alternatively, when the species has a well-defined normal or wild form, the normal or "wild-type" allelomorphs are indicated by a cross, to which, if necessary, the symbol of the locus is added (e.g. $+w$ is the wild allelomorph of the white eye locus in Drosophila). The mutant allelomorphs of the locus are written with small letters unless they are dominant over the wild (e.g. w^e eosin, w^{co} coral are allelomorphs of white, Bd beaded is dominant over $+^{Bd}$, the wild allelomorph). The cross of the wild type gene is now often written as an exponent, e.g. w^+.

Complex heterozygotes may be written $AaBBCc$, or the genes in one chromosome may be separated from those in the homologue by a dot (e.g. $ABC.aBc$) or a line (e.g. $+++/w^e + f$, where the two recessives w^e and f occur in the same chromosome).

3. *Mendel's Second Law: Independent Assortment*

Mendel's original experiments dealt with the inheritance of several characters simultaneously: tall or short plants, round or wrinkled seeds, etc. Each pair of allelomorphs was found to behave quite independently of every other pair. In a double heterozygote *AaBb* half the germ cells will of course contain *A*, and of these, it was pure chance which contained *B* and which *b*, so that the combinations *AB* and *Ab* were found in equal numbers.

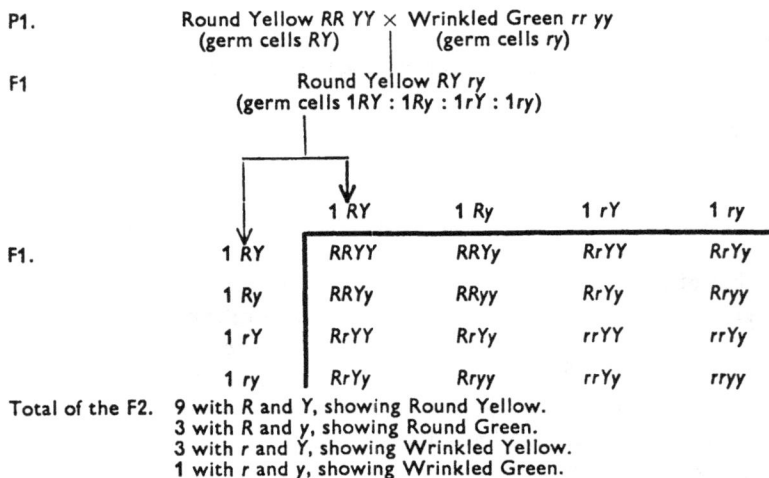

P1.   Round Yellow RR YY × Wrinkled Green rr yy
       (germ cells RY) │ (germ cells ry)

F1   Round Yellow RY ry
       (germ cells 1RY : 1Ry : 1rY : 1ry)

	1 RY	1 Ry	1 rY	1 ry
1 RY	RRYY	RRYy	RrYY	RrYy
1 Ry	RRYy	RRyy	RrYy	Rryy
1 rY	RrYY	RrYy	rrYY	rrYy
1 ry	RrYy	Rryy	rrYy	rryy

Total of the F2.   9 with R and Y, showing Round Yellow.
3 with R and y, showing Round Green.
3 with r and Y, showing Wrinkled Yellow.
1 with r and y, showing Wrinkled Green.

This is known as the "chequerboard" method of finding the progeny of a cross. An alternative is to multiply the two series of gametes together algebraically, e.g. to get the F2 above we have $(RY + Ry + rY + ry)(RY + Ry + rY + ry)$ $= RRYY + 2RRYy + 2RrYY + 4RrYy + RRyy + 2Rryy + rrYY + 2rrYy + rryy.$

Fig. 2. **Independent Assortment of Factors** in a cross between round yellow and green wrinkled peas (Mendel).

This "independent assortment" was later found not to be a general rule, although it holds for many factors. Bateson and Punnett[1] showed that with some loci in the sweet pea a heterozygote *AB.ab* (*A* affects flower colour, *B* shape of pollen grains), made by crossing *AABB* with *aabb*, forms more gametes of the sorts *AB* and *ab* from which it was itself derived than of the other two possible kinds *Ab* and *aB*. Thus there were fewer recombinations of factors than would result from pure chance. If the two dominants *A* and *B* were originally together, they tended to remain together ("coupling"), while if they were originally apart they tended to remain apart ("repulsion"). The general name for the phenomenon is linkage; and the strength of linkage between two factors is measured by the percentage of recombinations among the total number of gametes. Thus a strong linkage gives a low percentage of recombinations, while no linkage at all gives the chance expectation of recombinations, which is 50 per cent.

The progeny of any cross can be worked out by determining the ratio of the various sorts of gametes formed by the parents and multiplying together the two gametic series. This gives the ratio of the various classes of fertilized eggs on the assumption that fertilization is at random; if there is any selective fertilization (p. 55), each class must be multiplied by a coefficient expressing the relative probability of the type of fertilization involved. The ratio of the classes of adult organisms depends on the way in which the genes concerned interact in development; on relations of dominance and recessiveness, on the appearance of double heterozygotes, etc., and on whether any of the genes lower the viability of the organisms which thus tend to die off before attaining adult age.

4. *The Chromosome Theory*

Mendel's postulated factors are present in pairs in the zygotes, but are single in the gametes. The parallel between this behaviour, and the behaviour of the chromosomes to which attention had been drawn by Weissmann, was noticed soon after Mendel's work was rediscovered.[2] It was suggested that the factors might be chromosomes or parts of chromosomes. This suggestion has turned out to be correct, and we must examine the behaviour of the chromosomes to see how it explains the behaviour of Mendel's postulated factors.

---

[1] Cf. Bateson 1930.

[2] Montgomery 1901, Sutton 1902. For a full history of these discoveries, see Wilson 1928.

5. Behaviour of Chromosomes[1]

The chromosomes are normally visible only during cell division, when they appear in stained preparations as darkly staining bodies in the nucleus. In life they can be distinguished with some difficulty with ordinary light, but can be photographed with good contrast by ultraviolet light, which they absorb more strongly than does the cytoplasm.[2] There are two types of cell division to be considered.

5a. Mitosis: the normal type of cell division

The nucleus of a cell in the resting stage (or interphase) is more or less optically homogeneous in life, and shows only an irregular network when fixed and stained (see p. 39). The first structures which appear at the beginning of a division (the prophase stage) are thin threads which can be seen to be double. The members of each doublet are similar and lie closely side by side, but in certain organisms there may be several obviously different types of doublet. The double threads contract in length and become thicker, and gradually develop a definite shape. By the time their shapes can be clearly recognized, it can be seen that there are an even number of doublets and always two of each type (with a few exceptions, see sex chromosomes, p. 75). Each double body is one chromosome split longitudinally into two half-chromosomes or chromatids. Thus there are two of each sort of chromosome (four of each sort of chromatid), and the total number of chromosomes, which is normally even, is spoken of as the diploid number.

The chromatids are held together at one point, which can be seen as a non-staining gap in the chromosome (the attachment constriction or centromere). It was at one time thought that a fibre of the spindle was joined on to the chromosome at this point, but it is now believed that the spindle fibres are artefacts of fixation, corresponding indeed to some latent structure in the spindle cytoplasm but not actually existing as definite fibres until coagulated by the fixative (p. 363). By the time the contraction of the chromosomes is complete, the nuclear membrane has disappeared and the spindle has been formed; there may be acrosomes or centrosomes at the poles of the spindle in animals or lower plants but they are absent in higher plants, where the whole spindle mechanism is less highly developed. The chromosomes become arranged with

---

[1] For general accounts of the cytology of the nucleus, see Belar 1928, Darlington 1937, Geitler 1934, Sharp 1934, White 1937, Wilson 1928.
[2] Lucas and Stark 1931, Caspersson 1936.

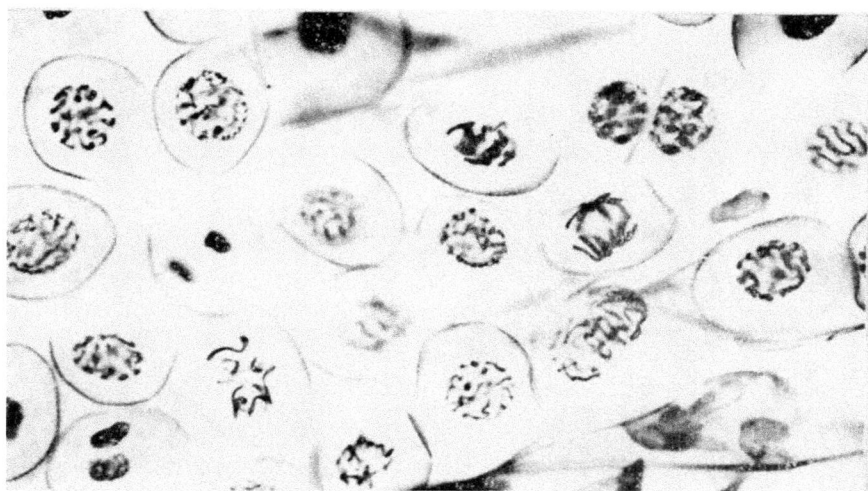

Plate 1. Above is mitosis in the pollen grains of *Frittillaria sp.* Notice the polar view of metaphase to the lower left, and the anaphase to the right, with an early telophase just above it. Most of the other grains are in prophase stages. Below is meiosis in pollen mother cells of *Frittillaria sp.* The nuclei are in diplotene and diakinesis stages and show many unterminalized chiasmata. (Courtesy of C. D. Darlington.)

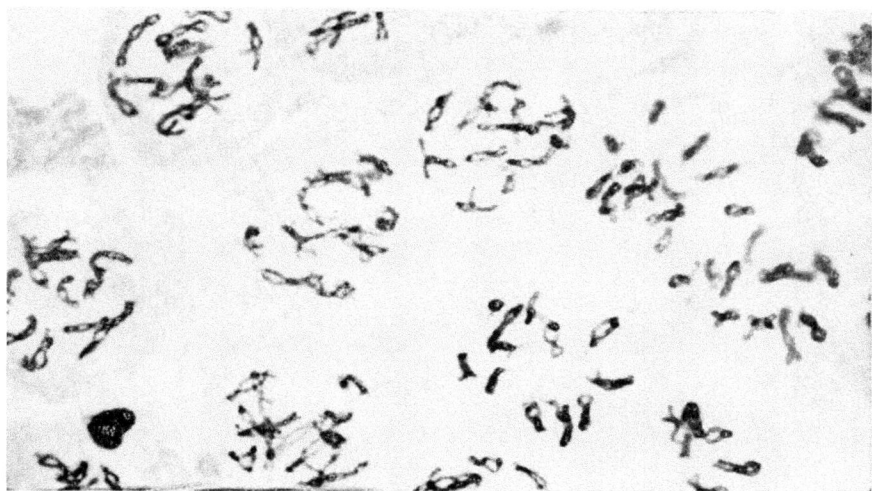

their centromeres lying in a plane across the equator of the spindle (the metaphase stage); the centromeres clearly repel one another and so do the chromosomes as a whole, so that the centromeres lie as far apart as possible consistent with being within the spindle, while the repulsion between the chromosomes causes the larger ones to lie on the edges of the equatorial plate, with the smaller ones more centrally placed.

The centromeres then divide into two, and, the daughter centromeres repelling one another, the chromatids are pulled apart and begin to move towards the poles (the anaphase stage). The repulsion of the centromeres seems insufficient to move the chromatids far apart and the later stages of their movement is caused by the narrowing and elongation of the material situated between the two groups of chromosomes, which forms the so-called "stem-body."[1]

When the chromatids reach the two poles (the telophase stage) the cell divides into two[2] and two new resting nuclei are reconstituted. The individual chromatids of this division are now to be reckoned as chromosomes, and in the ensuing interphase each one will manufacture a partner, or, if one likes to put it so, each one will become split longitudinally, so as to appear in the next prophase as a double body.

5b. Meiosis

Meiosis consists of two successive cell divisions and takes place only in the formation of the gametes in animals or spores in plants. The daughter cells which are formed from it contain only one of each kind of chromosome. They are said to contain the haploid number, which is half the diploid number. This reduction in number, in consequence of which the first division of meiosis is often called the reduction division, is brought about by the peculiarities of the prophase, which is more complicated than that of mitosis and is divided into several sub-stages.

(a) Leptotene.—The chromosomes appear as single unpaired threads, in the diploid number.

(b) Zygotene.—Similar chromosomes come together in pairs side by side.

(c) Pachytene.—Each chromosome splits longitudinally into two half-chromosomes or chromatids.

(d) Diplotene.—The four chromatids which were closely associated at pachytene fall apart in pairs, the two pairs being held together by

---

[1] Belar 1927.   [2] For a discussion of cell-division, see Gray 1931.

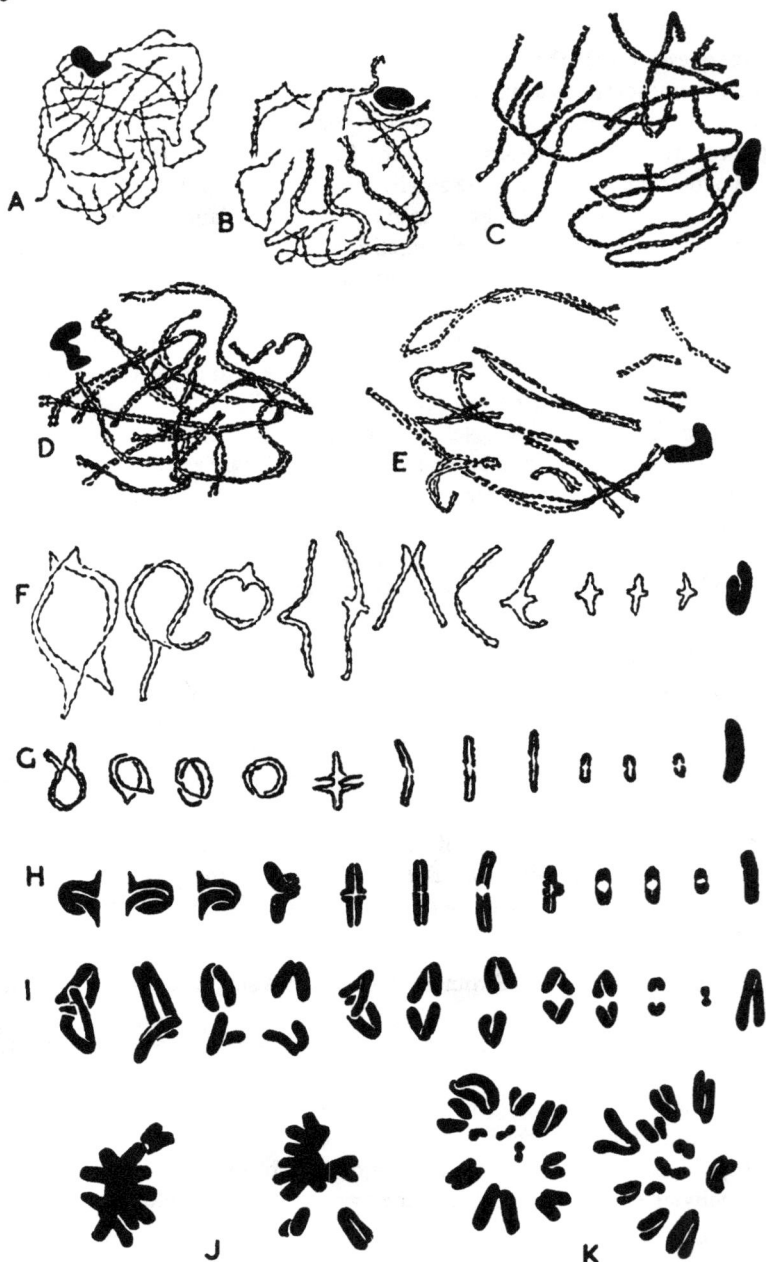

interchanges of partner among the threads. Each group of two chromosomes (four chromatids) is known as a bivalent and the interchanges of partner are chiasmata.

(e) Diakinesis.—The final stage before the chromosomes arrange themselves on the metaphase plate. The chromosomes have been contracting during the whole prophase and are now quite short and thick; as the prophase is more complicated and lasts longer than in mitosis, the contraction has been going on longer, and the chromosomes are shorter, perhaps only a third as long as in mitosis.

At metaphase each bivalent has two centromeres, one from each chromosome, and these become arranged so as to lie symmetrically on either side of the equatorial plane of the spindle. Their mutual repulsion gradually forces them apart (anaphase), unravelling the chiasmata also, each centromere carrying with it towards the poles two chromatids which make up one whole chromosome. After a telophase, resting stage nuclei are formed.

The first division is normally followed by a second division in which the chromosomes appear again still showing the split which they developed in pachytene, and still with undivided centromeres. A normal mitosis follows, the anaphase separation beginning when the centromeres eventually divide. This is sometimes spoken of as the equational division as opposed to the reduction division, but we shall see (p. 103) that this is misleading; another pair of names is heterotype division = first division, homotype division = second division.

The daughter cells of the second division are transformed directly into gametes in most animals. The pairing of two whole chromosomes together in zygotene, followed by their separation to different poles in anaphase, has brought it about that the daughter cells each contain only one of each kind of chromosome and has accomplished the reduction in number which we spoke of above. The relation between the mechanisms of meiosis and mitosis is discussed on page 113.

Fig. 3. **Meiosis** in the grasshopper *Melanoplus femur-rubrum.*—A–E, Optical sections of prophase stages; A Leptotene, with unsplit chromosomes. B, Zygotene, pairing in progress, C Pachytene, pairing complete, D and E stages of diplotene, the chromosomes splitting and falling apart into loops. F the chromosome complement drawn separately at early diplotene; note the exchanges of partner or chiasmata by which the threads are held together. The X chromosome on the right shows precocious condensation. G the chromosomes further contracted at diakinesis. H metaphase chromosomes seen in side view. I Early anaphase in side view. J Telophase of first meiotic division. K Metaphases (polar view) of second meiotic division. All magnified about 2,000 times.

(From Hearne and Huskins.)

6. *Chromosomes and Factors*

In order to show that the chromosomes behave like Mendelian factors we must demonstrate (1) that they have an individuality, that is to say that each one produces a specific effect on development and thus on the kind of adult which is produced; (2) that they retain this individuality throughout many cell divisions; and (3) that in a hybrid they segregate like Mendelian factors.

6a. *Evidence that the chromosomes have individual characters*

The classical evidence that the chromosomes have qualitatively different effects on development is that of Boveri (1907). By fertilizing

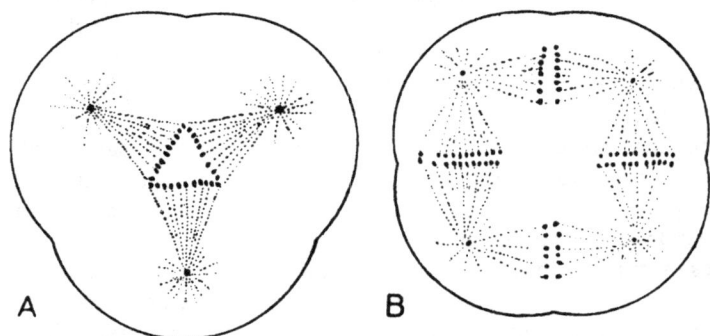

Fig. 4. **Diagram of Triaster and Tetraster Eggs in Echinoderms.**—In *A* a diploid number of chromosomes is being divided to give three daughter nuclei, in *B* a triploid number giving four nuclei.

(After Boveri.)

sea urchin eggs with an excess of sperm, he obtained double fertilization (two sperm entering one egg) which resulted in the formation of tetra-polar asters; and by shaking normal monospermic eggs he got triasters. The first four cells formed in the tetraster eggs, or the first three in the triaster eggs, could be separated by using Ca-free sea water. The percentage of the isolated cells which developed normally to the pluteus stage agreed approximately with the probability that the cell contained at least one of each kind of chromosome, remembering that in the first case 3*n* chromosomes had to divide and be distributed among four cells and in the second case 2*n* chromosomes among three cells. Moreover, Boveri could show that mere variation in number of chromosomes was not in itself enough to stop development, which was possible with either the haploid number (one of each kind), or with the normal or diploid number, or the tetraploid number (double the diploid number). Boveri concluded that development could only go on if there was one of each kind of chromosome, and that there were the haploid number of different kinds each with a specific role in development.

Of the recent evidence on this point, the clearest perhaps is that relating to organisms with one chromosome more than is normal for the species (trisomics, p. 83). Thus in Datura stramonium the haploid number is 12 and there are 12 kinds of primary trisomics known, each with one extra chromosome which has a specific effect on the characters of the plant. We shall see later that there are qualitative differences even between different parts of the same chromosome.

6b. *Evidence that the chromosomes retain their identity*

The difficulty in proving that each chromosome retains its identity through many cell divisions arises from the fact that its appearances are only intermittent. As a rule no satisfactory demonstration of the chromosomes in the interphase can be made. In life the interphase nucleus appears optically homogeneous or granular, and on fixation it shows either a granular structure, the graininess depending on the kind of fixation employed, or a structure of anastomosing threads which are swollen into nodes at the points of junction. Microdissection studies suggest that the nucleus is homogeneous and liquid, and measurements of viscosity give a surprisingly low value about equal to that of glycerine.[1]

There are some cases in which a resting phase is very short and the chromosomes remain more or less distinct as separate vesicles throughout it (premeiotic resting stage of spermatocytes, segmentation divisions of eggs). In other cases parts of the chromosomes, particularly the sex chromosomes (p. 78), may remain condensed and deeply staining and thus distinguishable from the rest of the nucleus. Similarly, parts of the chromosomes may remain condensed in the small nuclei of some plants, when the diploid number of deeply staining bodies may exist throughout the whole interphase; they are known as prochromosomes or chromocentres. The connection of true nucleoli with these bodies and with the rest of the nuclear apparatus is, in most organisms, unknown, but in some plants (e.g. *Vicia, Zea*)[2] and insects it has been shown that the nucleoli arise after division in close connection with the unstaining parts of certain chromosomes (the secondary constrictions; the chromosomes concerned have been called *SAT* chromosomes) and that when

[1] Cf. Gray 1931, Heilbrunn 1928.   [2] Heitz 1931, McClintock 1934.

the chromosomes appear again they are still in contact with the nucleoli.

All the parts of the chromosomes which remain deeply stained in the interphase are considered to be made of a material known as heterochromatin, as opposed to the rest of the chromosomes which are made of euchromatin. It is doubtful, however, how far all heterochromatin is really the same. The genetic properties of heterochromatin are only known in a few cases. In Drosophila the "inert" regions of the X and Y chromosomes, and the smaller inert regions near the centro-

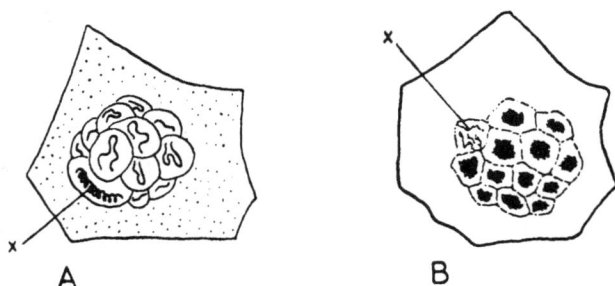

Fig. 5. **Semi-diagrammatic Drawings of Vesicular Nuclei.**—A in spermatogonium of grasshopper Aularches, note the more condensed condition of the X; B in spermatogonium of Phrynotettix, with a less condensed X.
(A after White; B after Wenrich.)

meres of the IInd and IIIrd chromosomes, seem to contain no hereditary factors; they are heterochromatic to the extent that they show differential staining in mitosis[1] and the permanent prophase of the salivary glands nuclei (p. 98), but they are not all concerned in the production of prochromosomes or nucleoli.[2] The centromeres themselves are not considered to be made of heterochromatin: they are somewhat similar in staining properties to the centrosomes.

In most resting nuclei we cannot see even the feeble traces of persisting chromosomes which have been described above. The evidence that they do persist is of a different kind and is overwhelming. Individual chromosomes can often be recognized by their shape, size, and peculiarities such as the distribution of constrictions, trabants, etc. Cells which are derived from a single ancestral cell by mitotic division are usually characterized not only by possessing the same number of chromosomes, which might be merely a function of the amount of

[1] *Rev.* Heitz 1935.   [2] Cf. Kaufman 1938.

chromatin, but by possessing the same number of chromosomes with the same sizes and shapes. This is most strikingly shown in organisms in which the chromosome complement has become altered from that

Fig. 6. **Nucleoli and Chromocentres** in *Vicia Faba.*—*A* One daughter nucleus in a late anaphase stage; the nucleolus-bearing chromosomes are drawn black, and the nucleolus-organizer is visible as a non-staining gap in the chromosome. *B* The resting stage nucleus with two nucleoli and (smaller) chromocentres.
(After Heitz.)

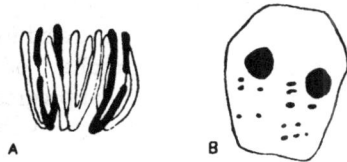

normal to the species, when the alterations are found to be common to all the cells of the organism and can indeed be inherited.

As well as this general evidence of repetition of shape, there is

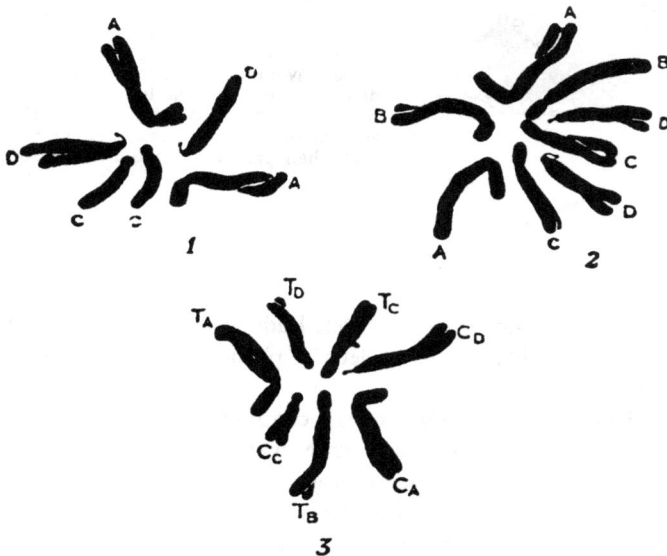

Fig. 7. **Constancy of Chromosome Shape.**—(1) Mitotic metaphase of *Crepis capillaris*; (2) of *C. tectorum*; and (3) of the hybrid between them, the individual chromosomes being recognisable, *Ta, Tb, Tc, Td* from *tectorum* and *Ca, Cc, Cd* from *capillaris*; the only change which has occurred is that *Td* has lost its trabant (terminal knob).
(From Darlington, after Hollingshead.)

evidence that each chromosome arises in a prophase in the same position and with the same peculiarities as it had when it disappeared in the preceding telophase. Thus shortly before the chromosomes disappear

at the end of a division they are orientated with their attachment constrictions towards one side of the nucleus (the so-called pole-field)
which may often be recognized either by its position within an asymmetrical nucleus or by the position of the nucleus in the cell. When the
chromosomes appear again in prophase they are still polarized, with
their centromeres concentrated in this region. A similar polarization is
often found in the prophase of meiosis, when the zygotene pairing
starts in the pole-field and progresses over the chromosomes from there
(this is the so-called bouquet stage, an exaggeration of which, particu-

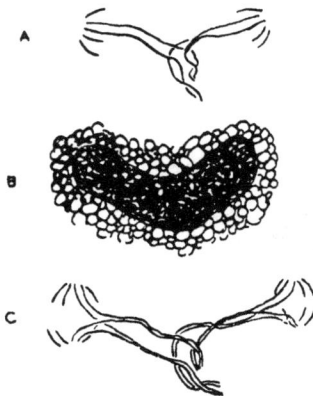

Fig. 8. **Persistence of Chromosome Arrangement through Interphase.**—A Anaphase in the Coccidian Aggregata; only some of the chromosomes are indicated. B interphase supervening shortly after the stage drawn in A. C the anaphase of the next division, showing the long chromosomes, which are now separating towards the top and bottom of the page, still not entirely free from their partners in the division before.
(After Belar.)

larly under conditions of bad fixation, leads to a collapse of the chromosomes to one side of the nucleus, a phenomenon which is called
synizesis).

A particularly clear example of the continuity of the chromosomes
was given by Boveri (1909), who investigated the nuclei of the first
blastomeres of the eggs of *Ascaris megalocephala var. univalens*, which
contains only one pair of chromosomes. At a telophase the four ends of
the chromosomes project from the surface of the two daughter nuclei,
forming cylindrical protruberances; the two nuclei are mirror images
of one another. The characteristic patterns of protuberances and
associated chromosomes which are found in telophases are also found
in prophase, suggesting that the chromosomes have persisted in some
way within the protruberances and have appeared again in the same
positions as they occupied when they disappeared from view. In
Aggregata eberthi (Protozoa) Belar[1] found that the interphases in the

[1] Belar 1926.

gametophytes might occur before the separation of the chromosomes was complete, and that when the chromosomes appeared again for the second division they still showed the interlocking left over from the first division.

A solution of the paradox of the continutity of existence of the

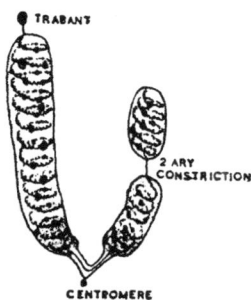

Fig. 9. **Diagram of the Structure of a Chromosome** at anaphase of mitosis.—The centromere is located at the bottom, in the angle of the V, in a non-staining region (the primary constriction). There are two secondary constrictions, one in the right arm and one near the end of the left arm, which cuts off a small lump of chromatin (known as a trabant). In the main body of the chromosome, the chromonema (central thread) is coiled in a spiral; in meiosis this small-scale spiral (minor spiral) may itself be coiled in a larger-scale spiral (major spiral). Some authors maintain that the chromonema is split longitudinally into two threads (see p. 116); it appears to be embedded in a "matrix." The chromomeres, which are best seen in prophases, may persist as swellings on the chromonema. The main part of the chromosome becomes less darkly staining in telophase; there may also be a part, the "inert" or "heterochromatic" part, which remains darkly stained; it is here drawn near the centromere. (After Heitz.)

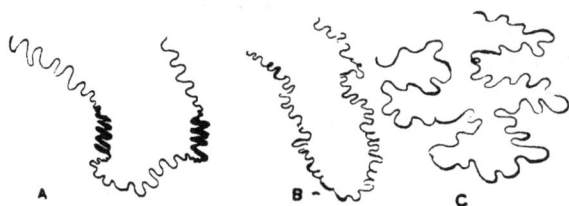

Fig. 10. **Persistence of Spiral Structure through the Resting Stage.**—A late telophase, the chromosome thread, which has been tightly wound in a spiral at metaphase, is becoming uncoiled. B earliest prophase, the chromosome is coiled to about the same extent as it was at the end of the last telophase. C slightly later prophase, the uncoiling is proceeding, and the thread (which is double) is being thrown into loose "super-spirals" owing to being confined within the nuclear membrane.

(After Darlington.)

chromosomes and the homogeneity of the interphase nucleus may perhaps be reached through a study of the spiral structure of the chromosomes (p. 364). Kuwada and Nakamura[1] showed that on treatment with ammonia vapour the spiral thread or chromonema of which the metaphase chromosome is made becomes unwound and that thereafter the

[1] Kuwada and Nakamura 1934a.

nucleus appears very similar to an interphase nucleus. Darlington[1] has shown that a similar uncoiling of the chromonema occurs in telophase, and further provides a very elegant demonstration of the continuity of the chromosomes, similar in principle to Boveri's above, by showing that the spirals of one telophase can be seen in the following prophase.

It is possible that the thread-like chromonema may be so flexible as to make little difference to the apparent viscosity of the nuclear contents and be insufficiently rigid to be demonstrated by the ordinary methods of fixation. But the structure of the interphase chromosomes is of more or less molecular dimensions (p. 377) and the model of a coherent thread, appropriate at larger dimensions, may be very misleading here. Moreover no description of the structure can be complete which ignores the fact that during the interphase the chromosomes become duplicated or split, and this is a process about which we know very little more than that it occurs.

6c. The Mechanism of Segregation

The chromosomal mechanism which is invoked to explain the segregation of genes is the pairing of like chromosomes in zygotene and their separation one to each daughter cell; the chromosomes are assumed to be in so far dissimilar that each contains a different allelomorph. The evidence as regards separation to the two poles is obvious, since it follows directly from observation, though it is not immediately obvious whether the two allelomorphs are separated at the first division or the second (p. 104). The crucial evidence we require to provide a basis for the whole theory is evidence that the two chromosomes which pair are similar or homologous ones.

The most direct evidence is derived from the observation of the similarity of pairing chromosomes in those organisms in which each chromosome is individually recognizable (e.g. Crepis). However it is usually difficult to make out the shape of chromosomes at the pairing stage (zygotene) because they are still uncontracted and tangled with one another. In some organisms the chromosomes show a row of small swellings or chromomeres, and in prophase the pairing chromosomes can be seen to correspond, not only in a general way but actually point for point. A similar detailed correspondence is seen between the two pairing members of the salivary gland nuclei in Diptera (p. 100), and although this pairing cannot be the same in every way as that found in

[1] Darlington 1935a.

zygotene, the point to point correspondence which it shows is probably a special case of a general phenomenon.

Less direct evidence that pairing is only between homologues is the

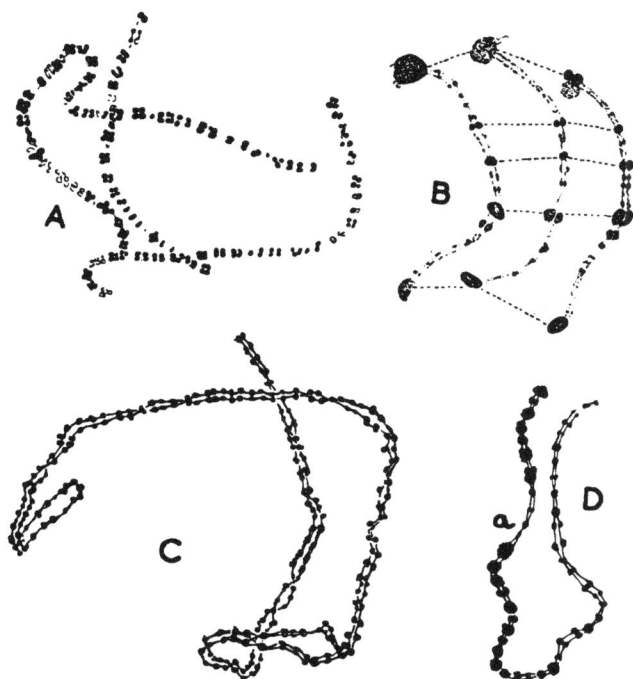

Fig. 11. **Zygotene Pairing.**—A in a lily, the chromosomes showing a structure of numerous small chromomeres ("ultimate chromomeres") (from Metz, after Belling). B Three drawings of one pair of chromosomes from a grasshopper, showing constancy of structure and homology of pairing elements; the chromosomes are shown at a lower magnification than A, and have a structure of larger and less numerous chromomeres. (From Metz, after Wenrich.) C a pair of chromosomes in *Trillium* (from Huskins and Smith). D. *Fritillaria Eggeri*, notice that the contraction and condensation of the chromosome has progressed farther near the centromeres, which are very closely paired at *a*. (From Darlington.)

fact that pairing normally fails completely in haploids (but see p. 72), while in triploids pairing is in threes, in tetraploids in fours, and so on (p. 68). Evidence of a similar kind was obtained by Gelei;[1] in Dendrocoelum the premeiotic division was sometimes irregular, leaving one of the daughter nuclei in two parts, and in some cases the smaller part

[1] Gelei 1913, 1921, 1922.

contained two chromosomes which failed to pair; the obvious explanation is that these two were not of the same kind.

General evidence of the parallelism between the phenomena of segregation of genes and chromosome pairing and reduction is found in the genetic behaviour of organisms with chromosome cycles different from the normal, in polypoids, and in cases of abnormal meiosis (Chap. 2.)

In some organisms pairs of chromosomes have been found in which the two members, although homologous, are yet recognizably different,

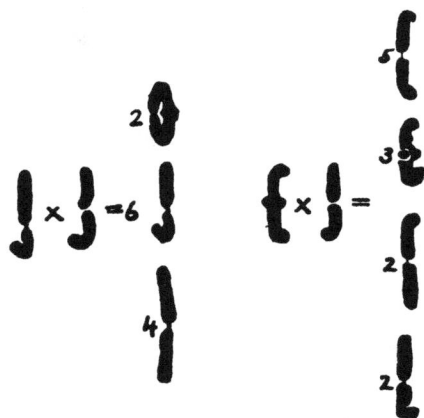

Fig. 12. "Mendelizing" Chromosomes. — Crosses between races of *Circotettix verruculatus* (Orthoptera) with heteromorphic pairs of chromosomes. On the left is a cross between two animals each with an *IJ* pair; it gave offspring with *JJ*, *IJ*, and *II* pairs. On the right is a cross between *JJ* and *IJ* animals, which gave *JJ* (two upper figures) and *IJ* (two lower figures) pairs in the offspring. The numbers of the different kinds of offspring are indicated.

(From Stern, after Carothers.)

one being larger than the other; they are probably cases of translocation or deletion (Chap. 4) such as are well known genetically in Drosophila. Thus in the Orthopteran *Circotettix*,[1] in crosses between a race with an unequal pair and a race with an equal pair, the unlike chromosomes were inherited exactly like a Mendelian factor in a cross between a heterozygote and a homozygote. This provides general evidence for the continuity of the chromosomes, their segregation, and their random recombination in fertilization.

7. *Independent Assortment and Linkage*

The independence which is sometimes found between different pairs of genes, and is shown by the formation of equal numbers of new and old combinations in the gametes of a double heterozygote, is explained as a consequence of the independent assortment of chromosomes to the

[1] *Rev.* Carothers 1926.

two poles of the meiotic spindle. Direct evidence of this independence was found by Carothers[1] in the Orthopteran *Trimerotropis* where there was a second unequal pair in addition to an unequal sex pair. In nearly

Fig. 13. Random Assortment of Chromosomes.—In animals with a heteromorphic pair of chromosomes (*IJ*) either the *I* or the *J*-shaped chromosome may go to the same pole as the unpaired *X* in the reduction division, and these two alternatives happen with equal frequencies.

(After Carothers.)

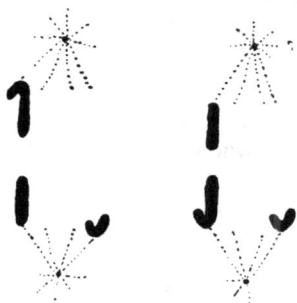

exactly half the cases (51·3 per cent) the larger member of the unequal pair segregated to the same pole as the *X*, as would be expected if the chromosomes were independent.

Independence of characters is not found in the exceptional cases

Fig. 14. Linkage.—The gametes produced by a heterozygote are best identified by crossing the heterozygote to a double recessive (the "backcross").

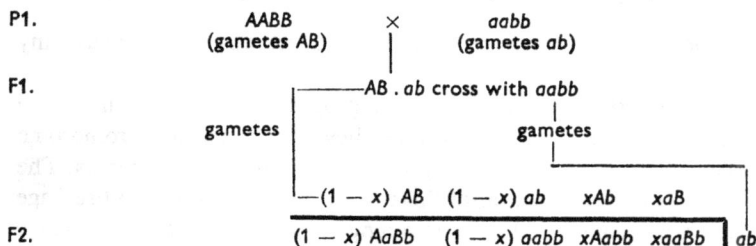

P1.   AABB   ×   *aabb*
           (gametes AB)   (gametes *ab*)

F1.   —AB . *ab* cross with *aabb*

       gametes   gametes

       $(1 - x)$ AB $(1 - x)$ *ab*   x*Ab*   x*aB*

F2. $(1 - x)$ A*aBb* $(1 - x)$ *aabb*   x*Aabb*   x*aaBb* | *ab*

When there is no linkage, $x = \frac{1}{2}$ and AB, *Ab*, *aB* and *ab* gametes are formed in equal numbers, which gives independent assortment.

When linkage is complete, $x = 0$ and only AB and *ab* gametes are formed.

When there is some linkage, x is between 0 and $\frac{1}{2}$ and more gametes are formed of the original kinds AB and *ab* than of the new combinations *Ab* and *aB*.

in which the chromosomes do not behave independently of one another (cf. ring formation, p. 109.)

Limited independence of factors, or linkage, was at first explained as a consequence of differential multiplication of the cells containing the different combinations of factors, which were supposed to be segregated

---

[1] Cf. Carothers 1926.

some considerable time before gamete formation (Bateson 1930). This theory had, however, to be abandoned, both on account of the evidence that segregation takes place at meiosis, after which only a single division occurs before the gametes are formed, and because it could not be made to fit the numerical data; if the different frequencies of the new and the old combinations are due to differential multiplication, the ratio of the two sorts of gametes must be $2^a : 2^b$, where a and b are small integers (some modifications of the theory would allow the ratio to be $a : b$ where a and b are small integers such that $a + b = a$ power of 2).

Fig. 15. **Theories of Linkage.**—The old theory (on the left) supposed that after segregation had occurred one set of gametes, in this case ab and AB, multiplied faster than the other aB, Ab set. The new theory supposes that A and B are linked if they lie in the same chromosome, but that they sometimes separate when the chromosomes break and rejoin in a different way.

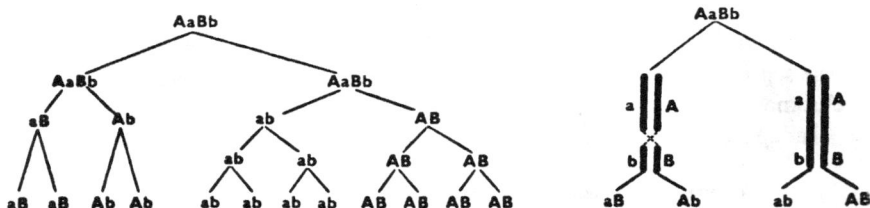

The ratios actually found are not limited in this way and may have any value.

The alternative theory of linkage, proposed by Morgan in 1911,[1] supposes that factors are linked when they lie in the same chromosome and behave independently when they lie in different chromosomes. The possibility of recombination of linked factors depends on the breakage of the chromosomes and their rejoining in such a way as to exchange parts. On this theory the genes in any organism should fall into a certain number (= to the haploid number) of linkage groups. This is true in all cases which have been sufficiently tested. Usually we know fewer linkage groups than the haploid number of chromosomes, but this is clearly due to a mere lack of knowledge of sufficient genes. In a few cases (e.g. the pea) it has at various times been claimed that although a large number of factors were known, they fell into more linkage groups than should be expected; but later work has always shown that two of the supposed linkage groups were not in fact independent but were very weakly linked. When, in *Drosophila melanogaster*, a factor

[1] The theory had been put forward on teleological grounds by Janssens 1909.

(Tubby) was found which could not be placed in any of the haploid number of linkage groups already known, it appeared to provide an exception to this rule until it was shown by cytological examination that these particular flies contained an extra fragment of chromosome.[1]

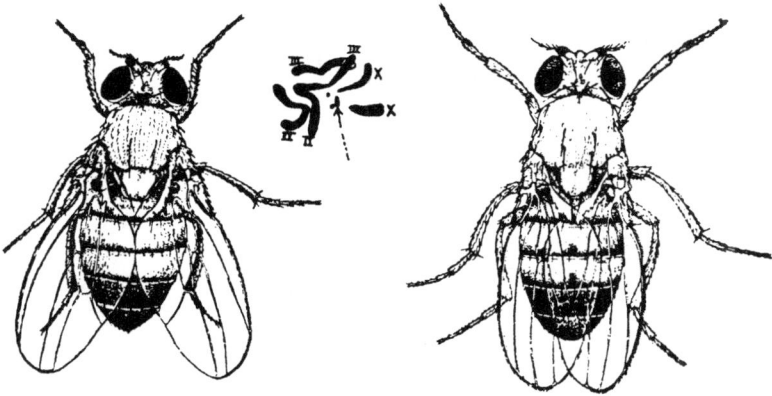

Fig. 16. **The Equality of the Number of Chromosomes and Linkage Groups.**—There are four chromosomes in the haploid set of *Drosophila melanogaster*, and four groups of linked factors. A fly (on the left) was found in X-rayed stock which showed several deviations from the normal (on right). It was called Tubby; but the gene determining its appearance was not linked in any of the known groups. Cytological examination was able to explain this apparent fifth linkage group by showing that the fly contained an extra chromosome fragment. which is indicated by the arrow in the metaphase figure shown.

(From Muller and Painter.)

8. *The Additive Theorem of Linkage Values*[2]

The further development of the chromosome theory of linkage was based on the discovery of the additive theorem of linkage values. If the percentage of recombinations found in a double heterozygote $AaBb$ is defined as the recombination value (or linkage value) between the loci A and B, then the additive theorem states that if A, B, and C are three linked genes, the recombination value AC is the sum of the values AB and BC (provided that the values AB and BC are small, cf. p. 89). Morgan and his associates saw that this fundamental relation could easily be accounted for by the hypothesis that the genes are arranged in a single linear series along the chromosome, and that a recombination is due to a chance breakage and rejoining between the

[1] Muller 1930.   [2] Morgan 1911, Sturtevant 1913.

loci in question. If the chance of a breakage is the same at all points in the chromosome, then the chance of a breakage between loci A and B simply depends on the distance between A and B, and the additive theorem of linkage values is explained by the additive theorem of intervals of length. We shall see later how far the chance

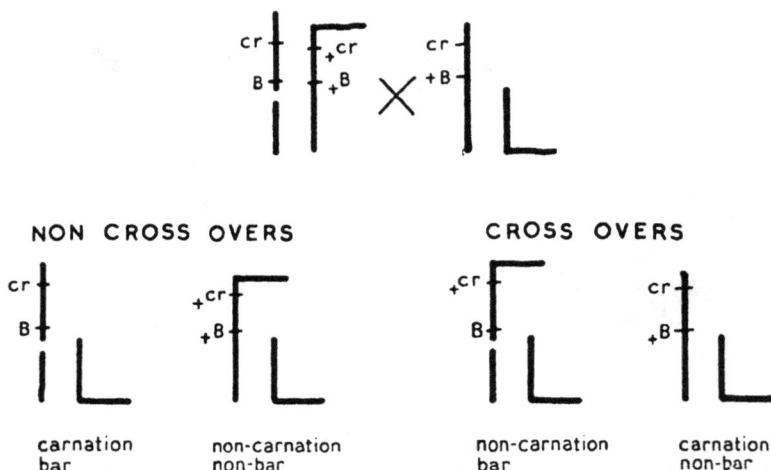

NON CROSS OVERS

carnation
bar

non-carnation
non-bar

CROSS OVERS

non-carnation
bar

carnation
non-bar

Fig. 17. Stern's Proof that Recombination involves Chromosome Breakage. —A fly (*D.melanogaster*) was obtained with one *X* chromosome (carrying carnation and Bar) fragmented into two pieces and the other *X* (carrying the normal allelomorphs) with a piece of the *Y* attached to it. This fly was crossed to a carnation non-Bar male. The diagram shows the males obtained in the *F*1; where crossing-over has occurred, and carnation becomes separated from Bar, the attached piece of *Y* has been broken off from the whole *X* and attached to the fragmented *X*. The same conclusion could be drawn from the *F*1 females, which are not shown. The *Y* chromosome is drawn L-shaped.

of a break is really the same at all points (p. 97). Here we may note that Stern[1] has conclusively demonstrated that the formation of a new combination does actually involve a breakage and rejoining, which is called a crossing-over; his proof depends on being able to label both ends of two chromosomes both cytologically and genetically, and showing that when a cross-over occurs the chromosomes exchange ends.

The additive theorem of linkage values is thus explained on the hypothesis that the genes are arranged in a linear order along the

[1] Stern 1931. Similar evidence was found in maize by Creighton and McClintock 1931.

chromosomes, a hypothesis which is the basis on which the whole subsequent developments of genetics are founded. By determining the linkage values of a set of linked genes we can determine their distance apart (measured in units of length in which the chance of a break is equal) and thus their order. Linkage maps can be prepared in this way and will be discussed in Chap. 4.

CHAPTER 2

The Modifications of the Chromosome Cycle

Mendel's original experiments were concerned with factors which are expressed in that part of the life cycle of the plant in which the nucleus is diploid, containing two of each type of chromosome. Predictions can be made from the chromosome theory as to the ratios to be expected for characters which are apparent in other phases of the life cycle, and also as to the course of inheritance in organisms with atypical nuclear cycles. These predictions form a valuable test of the theory. We shall consider first the principles which apply to factors which are expressed in various phases of the life cycles of plants, and then the results of some different chromosomes cycles. The special case of an abnormal chromosome cycle which is found in polyploids, i.e. in organisms which possess three, four, or more of each kind of chromosome instead of the usual two of each kind, is of particular importance and is considered in the last section of this chapter.

A. CHROMOSOME CYCLES[1]

In animals the reduction divisions take place immediately before the differentiation of the germ-cells, so that it is only in the gametes themselves that the nucleus is haploid. The haploid phase may, however, assume much greater importance in plants. In some lower plants, e.g. most fungi and algae, the reduction division follows immediately after fertilization, so that the nucleus is haploid throughout the life of the organism with the exception of the just fertilized egg-cell; such organisms are known as haplonts. In other plants the reduction division takes place at some time between fertilization and gametogenesis; in all higher plants and some lower ones (e.g. Fucus) the interval between reduction and gametogenesis is short, so that the haploid (gametophyte) generation is relatively unimportant; but in many lower plants the haploid phase which bears the gametes may be equal in duration and morphological complexity to the diploid phase which occurs after fertilization and before reduction. The haploid cells produced by the reduction division are known as spores (microspores in the male, macro-

[1] General references: Darlington 1937, 1939, Hartman 1929a, Stern 1928, Wilson 1928.

spores in the female) and the diploid plant which bears them is the sporophyte. They develop into a haploid plant, which, since it bears the gametes, is known as the gametophyte. The factors which affect the

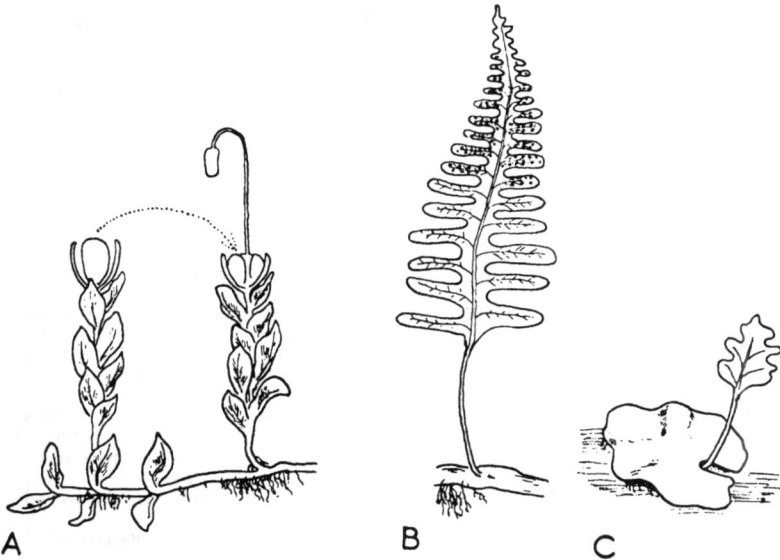

Fig. 18. **Life Cycles of Mosses and Ferns.**—A. In a moss the main plant is the haploid gametophyte phase. The gametes are borne on sexual organs among the leaves, sometimes both sexes on the same plant, sometimes on different plants. Fertilization takes place in these organs, and from the female organ the zygote grows into the diploid sporophyte, which remains attached to the gametophyte. The reduction division occurs in the spore capsule, and the haploid spores, when released, grow into new haploid plants. B. In ferns the main plant is the diploid sporophyte. The reduction division takes place in the "sori" on the under side of the leaves, and the spores, after being set free, grow into an inconspicuous flat plant, C, which is separate from the sporophyte. This is the gametophyte, known as the prothallus; it bears the sexual organs in which the gametes are formed, and after fertilization the new diploid sporophyte grows out from it and absorbs it. In the drawing on the right it is more enlarged than the fern B, and the first leaf of a sporophyte is just growing out from it.

characters of the diploid phase are inherited according to Mendel's original laws, and need not be further discussed here.

1. *Inheritance in the Haploid Phase*

The chromosomes of the haploid phase are in the reduced condition; there is only one of each kind. If the chromosomes contain the genes, there can be only one gene of each locus. Heterozygosity should therefore be impossible. Thus in the haploid phase we can observe the effects of segregation uncomplicated by those of fertilization. From a cross between two different gametophytes of, for instance, a moss, in which the haploid and diploid phases are of more or less equal importance, we may expect to, and do, obtain a hybrid sporophyte; but F1 gametophytes which are formed from this hybrid are each of them "pure," that is, of each pair of contrasting characters they show one only. This is a demonstration that heredity is particulate and not blending, as the pre-Mendelians thought; each gene retains its individuality without contamination by any other character which may happen to be associ-

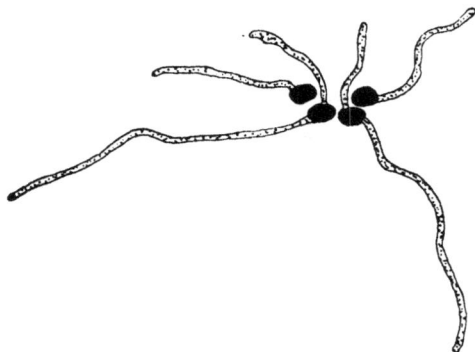

Fig. 19. **Germination of a Spore Tetrad** in the moss *Funaria hygrometrica.*—The haploid spores were derived from a sporophyte heterozygous for the factor *Gg* which causes rapid growth. Two spores are fast-growing, and have already sent out a long and a short thread, while the other two slow-growing ones have only formed one medium-sized thread each.
(After Wettstein.)

ated with it in a hybrid. New combinations of factors may of course be formed in the gametophytes, and give assemblages of characters which, however, are not blends but are mixtures of units from one parent with units from another.

The inheritance of haploid phase characters has been followed in most of the orders of plants where it is to be expected.[1] Particularly interesting results have been obtained in mosses and some fungi, where it has been possible to isolate the four spores formed by the reduction of one diploid cell in a sporophyte and to show that, as regards any one character, they germinate into two gametophytes of one allelomorphic type and two of another. Similar tetrad-analysis of spores from doubly heterozygous sporophytes is dealt with later.

It should be noted that all the factors known in plants with a more or less equal alternation of generations affect only one of the two

[1] *General,* Sansome and Philp 1932; *Ferns,* Anderson-Köttö 1931; *Bryophytes,* Allen 1935; *Fungi,* Dodge 1936, Kniep 1929.

generations, either the haploid phase or the diploid phase but not both. This is not true in higher plants and exceptions may be expected in the comparatively few organisms (some algae) in which the haploid and diploid phases are nearly identical in appearance.

2. Haploid Heredity of Higher Plants

In the higher plants the haploid phase is considerably reduced. In the male the microspore forms only the pollen tube, in which one or two nuclear divisions, unaccompanied by cell division, may occur, and in the female the macrospore undergoes a variable number of divisions giving some hundreds of nuclei in gymnosperms, but usually only about eight nuclei in angiosperms. In spite of its short duration, even the gametophyte of higher plants seems to be under the control of its own special set of genetic factors. Most of the factors which have been discovered affect the male. Two of the clearest examples are factors which affect the reserved carbohydrate of the pollen grain. Parnell[1] showed that two races of rice exist, differing in a factor of which the dominant allelomorph *Gl* produces starchy pollen which stains blue in iodine, the recessive allelomorph *gl* a glutinous carbohydrate which stains brown. An exactly analogous factor is known in maize, where the waxy gene (recessive) gives a pollen grain containing a carbohydrate which stains red in iodine.[2] In both these cases, hybrid plants form equal numbers of the two types of pollen.

Many factors have been found which affect the growth of pollen-tubes. The effect may sometimes be a direct influence on the pollen; in extreme cases factors are known which make it impossible for the pollen to function (pollen lethals), as it seems to be in the well-known case of Matthiola,[3] where, however, this may be due to a chromosomal deficiency.[4]

The effect may be less extreme, merely causing one type of pollen grain to grow faster than another.[5] This phenomenon, so called certation, leads to aberrant ratios when a heterozygous plant is used as the male parent. An example may be given from the work of Correns on Melandrium in which it was shown that the female determining pollen tubes grow faster and achieve fertilization more often than the male-determining.[6]

Often, however, the expression of the factor depends on an inter-

[1] Parnell 1921.   [2] *Rev.* Brink 1929.
[3] Saunders 1928, Waddington 1929.
[4] Philp and Huskins 1931, but see Kuhn 1938.   [5] *Rev.* Jones 1928.
[6] Correns 1921.

action between the pollen tube and the tissues of the style in which it is growing. Factors of this kind give rise to the phenomenon of self-sterility, in which a plant cannot be fertilized by its own pollen, a state of affairs which is found in many different genera (Nicotiana, Veronica, Verbascum, Prunus, etc.).[1] Usually the incompatibility extends to a group of plants, so that the species is divided into a set of inter-fertile groups within each of which fertilization is impossible. The genetic

Fig. 20. **Pollen Lethals in Stocks** (*Matthiola incana*).—The "eversporting" stock, when selfed, gives about equal numbers of single flowered stocks and plants with sterile double flowers. It must, therefore, be heterozygous for singleness; but it transmits only doubleness through the pollen, since the *F*1 from a cross of pure single by eversporting male consists entirely of heterozygotes. The factorial interpretation is that in the eversporting race a recessive pollen lethal factor *l* is closely linked to the dominant factor *S* for singleness. The crosses are, then:

Eversporting *Sl sL* selfed gives *Sl sL* (eversporting singles) and *sLsL* (doubles), the *Sl* pollen not functioning.
Pure single *SL SL* by eversporting *Sl sL* gives only *SL sL* (heterozygous singles), which give a normal 3 : 1 ratio in the *F*2.

The lethal factor *l* can be identified as the absence of a small fragment (trabant) which is normally attached to the chromosome bearing the singleness factor. The appearance of this chromosome in the three types is shown below.

(After Philp and Huskins.)

PURE   EVERSPORTING   DOUBLE
SINGLE   SINGLE

basis of the infertility is a group of allelomorphic factors $S_1, S_2, S_3 \ldots S_n$, such that a pollen grain containing a gene S_x cannot grow satisfactorily on a style which contains S_x. Thus self-fertilization is always impossible but cross-fertilization may be possible giving either two classes of progeny in equal numbers ($S_1S_2 \times S_1S_3$ gives S_1S_3 and S_2S_3) or four classes in equal numbers ($S_1S_2 \times S_3S_4$ gives S_1S_3, S_1S_4, S_2S_3, and S_2S_4). Analysis of self-infertility has considerable economic importance in those plants where the crop depends on the setting of fruit and thus on fertilization as in apples. Unfortunately in many of the important cases the plants are polyploid, so that very complicated possibilities of interaction between stylar tissue and pollen tubes must be envisaged.

[1] *Revs*. Brieger 1930, East 1929.

3. *Endosperm Characters*

Few genes, except simple lethals, are known affecting the female gametophyte of higher plants. A peculiar situation arises, however, in the formation of endosperm, and associated tissues such as the aleurone, which are produced (by "double fertilization") by fusion of one of the

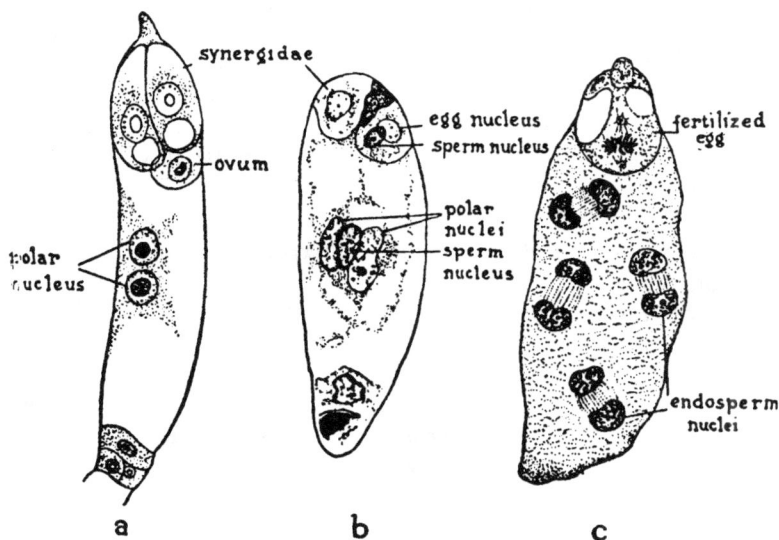

Fig. 21. Double Fertilization and the Formation of Endosperm in higher plants. In *b* two sperm nuclei have entered the embryo sac from the pollen tube. One unites with the egg nucleus to give the diploid embryo, while the other unites with two embryo sac nuclei ("polar nuclei") to give the endosperm, the nuclei of which (shown in telophase in *c*) are therefore triploid.

(From Wilson.)

pollen tube's haploid nuclei with two of the haploid embryo-sac nuclei and are therefore triploid in constitution. Several endosperm characters are known; they appear to the unsophisticated eye as characters of the seed. Their inheritance depends on the dominance relations in the triploid tissue; in some cases an endosperm with one dose of the dominant and two of the recessive shows the dominant character (e.g. *Su* starchy dominates over two doses of *su* sugary in maize grains), in others the recessive (e.g. two doses of *f* floury show in the presence of *F* flinty in maize). The immediate apparent effect of pollen on the characters of the seed in a cross between sugary female and starchy male, for instance, is spoken of as xenia, a phenomenon which may also occur in straightforward diploid inheritance if the character brought in by the male shows early enough.

4. Haploid Phase Characters in Animals

Few genes are known which affect the gametes of animals.[1] Self-incompatibility, apparently similar to that described in plants, is known in some hermaphrodite animals (e.g. the ascidian Ciona)[2] but nothing is known of its genetic basis. In some animals, it has been stated that the sperm fall into two size classes, or have a bimodal size-frequency curve.[3] The two sizes are usually supposed to be associated with the two classes of sex-determining sperms, since bimodal curves are only known in forms with male heterogamety.

Cases in which the eggs fall into two classes in size have also been reported and in some of these the size differences have been supposed to be the causes of sex determination (p. 230); the genetic basis of the dimegaly is, however, unclear. The best known cases of variation in the egg characters of animals have been shown to be due to factors which act in the diploid tissue of the mother causing all her eggs to be of one specific kind, e.g. egg-shape and colour in silk worms, type of coiling in Limnea (p. 143). These genes must be clearly distinguished from genes which act during the haploid phase itself.

B. APOMIXIS[4]

The normal process of sexual reproduction, on which Mendelian inheritance depends, has two essential features: reduction of chromosome number and fertilization. In the reproduction of some organisms one or other or both of these features may be absent. In cases where reproduction occurs simply by the development of some of the somatic cells into a new organism we speak of budding, or vegetative propagation; the genetic result is of course that the offspring exactly resemble the parent, with rare exceptions due to somatic mutation. Genetic series of organisms derived vegetatively from one parent constitute a clone. Reproduction of this kind occurs in few animal species but is frequent in plants where it is of great economic importance in the propagation of varieties (hybrids, unbalanced polyploids, etc.) which do not breed true. Many of the best varieties of cultivated apples, oranges,

[1] Cf. Muller and Settles 1927   [2] Morgan 1938.
[3] Wodsedalek 1913, Zeleny and Faust 1915, but see Krallinger 1928.
[4] *General references:* Darlington 1937, chap. 11, Peacock 1925, Rosenberg 1930, Stern 1928.

roses, etc., have been obtained by the vegetative propagation of somatic mutations, which are often known as bud-sports.

In other cases, reproduction, although not fully sexual, involves the sexual organs or related structures. Such phenomena are known as apomixis. In cases where the sexual cells (gametes) are concerned, we speak of parthenogenesis; when the gametes fail and their functions are taken over by other (gametophyte) cells we have apogamy, which occurs in plants; while finally examples are known, also in plants, in which the gametophyte is developed from cells other than the spore mother cell and these are referred to as apospory. The nomenclature adopted here is that proposed by Darlington.

1. Parthenogenesis[1]

Parthenogenesis is of two kinds: diploid parthenogenesis, in which the reduction is abortive and diploid eggs are formed which develop

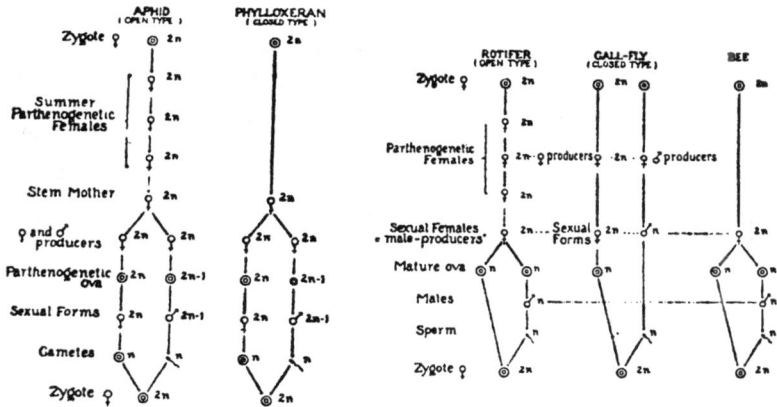

Fig. 22. **Parthenogenesis in Animals.**—Two diploid parthenogenetic life cycles are shown on the left, the aphid with a succession of females reproducing by diploid parthenogenetic eggs, the Phylloxeran with only two such females before sexual forms are again produced. On the right are three cycles involving haploid parthenogenesis. In the rotifers there is a series of diploid parthenogenetic females, in the gall flies one, in the bees none. In all cases the males are haploid, developed from unfertilized eggs.   (From Wilson.)

without fertilization, and haploid parthenogenesis in which normally reduced eggs develop without fertilization. Either of these forms may be found constantly in a species (obligatory) or only under certain conditions (facultative).

---

[1] *General references:* Ankel 1927, 1929, Vandel 1931, For genetics of Protozoa see Jennings 1930, Rattel 1932, Sonneborn and Lynch 1934.

1a. Diploid Parthenogenesis, Apospory and Apogamy

In some animals diploid parthenogenesis occurs under the influence of rather unspecific external conditions, e.g. diet, temperature, crowding, etc.[1] In other forms, e.g. Phylloxera, there is a regular alternation of diploid females which produce parthogenetic diploid eggs and sexual males and females which develop from these eggs. In still other forms (some nematodes, some insects) diploid parthenogenesis is obligatory and males are very rare or even unknown. Among plants, diploid parthenogenesis occurs in ferns and in many flowering plants.

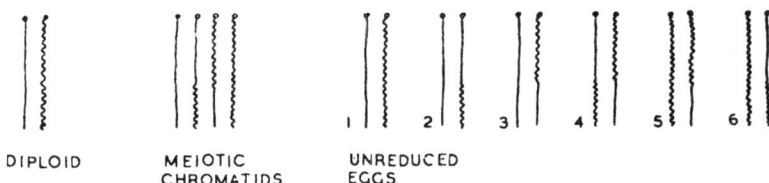

DIPLOID   MEIOTIC   UNREDUCED
                 CHROMATIDS   EGGS

Fig. 23. Segregation in Diploid Parthenogenesis.—In a heterozygous diploid, four types of chromatids must result from meiosis, since there must be crossing over of factors far removed from the centromeres (p. 121). If diploid parthenogenetic eggs are formed, there are six ways in which two of these chromatids may come together in a single egg. Only one of these (1) gives a diploid of the same constitution as its parent; in all the others segregation has occurred. Types 1, 3, 4 and 6 are formed if the two daughter nuclei come together after the first metaphase and form a restitution nucleus; there is segregation only of factors distal to the crossing over. Types 2 and 5 occur when an egg unites with the second polar body; segregation occurs for factors between the centromere and the crossing over.

Diploid parthenogenetic eggs may develop spontaneously or their development may be stimulated by sperm or pollen of which the nucleus degenerates and takes no part in the subsequent formation of the embryo. This process is known as pseudogamy and is found as a naturally occurring process in some nematodes and plants (Potentilla and blackberry).

Apospory may, in its genetic effects, be considered as a variety of diploid parthenogenesis, since it consists in the development of the gametophyte generation from an unreduced (diploid) cell. It is known in flowering plants and also in ferns. Apogamy is not so well authenticated a phenomenon, but when it occurs it also gives rise to diploid offspring by the development of two fused embryo-sac cells. Combinations of apospory and apogamy also occur.

All these forms of diploid apomixis give the same genetical results:

[1] Cf. Shull 1929, Mortimer 1935.

offspring whose genetic characters are identical with those of their parent. An example may be taken from the work of Agar[1] on Daphnia, in which hybrids between *D. obtusa* and *D. pulex* bred true through ten generations of diploid parthenogenesis. Obligatory diploid apomictic plants form closed series of plants with almost perfectly constant characters; they may be compared with clones of vegetatively propagated organisms. It is surprising, then, to find that groups of plants reproducing in this way are often extremely polymorphic (e.g. the group of Eu-hieracium). How is this divergence in evolution to be explained if segregation and recombination have been abolished? The answer appears to be that segregation, although hindered, is not entirely abolished; it is still possible owing to crossing over, while on the other hand the suppression of hybridization prevents the blending of slightly different races into a uniform population. (Fig. 23.)

1b. Haploid Parthenogenesis

The development of haploid organisms from nuclei with the reduced number of chromosomes can occur only by haploid parthenogenesis or rarely by apogamy. Haploid parthenogenesis, may, however, yield diploid organisms, since in some forms the development begins by doubling of the original reduced chromosome number; such organisms will clearly be homozygous.

Facultative haploid parthenogenesis occurs in many animals. The offspring show segregation of the factors present in the parents, and are homozygous for factors for which the parents were heterozygous. Thus Fryer[2] isolated a grasshopper (*Clitumnus*) heterozygous for a dominant factor producing a process on the head, and from one female obtained twelve horned females and ten unhorned females. Similarly Nabours[3] describes experiments with the grouse locust Apotettix, in which an isolated female of the constitution $\dfrac{YZk}{yzK}$ (colour pattern factors) gave offspring $9YZk : 18yzK : 1yZk : 1YzK$ by haploid parthenogenesis with subsequent doubling of the chromosome number. This case shows coupling as well as segregation. Similar haploid parthenogenesis, with segregation, may occur in hybrids in Lepidoptera even between strongly sexual races.[4]

The most important examples of haploid parthenogenesis in animals

[1] Agar 1920.   [2] Fryer 1913.   [3] Nabours 1919.   [4] Peacock 1925.

are in those organisms which show the so-called Hymenopteran method of sex determination. In the bees and wasps themselves fertilized eggs develop into females, unfertilized eggs develop parthenegetically into males. A similar type of sex determination is combined with a series of diploid parthenogetic generations in Rotifers and alternates with one such generation in Gall-flies. The inheritance of factors in Hymenoptera has been extensively studied in the parasitic wasp Habrobracon by Whiting.[1] He has shown that, as might be expected, the characters of a male are dependent solely on those of his mother, and that he is homozygous or "pure"[2] for characters for which she is heterozygous. A father's genes pass only into his daughters, in fact his sons can only be considered his by courtesy. Rare exceptions to this behaviour occur, since diploid males have been found (p. 230).

Haploid parthenogenesis in plants is known only in a few flowering plants (e.g. Solanum), although the germination of haploid spores into the gameteophyte generation may be regarded as a somewhat similar phenomenon.

2. *Pseudogamy and Male Parthenogenesis*

The parthenogenetically developing eggs of an animal or plant, whether diploid or haploid, may develop spontaneously or may require activation by a definite stimulus. This stimulus may be artificial,[3] as in the haploid parthenogenesis of animal eggs which has been performed by many authors (e.g. pricking frogs' eggs, treatment with chemicals, etc.) or in the apospory provoked by pricking in the alga Vaucheria.[4] On the other hand, the stimulus required may be that of fertilization by a sperm, which may be of the same or another species, but which in either case may play no further part in development after activating the egg. Such cases are spoken of as pseudogamy. They are well known in animals (*echinoderm* by *annelid* crosses); and in the higher plants haploid parthenogenesis is probably impossible without the stimulus of homologous or foreign pollen, while the phenomenon is also common in diploid apomixis. The offspring of such pseudogamous modes of reproduction are of course entirely maternal in character since the male gamete is responsible only for activation of the egg.

In some plants it has been found that haploid organisms derived

---

[1] Whiting 1932.

[2] Muller has suggested that it would be better to speak of factors in haploid organisms such as male bees as hemizygous; and this word can also be used for factors (e.g. in an unpaired X chromosome) for which there is no allelomorph in a normal diploid.   [3] *Rev.* Just 1937.   [4] Wettstein 1920.

from a cross fertilization are entirely paternal in character (e.g. *Nicotiana*). In these "hybrids" it must be assumed that the reduced egg cell has been fertilized by the male gamete, and that the female nucleus has subsequently degenerated. Similar phenomena have been produced artificially in animals (sea urchins, newts), in which it has been possible to fertilize enucleate egg fragments with sperm either of the same species (*merogony*) or of a different species (*bastard merogony*). We shall return to a consideration of these experiments in connection with the relations of the nucleus and cytoplasm during development (Chap. 6).

3. *The Production of Unreduced Gametes*

The production of unreduced gametes, regularly in diploid parthenogenesis, and occasionally in sexual organisms where they lead to the formation of polyploids, may be due to (1) a failure of pairing, or (2) a failure of the mechanism for separating the chromatids.

(1) A failure of pairing in regularly diploid-parthenogenetic organisms may be due to a genetically controlled failure of the conditions which normally cause meiosis, in particular to a failure of the precocious chromosome contraction. In other cases it is due to too great a dissimilarity between the chromosomes. This may occur in true haploids of non-polyploid species, e.g. *Datura*, or in hybrids, e.g. *Raphanus-Brassica*. In such cases there is no pairing and therefore no chiasma formation. The chromosomes are usually scattered over the spindle in the first division and pass at random to the two poles; lagging chromosomes may form subsidiary nuclei. Unreduced gametes are only formed if these groups of chromosomes subsequently unite, forming a sort of restitution nucleus. Some chromosomes may pass on to the metaphase plate and divide at the first division; if many of them do so, the second division is often suppressed, but otherwise the chromosomes which fail to divide at the first division divide at the second. The process is clearly a very irregular one, and will give at best a very few haploid gametes from a haploid organism or diploid ones from a diploid hybrid.

Meiosis in haploid cells leads to a regular result only in males produced by haploid parthenogenesis. Thus, in male bees no chromosome pairing takes place but all the chromosomes pass to one pole and the division merely results in the pinching off of an enucleate bud of cytoplasm. In rotifers and coccids the formation of unreduced gametes is rather less regularly ensured by the failure of one division.

C

(2) Failure of the spindle mechanism, although pairing is regular, occurs in plants, e.g. Hieracium. The first division comes to an end at the diakinesis stage and a spindle is not formed; the diploid nucleus resulting is said to be a restitution nucleus. A second division follows in which the bivalents divide, so that two diploid gametes are formed. In animals the failure may affect the second division, which is either omitted altogether or gives rise to a second polar body which then fuses again with the egg-nucleus. A more refined failure of the spindle mechanism is probably responsible for the differential elimination of one chromosome in the production of males in diploid parthenogenesis of Aphids. Failures of the spindle may also occur in mitosis, either regularly as in the first cleavage division of some haploid partheno-genetic eggs, or sporadically. It leads to a doubling of the chromosome number. In the eggs this merely restores the normal diploid number; but if a failure occurs in germinal cells shortly before meiosis it leads to the formation of tetraploid cells which on reduction give diploid gametes. The phenomenon is known as syndiploidy and provides another mechanism by which occasional diploid gametes may be formed.

A failure of the spindle in meiosis has been provoked artificially by abnormal temperatures and narcotics (see p. 255). A similar failure in mitosis has been brought about in the cleavage divisions of echinoderms by shaking and in plants by various treatments. In a few plants, particularly Solanaceae, a fairly high proportion of tetraploid cells occurs in the callus tissue which grows over wound surfaces, and these cells may grow into tetraploid shoots.

C. THE THEORY OF POLYPLOIDS[1]

The normal diploid organism contains two of each kind of chromosome. Chromosome complements are also found, particularly in plants, in which each chromosome is represented three, four, or more times. Such organisms are spoken of as polyploids, and we may find polyploid series of related plants in which the chromosome numbers form an arithmetical progression. Thus there are different species of the genus *Solanum* with the chromosome numbers 24, 36, 48, 60, 72, 96, 108, 120, and 144, all the numbers being multiples of 12. This can be taken as the basic number of the series, the members of which can be represented as $2x$ diploid, $3x$ triploid, $4x$ tetraploid, $5x$ pentaploid, $6x$ hexa-

[1] *General references:* Darlington 1937, chap. 6, Müntzing 1936, Sansome and Philp 1932

ploid (7x heptaploid is missing), 8x octoploid, 9x nonaploid, 10x decaploid, etc. If any of these types regularly forms gametes with half the somatic number of chromosomes, it can also be written as 2n, as is

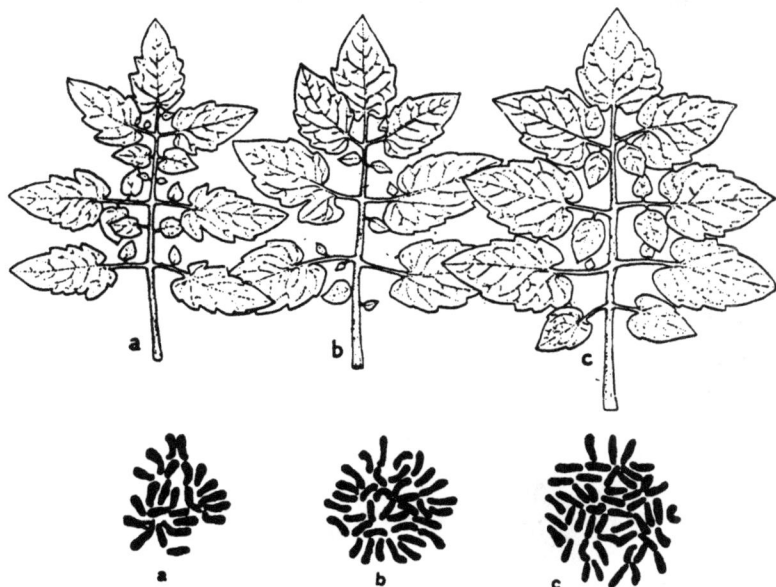

Fig. 24. **Autopolyploid Tomatoes.**—Mitotic chromosomes below and leaves above, A diploid 2n = 24, B triploid 2n = 36, C tetraploid 2n = 48.
(From Hurst, after Jørgensen.)

usual for the diploid; but we shall see that gametes are not always regularly formed, and a notation which gives the multiple of the basic number, such as 6x, conveys more important information than one which merely gives the somatic number.

Fig. 25. **Polyploid Series in Crepis.**—Mitotic metaphases in haploid, diploid, trisomic diploid, triploid and pentaploid *Crepis capillaris* (n = 3).
(From Darlington, after Hollingshead and Navashin.)

Polyploids fall into two groups, those with an even multiple of the basic number, in which the formation of gametes with half the somatic number of chromosomes is theoretically possible; and those with an uneven multiple in which a regular reduction division is impossible.

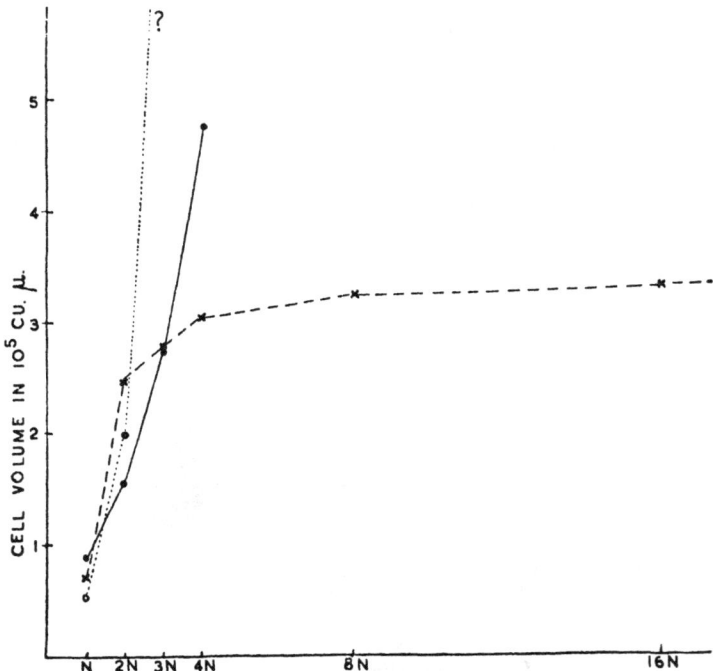

Fig. 26. **Polyploidy and Cell Size.**—Cell volume (in 100,000 cu. μ) in polyploids of the mosses *Physcomitrella patens* (dotted), *Funaria hygrometrica* (plain line), and in allopolyploid hybrids between them (dashes).

(From Wettstein.)

The term "polysomics" or "aneuploids" is reserved for organisms with one or a few chromosomes more or less than a multiple of the basic number; e.g. $4x + 1$ is a pentasomic tetraploid, $6x - 2$ a doubly pentasomic hexaploid.

The possibility also arises that the different basic sets of chromosomes may not be identically alike. A tetraploid, for example, may arise by the doubling of the chromosome number in a hybrid between the species A and B (p. 254) and the A and B sets of chromosomes will be to some extent different from one another, the difference depending on the divergence which has taken place during the evolution of the two species. The diploid hybrid may be represented as *AB*, and the tetraploid derived from it as *AABB*; it is spoken of as an allotetraploid, in distinction from an autotetraploid *AAAA* in which all four sets of chromosomes are identical.

1. *The Appearance of Polyploids*

A mere doubling of the chromosome number, to give an autopolyploid, does not usually have any great effect on the phenotype. In a

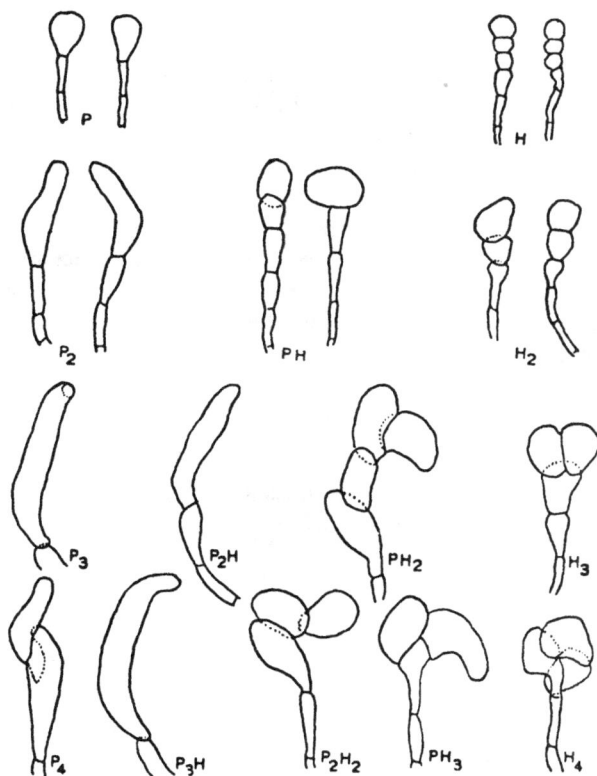

Fig. 27. **Allopolyploids in Mosses.**—The drawings show the shape of the paraphyses in polyploid hybrids (gametophytes obtained by regeneration) between *Physcomitrium piriforme* and *Funaria hygrometrica*. The chromosome sets are indicated by *P* and *H*; thus P_3H is a tetraploid with three *piriforme* sets and one *hygrometrica* set. They have all, except the pure *H* polyploids, cytoplasm derived from *piriforme*. Note the "balance" between the *P* and *H* sets, with some dominance of *H*.

(After Wettstein.)

fair number of cases, however, there is an increase in the size and general vigour of the organism parallel to the increase in chromosome number. This is probably dependent on an increase in the size of individual cells, which acts so as to keep a constant ratio between the size of nucleus and cytoplasm. The phenomenon is not always found. Wettstein[1] showed that in mosses, in which polyploids can be artificially produced by regeneration of sporophytes, the relations were very different in autopolyploids and allopolyploids. In the former, increase in chromosome number led to a rapid increase in cell size, each species having a characteristic rate of increase for a given change in number. In allopolyploids made from the hybrids between pure species which are not too nearly related, the cell size increased rapidly between the haploid and diploid, but much less in the higher multiples, eventually becoming more or less constant, so that very high polyploids (e.g. hexadekaploids 16x) could be prepared. The mechanism of this reaction is not fully understood.

In allopolyploids, the appearance of the organism is usually intermediate between that of the two ancestral species, the exact appearance depending on the balance between the chromosome sets and the dominance relations. Again the artificial moss polyploids provide a very clear example. (Cf. p. 171).

2. *Meiotic Behaviour and Segregation in Polyploids*

The behaviour of the chromosomes of a polyploid at meiosis, and therefore the segregation of genetic factors, depends on whether the organism is a auto- or allo-polyploid.

2a. *Autopolyploids*

In an autopolyploid, all the homologous chromosomes of any one kind may pair in a compound body at zygotene. The pairing, however, is always strictly in twos at any one place, the compound body being held together only by exchanges of partner among the threads, rather as the four diakinesis threads in a diploid are held together at a later stage. This pairing in twos can be directly observed in cytological preparations, and is also indicated by the proportions of recurrent and progressive cross-overs in triploid Drosophila (p. 107).

The pairing seems to take place by a more or less random coming together of the threads, but the randomness is modified by two factors.

[1] Wettstein 1926, 1927, 1937.

Firstly, if two chromosomes are paired at one point, they tend to be paired at nearby points, as might be expected if the chromosome has a certain rigidity. Darlington has described this as a tendency of the chromosomes to behave as if they were made of a small number of pairing blocks, but this phrase must not be taken to imply that the regions paired are the same in different nuclei, or that there are any non-pairing segments; it is only intended to describe the fact that the

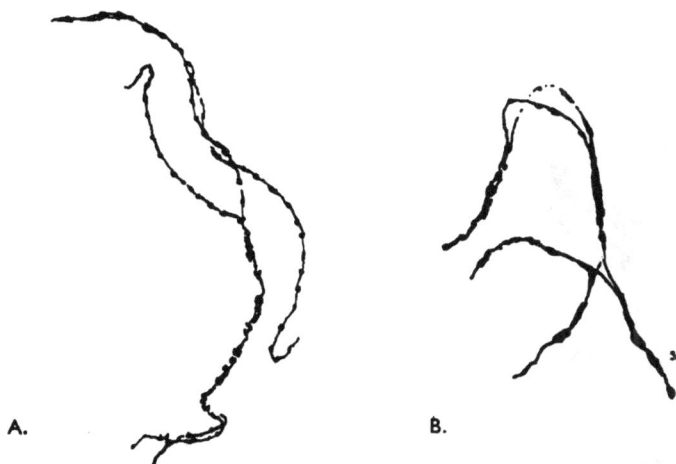

A.   B.

Fig. 28. **Zygotene Pairing in Autopolyploids.**— A in triploid *Tulipa*, B in tetraploid *Primula sinensis*. The figures show parts of the chromosomes, which are too long and coiled to be drawn completely. Notice the changes of partner, and that there are never more than two threads paired at any point.
(After Darlington.)

chromosomes do not zig-zag backwards and forwards from one association to another very often, but only rather seldom. Secondly, the zygotene pairing may be interfered with by the large number of chromosomes which are squeezed inside the nucleus. These may produce such a tangle of threads that some possible pairing fails to occur for purely mechanical reasons.

Both these factors tend to reduce the total amount of zygotene pairing, and since chromosomes not paired at zygotene cannot be held together at metaphase, unassociated univalent chromosomes appear. Even chromosomes which were paired at zygotene will only remain associated if chiasmata are formed between them. If the chiasma frequency is low, there may not be enough chiasmata to hold together all the chromosomes, and some of the zygotene pairings will not be preserved. Thus the metaphase associations, in an autotetraploid for example, usually include univalents, bivalents and trivalents as well as quadrivalents. It was in fact the discovery of incomplete association in autopolyploids which caused the abandonment of the theory that the metaphase association is due to a generalized attraction between homologues, and provided the basis of Darlington's theory that it is depen-

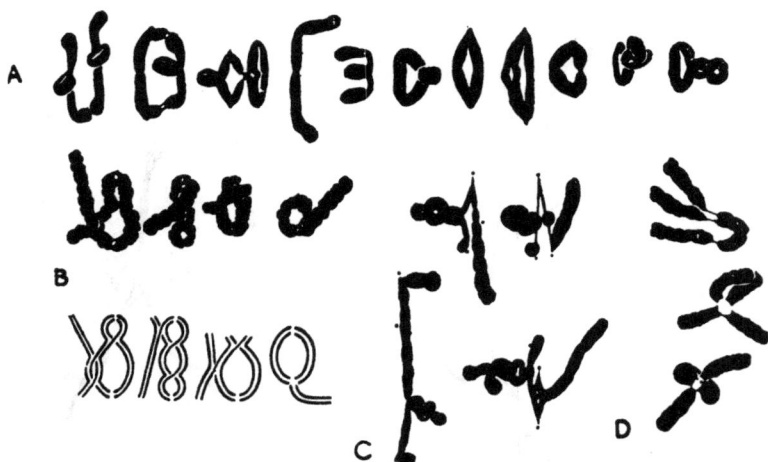

Fig. 29. **Metaphase Associations in Autopolyploids.**—A side view of metaphase chromosomes in tetraploid oats (*Avena*); the 28 chromosomes have formed three quadrivalents (at the left) and eight bivalents. B Diakinesis trivalents in Tulipa (little terminalization) with a diagram of the chromatid structure below. C Metaphase trivalents in Fritillaria showing orientation of centromeres. D some types of quadrivalents found in tetraploid *Primula sinensis*.

(After Darlington.)

dent on chiasma formation. Darlington found one of the main supports for his hypothesis in the fact that in organisms whose chromosomes are of different lengths, it is in the short ones that association most often fails (p. 121).

Incomplete metaphase association usually leads to very irregular segregation and the formation of gametes with variable and unbalanced chromosome sets. A similar irregularity is common even when association is complete, since, e.g., quadrivalents are not always separated two to one pole and two to the other. The gametes with unbalanced complements are usually inviable, and autotetraploids tend in consequence to be rather infertile. In some plants, however, the chiasma frequency is just sufficient to give nearly regular association of the chromosomes in twos, and these bivalents are separated normally at anaphase, giving diploid balanced gametes which function. The theoretical segregation[1] of genes in such a system can be calculated on various assumptions, of which the simplest, appropriate for a strict autotetraploid, is that the similar chromosomes pair at random and segregate as units. An auto-tetraploid duplex for a factor A (i.e. $AAaa$) will have association into two groups of Aa twice as often as into one AA and one aa. From the first association we should get equal numbers of AA, aa and Aa, Aa pairs of gametes, while the latter would give all Aa, Aa pairs. Thus the total gametic output would be $1\ AA : 4\ Aa : 1\ aa$. The ratios of off-spring actually observed in some plants (e.g. tetraploid *Primula sinensis*) agree fairly well with the predictions on this basis. Another theo-retical possibility, is, however, sometimes realized. It has been shown (p. 106) that it is only in the region of the centromere that the paired homologous chromosomes are regularly separated at the first division, whereas in a region distant enough from the centromere for crossing-over to occur between them, portions of two sister chromatids may become attached to different centromeres and thus eventually get into the same gamete. This gives the possibility of random chromatid segregation, which will occur in a region so far removed from the centromere that it becomes an equal chance which centromere a chromatid remains attached to after crossing-over had taken place. The consequences of this type of segregation have been discussed on p. 105.

2b. Allopolyploids

Any increase in the differentiation between the various basic sets in a polyploid decreases the chance that the zygotene pairing will be purely at random and increases the chance that it will be strictly in pairs of exactly similar homologues. If a tetraploid $AABB$ was ori-ginally derived from a hybrid AB, the pairing of A with A and B with B may be much more frequent than that of A with B, the actual rela-tive frequencies depending on how alike the A and B chromosomes are. Even if an A chromosome pairs to some extent with a B in zygo-tene the length through which it is paired with its A homologue may be so much greater that the chiasmata nearly always associate together the two A's and the two B's at metaphase. In such a case, we can speak of a differential affinity of the A chromosomes and the B chromosomes.

Metaphase association of the chromosomes of the two exactly

---

[1] Haldane 1930*a*, Mather 1933, 1936*b*.

C*

similar sets, as in the example above, leads to a condition which is genetically indistinguishable from that in a diploid. Balanced gametes of the type *AB* are formed regularly, and the tetraploid is comparatively, sometimes very, fertile. This fertility depends on there being a considerable difference between the "homologous" chromosomes of the two original species from which the tetraploid was derived; this same differentiation will tend to cause a failure of pairing in the diploid hybrid *AB*, and we therefore find an inverse correlation between the fertilities of the diploid and tetraploid hybrids.[1]

Association of the chromosomes derived from one of the original ancestors may be called homogenetic association, while the association of *A* with *B* chromosomes may be called heterogenetic association. Homogenetic association in an allotetraploid hybrid may lead to the complete disappearance of some of the characters of the ancestors. Thus if a character of one species is determined by a factor which is recessive to the homologous factor in the other species, it will never appear in the progeny of the hybrid, since the recessive factor, e.g. *w* from *A*, will always be accompanied by the homologous dominant *W* from *B*. This phenomenon is very noticeable in wheat crosses, where some of the forms dependent on the more extreme members of polymeric series of factors can never be recovered; it has been spoken of as "shift."

Heterogenetic association also occurs in allotetraploids, and can be detected cytologically by the appearance of tetravalents. An example is *Primula kewensis*, a tetraploid derived from doubling in a hybrid between *P. floribunda* and *P. verticillata*; metaphase association is normally in twos, but occasional quadrivalents can also be found. Similar association occurs in derivatives of allopolyploids, for instance, in the "haploids" (really diploids as regards the basic number) which can be derived by parthenogenesis in an allotetraploid. *Nicotiana tabacum* is itself an allotetraploid *AABB*, and from it a "haploid" *AB* has been formed; in this some bivalents are formed, which must be the result of pairing between *A* and *B* chromosomes.[2] Similarly, cases are known, e.g. in Prunus, in which a hybrid *AABC* between a diploid *AA* and a hexaploid *AABBCC* has complete association of the *BC* sets or even of all four sets together to give quadrivalents.

Regular heterogenetic association, to the exclusion of homogenetic, is not to be expected. If it were found, it would give a gametic output, from a tetraploid *AABB*, of 1 *AA* : 2*AB* : 1 *BB*. It probably usually occurs only in individual chromosomes, associated with more or less

[1] Darlington 1932*a*.   [2] Lammerts 1934.

homogenetic association, and even then it is probably never more frequent than would be expected on a basis of pure chance, which gives a gametic output from $AABB$ of $1\ AA : 4\ AB : 1\ BB$.

Polyploids of higher number than tetraploids behave in essentially the same way, showing failure of association of homologues as a consequence of an insufficient amount of chiasma formation, and homo- or hetero-genetic association according to the differentiation between the chromosome sets.

3. *Auto- and Allo-syndesis*

Homo- and hetero-genetic association are frequently discussed in terms of auto- and allo-syndesis. The usages of these words are, however, by no means uniform, and even authors as closely associated as Darlington and Philp and Huskins use them in quite different senses. Philp and Huskins use autosyndesis to mean the pairing of chromosomes which are phylogenetically derived from the same diploid species; that is, in the sense in which homogenetic is used above. Darlington, on the other hand, uses autosyndesis to mean the pairing of chromosomes derived from the same parental gamete. Thus the association of A with A and B with B in a tetraploid $AABB$ is autosyndesis for Philp and Huskins, but allosyndesis for Darlington.

Darlington's usage is in accordance with the original definition of Ljungdahl,[1] and when the word is used in this work it will be in the sense he defined. The words homo- and hetero-genetic association, or Philp and Huskin's auto- and allo-syndesis, can only be used if one knows or can deduce the phylogenetic relations of the chromosomes, but Ljungdahl's words may be used without implying any such speculation. Thus if complete association takes place in a hybrid between a diploid and a hexaploid, some of the chromosomes derived from the latter must have associated, and autosyndesis must have occurred. This fact may or may not enable one to make deductions as to the phylogenetic relations concerned, and thus to translate the facts into terms of homo- and hetero-genetic associations.

4. *Secondary Pairing*

Homologous bivalents in an allopolyploid, even if not associated by chiasmata, often lie near together in meiotic metaphase. This secondary pairing is probably an expression of an attraction due to homology balancing the normal repulsion between paired chromosomes. It may

[1] Ljungdahl 1924.

be regarded as analogous to the somatic pairing which causes homologous chromosomes to lie near each other in mitotic metaphase in organisms such as Diptera. Catcheside[1] has recently discussed the mechanical conditions involved. The main importance of the phenomenon is the clue it provides as to the constitution of secondary polyploids (p. 261).

Fig. 30. **Secondary Pairing.**—Polar views of metaphase. A, in an apple, a secondary polyploid with a haploid number of 17 made up of two sets of 7 (the usual basic number in Rosaceae) with three additional chromosomes (after Darlington). B, in *Brassica oleracea*, a secondary polyploid with haploid number 9 made up of a basic set of 6 plus 3 extra chromosomes. Each dot represents two closely associated homologous chromosomes.

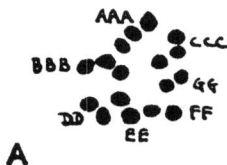

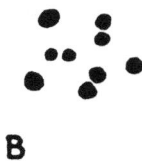

(After Catcheside.)

[1] Catcheside 1937.

CHAPTER 3

The Behaviour of Individual Chromosomes

A. SEX CHROMOSOMES[1]

The mating of a male and a female organism produces males and females in approximately equal numbers; in this equality of number it is analogous to the cross between a heterozygote and homozygous recessive. Correns[2] first showed that the analogy is more than superficial. In Bryonia the male is the heterozygous form and produces two sorts of pollen, a male-producing and a female-producing, while the eggs are all of one kind. Similar sex-determining mechanisms have since been found in most organisms. The action of the sex-determining factors is discussed in Chap. 10.

In 1902 McClung first discovered chromosomes which formed an unequal pair in one sex, while corresponding to them in the other sex there was an equal pair. These chromosomes therefore behave like the postulated sex factors and were called the sex chromosomes, the remaining chromosomes being called the autosomes. They have since been found in very many animals and plants. The sex with the unequal pair may be either the male (more usually) or the female (Lepidoptera, birds, some fishes, Fragaria as the only known plant); it is known as the heterogametic sex, and is the sex which is heterozygous.

1. *Types of Sex Chromosomes*

The pair of equal chromosomes are known as the XX pair, the unequal pair as the XY; in cases where the female is the heterogametic sex the nomenclature $ZZ\ \male\ WZ\ \female$ is sometimes used, but is probably unnecessary. The sex controlling mechanism depends originally on the difference between the X and Y chromosomes, but in the course of evolution the Y often loses its importance and the sex determination comes to depend solely on a balance between the X and the autosomes.

The Y chromosome may be bigger than the X, but is more usually smaller. A complete series of types can be found, ending with types in which the Y is totally absent and the sex determination depends on an

---

[1] *General references:* Darlington 1937, chap. 9, Schrader 1928, Wilson 1928.
[2] Correns 1907.

XX-XO chromosome mechanism. No cases are known of the opposite phenomenon, the disappearance of the *X*, which would leave sex determination dependent on the presence of a single supernumerary chromosome in one sex (*OO-OY* mechanism).

Either the *X* or the *Y* may become compound either by a reciprocal translocation with an autosome, or by fragmentation, which may perhaps always be a result of translocation (p. 95). A translocation may produce ring formation or chain formation of chromosomes in meiosis analogous to the phenomena found in Oenothera. The *XX-XY* mechanism may sometimes be secondarily produced from an *XX-XO* type by fusion of the *X*'s with an autosomal pair, giving an equal *A + X* pair in one sex and an unequal *A + X,A* pair in the other.

Fig. 31. **Some Types of Sex Chromosomes.**—The chromosomes are shown in side view of meiotic metaphase, the *Y* being above. *A* equal *X* and *Y* (e.g. many Hemiptera), *B* and *C* smaller *Y* (e.g. rat), *D* larger *Y* (e.g. Drosophila), *E* compound *Y* (e.g. Humulus), *F* compound *X* (e.g. Tenodera), *G* compound *X* (Blaps), *H* fusion of *X* to autosome (e.g. Mermiria).

(After Wilson, Darlington, etc.)

This state of affairs can sometimes be recognized by the occurrence of precocious condensation (heteropycnosis) of the *X*-chromosome part in the prophase of meiosis. In other cases the evidence is more indirect; e.g. *Drosophila melanogaster* has a rod-shaped *X* and *V*-shaped autosomes, *D. willistoni* has a *V*-shaped *X* (and one pair of rod-shaped and one of *V*-shaped autosomes) and part of the *X* is homologous, as is shown by comparison of the linkage maps with one of the autosomes of *melanogaster*.[1]

2. *Cytological Behaviour of the Sex Chromosomes*

The peculiarities in the cytological behaviour of the sex chromosomes can be regarded as consequences of, or adaptations to, the fact that the members of the *XY* pair are unlike and must be segregated by some mechanism which does not allow their differences to be annulled by crossing-over. The simplest case is of course that of the *XX-XO* mechanism, where in meiosis the isolated chromosome of the heterogametic sex either (1) passes undivided to one pole in the first division,

[1] Cf. Morgan, Bridges and Sturtevant 1925.

usually either before, or after, but not synchronously with, the autosomes, and divides equationally in two at the second division, or (2)
divides equationally at the first division and passes to one or other pole
at the second.

In *XY* organisms pairing between the *X* and *Y* is very variable in
intensity, probably correlated with the length of chromosome through
which the sex differential is spread, and which must therefore be
guarded from crossing-over. At one end of the series, in organisms with
diffuse sex differentials, e.g. Lygaeus, the chromosomes divide equationally in the first division, and pair transitorily before segregation in

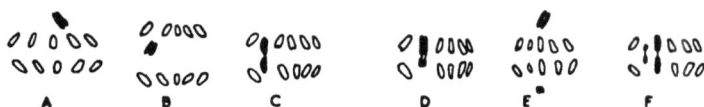

Fig. 32. **Segregation of Sex Chromosomes.**—The sex chromosomes are drawn
in black; the diagrams represent anaphases of the first division of meiosis seen in
side view. A, unpaired *X* passing to one pole before the rest of the set (e.g.
Stenobothrus). B, unpaired *X* passing to one pole after the rest of the set (e.g.
Aphis). C, unpaired *X* split into two chromatids, one of which goes to each pole
at the first division (equational separation) (e.g. Plotinus). D, unequal *XY* segregating with the other chromosomes (many animals). E, unequal *XY* segregating
before the other chromosomes (e.g. man). F, unequal *XY* separating equationally
at first division (e.g. Lygaeus).

(Modified from Darlington.)

the second division. At the other end, are organisms in which the sex
differential is localized at one point and pairing of the *X* and *Y* at the
prophase of the first division is apparently normal, and crossing-over
occurs between them (Lebistes). Drosophila occupies an intermediate
position, since pairing occurs, but there is less affinity between an *X*
and a *Y* than between two *X*s in *XXY* flies. This differential affinity
is due to the fact that only part of the *X* possesses a homologue in the
Y; the limits of the homologous section can be determined by studying
the pairing of *XY* flies containing small reduplicated sections of the *X*.
Within the homologous segment of the *X* and *Y* pairing occurs and
crossing-over takes place, as is necessary if the chromosomes are to
remain associated at metaphase (p. 121). This causes no difficulty unless,
as is the case in *D. melanogaster*, the pairing segment lies between two
sex-determining segments which might get separated by crossing-over
and thus lead to a disturbance of the sex-determining mechanism. In
such a case the crossing-over is probably by two reciprocal chiasmata,
both on the same side of the centromere, so as to preserve the relation

between the centromere and the ends of the chromosomes.[1] It has little observable consequence, as the homologous parts of the X and Y are nearly empty of active genes. This part of the chromosomes is therefore called the inert part of the X; it is made of heterochromatin (p. 40) and shows the characteristic differential condensation. Such condensation or heteropycnosis is in fact a fairly common property of sex chromosomes, and, as was mentioned above, may be found in XX-XO or derived mechanisms, where there is no apparent need for an inert

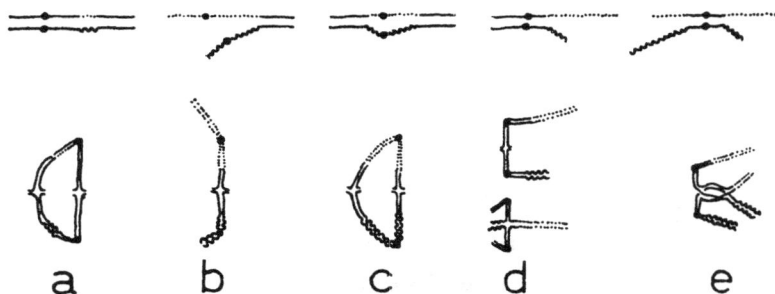

Fig. 33. **Homologous and Differential Parts of Sex Chromosomes.**—The homologous parts of the sex chromosomes are represented by simple lines, the differential, sex-determining parts by dotted lines (for X) and wavy lines (for Y). Centromeres black. The upper row shows the zygotene pairing, the lower the metaphase configurations. *A* a short differential segment in one arm (e.g. Lebistes). *B* pairing segments at one end (e.g. Melandrium). *C* pairing segments at both ends (e.g. Phragmatobia). *D* differential segments at one end, two alternate metaphase configurations (e.g. Mammalia). *E* differential segments at both ends (e.g. Drosophila).
(After Darlington.)

pairing part of the chromosome differentiated from the sex determining part.

The necessity to prevent the mingling of the sex-determining regions of the X and Y chromosomes, which is expressed in the development of inert pairing regions in forms such as Drosophila, is probably also responsible for the generally observed reduction in frequency of crossing-over in the heterogametic sex.[2] In Drosophila and other Diptera this goes so far as a complete suppression of crossing-over of all chromosomes in the male. This is correlated with the development of a completely abnormal mechanism of pairing and segregation (p. 123), a development which has brought about the peculiar paradox that in the male Drosophila the only chromosomes associated by crossing-over (chiasmata) in the ordinary way are the XY pair, although it

[1] Phillip 1935.   [2] Haldane 1922, Huxley 1929, Eloff 1932.

is because of the injurious effects of crossing-over in the differential segments of the X and Y that crossing-over has had to be suppressed in all the other chromosomes of the complement.

3. Sex Chromosomes in Polyploids

Polyploids arising by the doubling of all the chromosomes of a diploid set will be either XXXX or $XXYY$. These are probably sexually normal females and males in most organisms. The tetraploid XXXX female form is known in Drosophila but not the corresponding male. *Vallisneria spiralis*[1] is diploid and dioecious, and *V. gigantea*, which is tetraploid and may be derived from it, has two separate sexes which were probably formed by simple doubling occurring on two separate occasions, once in a male and once in a female. Probably, however, new polyploid forms have a low chance of survival unless they are hermaphrodites and can be self-fertilized (unless they are propagated vegetatively).[2] The sex chromosomes will probably pair in equal pairs XX,XX or XX,YY and differentiated sex chromosomes will therefore be difficult to detect. In fact, the only case known of sex chromosomes in a polyploid is in *Fragaria elatior*,[3] a hexaploid, where it is probable that sex differentiation has been evolved *de novo* subsequent to the evolution of the species; the female sex is heterogametic with a simple XY pair, although all other plants have the male sex heterogametic.

4. Sex-Linked Inheritance

Some characters appear to be associated in inheritance with sex. This association is of two kinds, (1) sex-limited inheritance, which is shown by some genes which have an effect in one sex only, because of some physiological connection with sex differentiation (p. 162); (2) sex-linked inheritance, where the association is due to the gene in question lying in the X or the Y chromosome.

A characteristic phenomenon of inheritance of genes which lie in the X chromosome is a criss-cross heredity by which, assuming male heterogamety, in the F1 from a cross between a homozygous recessive female and the dominant male, the sons show the recessive character from their mother and the daughters the dominant from their father. A similar phenomenon will occur, *mutatis mutandis*, in cases of female heterogamety, where the phenomenon was actually first observed.[4]

---

[1] Jørgensen 1927.   [2] Cf. Muller 1925.
[3] Kihara 1930.   [4] Doncaster 1908.

Factors in the Y will remain associated with the heterogametic sex, descending simply in the male line in male heterogamety, and may simulate sex limited inheritance unless abnormal organisms such as XO males or XXY females can be found which allow the Y chromosome factors to be dissociated from the sex-controlling mechanism. Examples of Y chromosome genes are comparatively rare, since the Y usually consists nearly entirely of inert heterochromatin. They are found chiefly in organisms in which the genetic sex-determining mechanism is not highly developed, e.g. in some fish (Lebistes),[1] and in these forms may show a fairly high percentage of crossing-over into the X. In organisms with more highly evolved sex chromosomes,

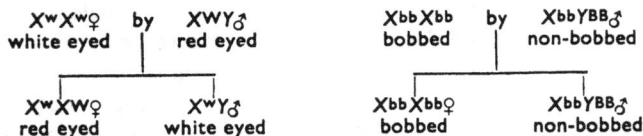

Fig. 34. **Sex Linkage in the X and Y.**—In cases of female homogamety, a daughter must get one of her X chromosomes from her father, a son must get his single one from his mother; thus if a recessive female is crossed to a dominant male, the sons are like their mother and the daughters like their father (on left, X^w = recessive white factor in X, etc.). Factors in the Y are handed on from father to son (on right, bb = bobbed).

between which crossing-over is more completely suppressed, few Y chromosome factors are known. In *Drosophila melanogaster* the bobbed locus, which causes short bristles, is the best known. There are also two regions of the Y which are both necessary for fertility in males; they probably contain factor complexes.

Crossing-over between the X and Y gives rise to partial sex-linkage, a phenomenon which is best seen in fish. It has also been shown to occur in *Drosophila* for the locus of bobbed, which may cross over from the Y into the X if a chiasma occurs farther than usual away from the centromere. Partial sex-linkage has recently been detected in man (p. 332).

B. NON-DISJUNCTION

Two chromosomes associated at meiotic metaphase may fail to separate ("disjoin") in the anaphase and may therefore pass to the same pole, and, dividing equationally in the second division, they may

[1] Cf. Winge 1923, 1927, 1931.

be included in the same germ cell. This phenomenon, which gives rise to germ cells containing one sort of chromosome represented twice over, is known as non-disjunction, and the name has now been used to cover other processes which lead to the same result.

1. *Non-disjunction of X*

Non-disjunction of the X leads to a reversal of the usual criss-cross sex-linked inheritance. In a type with female homogamety, *Drosophila* for example,[1] the non-disjunctional XX eggs with the extra chromosome

Fig. 35. Secondary Non-disjunction.—Start with a female, derived from non-disjunction, having the constitution XXY, and bearing the factor w for white eyes in both X chromosomes. The proportions of the different types of pairing are inferred from the observed ratios of offspring.

(After Bridges.)

		wwY♀		
Pairing	84 % ww Y		16 % wY w	
Segregation	84 % wY and w		8 % ww and Y 8 % wY and w	
Eggs	46 % wY	46 % w	4 % ww	4 % Y
Crossed with WY♂	23 % WwY red ♀	23 % Ww red ♀	2 % Www red super ♀ (dies)	2 % WY red ♂ (exception)
	23 % wYY white ♂	23 % wY white ♂	2 % wwY white ♀ (exception)	2 % YY (dies)

when fertilized by X sperm give super-females (p. 221) most of which fail to survive, and, with Y sperm, sexually normal females which show the recessive characters of their mothers and are thus exceptions to the general rule of criss-cross sex-linked inheritance. Similarly, the eggs lacking an X chromosome (O eggs) give with X sperm exceptional males showing their father's characters and with Y sperm inviable zygotes which fail to develop.

The non-disjoining XX pair probably lags behind the other chromosomes at anaphase and often fails to be included in either daughter nucleus, so that more O eggs than XX are formed and exceptional males are found more frequently than exceptional females.

The discovery of exceptional males and females of this kind was made by Bridges, who proposed the explanation depending on the occurrence of non-disjunction. This was confirmed by the prediction of the types to be expected from the breeding of the exceptional females

[1] Bridges 1914, 1916.

(secondary non-disjunction). The proportions in which the various types should occur could not be predicted theoretically since it depends on the differential affinity of the X and Y as expressed in the relative frequency of XY and XX pairing. The qualitative prediction of the types of the F1 was however a sufficiently striking proof of the correctness of the theory, which was further confirmed by cytological examination and demonstration of the predicted XXY and XO individuals. This was the first example of the successful prediction from genetical data that a definite cytological abnormality had occurred and was very important in persuading scientists that the chromosome theory of heredity was justified.

Primary non-disjunction of X occurs spontaneously in *D. melanogaster* with a frequency of about 1 in 2,000 and can be increased by selection, and some environmental agents such as X-rays, ammonia vapour, etc. L. V. Morgan[1] obtained a stock showing nearly 100 per cent primary non-disjunction, which turned out to have the two X chromosomes attached to each other near the attachment constriction. This is the famous attached-X stock, much used in work on mutation frequency (p. 382). Apparent non-disjunction of the X and Y sometimes occurs in males, both chromosomes being included in the same gamete. This is more probably due to a failure of pairing and chance assortment to the same pole, than to a failure of disjunction occurring after normal pairing. Non-disjunction of sex chromosomes is known in other organisms,[2] and has sometimes been invoked to explain abnormal inheritance in man.[3]

2. Non-disjunction of other Chromosomes in Drosophila

Non-disjunction of the autosomes in *Drosophila* would give types with one too many or one too few of the IInd, IIIrd, or IVth chromosomes. Of these only the types involving the IVth chromosome are viable.

The type with only one IVth chromosome, called haplo-IV, was discovered by chance.[4] A fly was found which appeared to contain a factor with a dominant phenotypic effect (smaller body, lighter colour, smaller bristles) which was named Diminished; the factor was lethal when homozygous. It showed free recombination with factors in the linkage groups of the X, IInd and IIIrd chromosomes. Crossed with flies homozygous for the fourth group recessive factors eyeless, shaven, and bent, a few Diminished flies appeared in the F1 and not only

---

[1] Morgan 1922.
[3] Gowen 1933, Haldane 1932*c*.
[2] E.g. fowl, cf. Crew 1933.
[4] Bridges 1921.

showed the characters of eyeless, shaven, and bent, for which they should have been heterozygous, but even showed them more strongly than they normally appear in the homozygote. This "pseudo-dominance" and exaggeration is characteristic of flies which contain only one dose of hypomorphic genes (p. 165), and since all the factors in the IVth show pseudo-dominance, all of them must be present in only one dose, that is, there must be only one IVth chromosome. Cytological examination confirmed this. "Diminished" is in fact haplo-IV.

The flies with three IVth chromosomes (triplo IV) which should

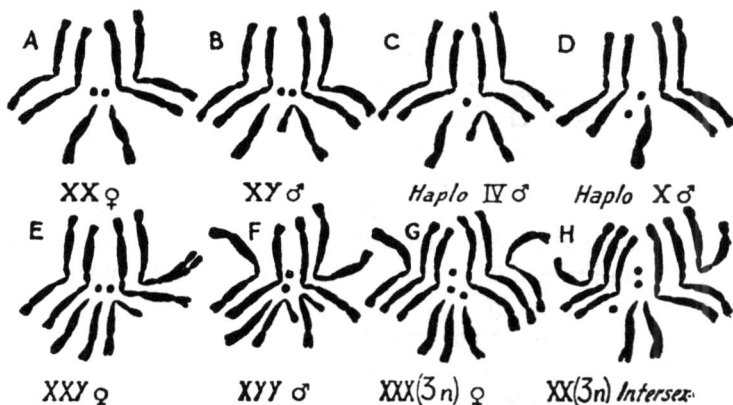

Fig. 36. **Chromosome Complements in** *Drosophila melanogaster.*
(From Darlington, after Bridges.)

result from fertilization of non-disjunctional IV.IV eggs were searched for as likely to show the opposite characters to haplo-IV flies and were eventually found in the progeny of triploids. Tetra-IV flies have also been obtained. These hypo- and hyper-ploid variations allow one to compare the effects of different dosages of factors in the IVth chromo-some (p. 165).

3. *Non-disjunction in Other Organisms*

Organisms with one chromosome represented three times are usually known as trisomics and are found in many species, sometimes as a result of primary non-disjunction, sometimes of secondary non-disjunction in polyploid variants. The most complete series are in plants. In *Datura stramonium* the haploid number of chromosomes is

12 and all 12 trisomic types are known.[1] In the tomato, also, with a haploid number of 12, 11 primary trisomics have been described,[2] and many other nearly as complete series are known, e.g. in *Matthiola*, *Crepis*, *Zea*. The simplest reduplication of a chromosome gives a primary trisomic; reduplications combined with various types of fragmentation and interchange give the secondary and tertiary types (p. 108). The phenotypic appearance of each trisomic is characteristic and shows the effect of an extra dose of the assemblage of genes in the reduplicated chromosomes (p. 170).

4. *Segregation in Trisomics: Secondary Non-disjunction*

The three similar chromosomes in a trisomic may all be associated as a trivalent at meiotic metaphase, as is usually the case in *Datura*, or in a certain percentage of cases the association may be only of two members, the other chromosome being free. In this respect the trisomic behaves exactly like a triploid, and follows the general rule for polyploid pairing (p. 68). Segregation will give various proportions of gametes with and without a duplicated chromosome. The gamete ratio for a factor A present "simplex" Aaa will be $1A : 2Aa : 2a : 1aa$ if there is no preferential pairing; the ratio may also be modified by chromatid segregation (p. 103).

The zygotic ratios are not immediately calculable from the gametic ratios because the presence of the extra chromosome usually has a very depressing effect on the viability both of the gametes and of the zygotes in which it is contained. In animals the effect is mainly on the viability of the trisomic zygotes, the gametes being apparently little affected, but in plants the trisomic chromosomes, although transmitted almost in the expected frequency through the ovules, may be eliminated nearly completely in the pollen, which even if viable may grow too slowly to be effective in competition with pollen tubes containing normal nuclei.

5. *Chromosome Mosaics*

Non-disjunction in the strict sense can only take place at meiosis, but somewhat similar abnormalities of chromosome separation can occur in mitoses, one of the sister halves of a chromosome failing to move away from the metaphase plate and thus not being included in either of the daughter nuclei. This elimination of a chromosome gives rise to a patch of tissue which has less than the normal quantity of (is

[1] *Rev.* Blakeslee 1928, 1934.　　　　　　　　[2] Lesley 1928, 1932.

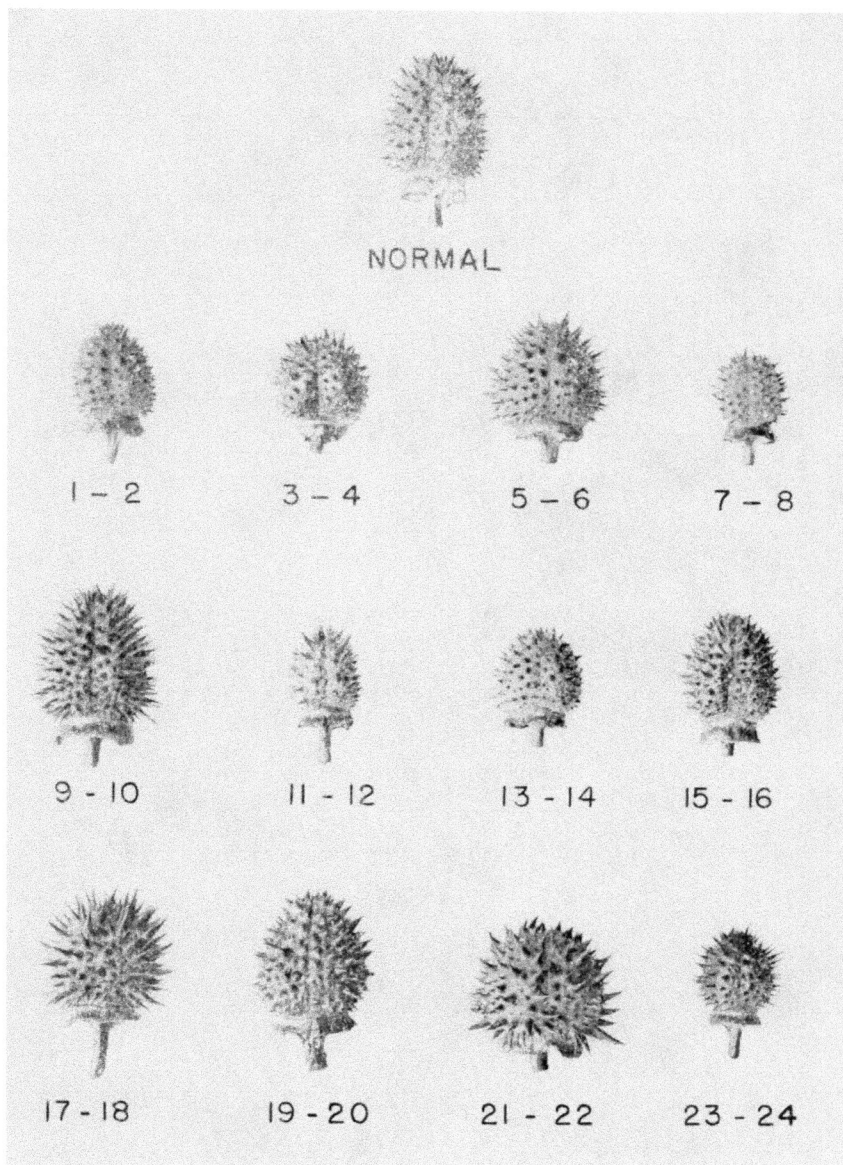

Plate 2. **Capsules of normal and trisomic Daturas.**—Above is the capsule of a normal diploid plant. Below are those from the twelve possible types of trisomic; the numbers are those conventionally used to designate the ends of the chromosomes, the capsule labelled 1–2, for instance, having three of the chromosomes whose ends are known as 1 and 2. (Courtesy of A. F. Blakeslee.)

"hypoploid for") factors contained in the chromosome which has been lost. If the lost chromosome contained dominants masking the presence of the corresponding recessives in the sister chromosome, these recessives will now be able to show, perhaps with exaggeration.

Mosaics in which an X chromosome has been eliminated from a female show male characters in the parts from which the second X is missing. They are known as gynandromorphs.[1] The eliminations seem to be most usually in one of the earliest segmentation-divisions of the egg, often at the first, and gives rise to a gynandromorph with approximately one half of the body male and one half female.[2] Animals mosaic for sex or other characters are known in birds and mammals[3] as well as insects. By bringing tissues of genetically different constitutions into intimate contact with one another they provide opportunities for investigating the mutual interaction of the tissues during development, a subject which will be discussed more fully later (p. 177). Here it is sufficient to remark that in insects the remarkable thing is not the interaction of the different tissues but their lack of interaction, each cell developing the characters corresponding to its own nucleus with little reference to the genetic constitution of its surroundings. For instance, gynandromorphs in the parasitic wasp Habrobracon, with a head of one sex and gonads of another, are said to behave purely according to the type of their cephalic ganglion;[4] but this praiseworthy effort to make the sex life subservient to the dictates of reason seems to be even less likely than usual to succeed in this case.

If the mosaic is of such a kind that each cell lacking the eliminated chromosome can be recognized, it may enable one to distinguish a certain group of cells as necessarily derived from a single cell in which the elimination took place. This would make it possible to trace the cell lineages of insect development, which would be a considerable refinement of the description of insect development as it is known at present. So far only the explanation of gynandromorphs in Drosophila has been fully worked out on these lines[5] but factors have been found which give a high percentage of chromosome elimination at late stages of development,[6] and the resulting small mosaic patches may enable cell lineages in insects to be described in greater detail.[7] In some plants ring-shaped chromosomes are known which suffer frequent

---

[1] Cf. Morgan and Bridges 1919.   [2] Cf. Sturtevant 1929b.
[3] E.g. mouse, see Dunn 1934.   [4] Whiting 1932a.   [5] Parks 1936.
[6] It is probable that these should be re-interpreted as due to somatic crossing-over (p. 373).   [7] Bridges 1925b, Noujdin 1936, Sturtevant 1929b.

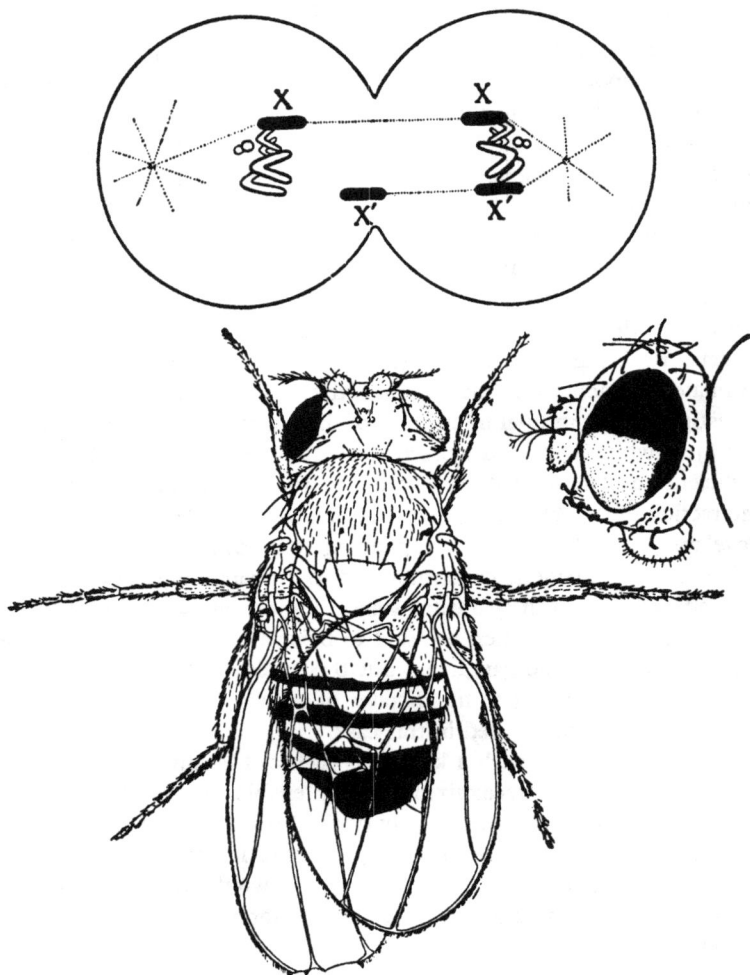

Fig. 37. Gynandromorphs.—On the left above is a diagram of the first cleavage of a female egg, showing one of the X chromosomes being eliminated by lagging on the equator of the spindle. Below is the resulting gynandromorph, whose left side is female (XX) and shows the dominant character Notch, while the right side is male (XO), having lost the Notch X-chromosome and showing the recessives scute (no bristles on scutellum), broad (wing), echinus (rough eye), ruby (eye colour), and tan (body colour) from its remaining X.

(From Morgan, Bridges, and Sturtevant.)

On the right is the head of a fly from which an X-chromosome has been eliminated from a late somatic mitosis. The mosaic spot shows the eye colour eosin and singed bristles, for which the fly as a whole was heterozygous.

(From Bridges.)

somatic elimination and thus give rise to mosaic patches.[1] Some mosaics in insects have arisen in a way other than by chromosome elimination. Doncaster showed that in Lepidoptera there are gynandromorphs which are formed following fertilization of binucleate eggs (i.e. eggs in which the second polar body is retained). IInd and IIIrd chromosome mosaics in Drosophila are probably not formed by elimination, though IVth chromosome mosaics are. Stern[2] has described a case of a mosaic found in the F_I of a cross between Stubble (III) and Haplo-IV Curly (II); the mosaic was haplo-IV $+$ ^{Cy}Sb in the head and left part of the

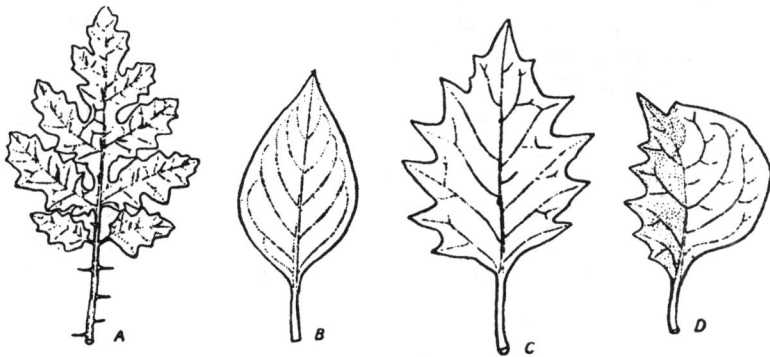

Fig. 38. **Plant Chimaeras.**—A leaf of *Solanum sisymbrifolium*, B of *S. nigrum*, C a periclinal chimaera with an interior of *S. nigrum* above which is one layer of tissue from *S. sisymbrifolium*. D a partial periclinal ("mericlinal") in which the superficial layer of *S. sisymbrifolium* tissue is present only in the shaded region. Note the interaction of the tissues in determining the shape of the leaf.

(From Jørgensen and Crane.)

thorax and haplo-IV $+$ ^{Sb}Cy in the right part of the thorax and abdomen. The first constitution must have originated by the fertilization of an Sb egg-nucleus by $+$ ^{Cy} haplo-IV sperm, the second from $+$ ^{Sb} egg-nucleus by haplo-IV Cy sperm. There must in fact have been double fertilization of a binucleate egg.

Mosaics due to somatic elimination of chromosomes are also found in plants (Crepis, etc.). The most interesting type of plant mosaic arises in a different way and is spoken of as a graft hybrid or plant chimaera.[3] If a graft is made of a shoot of one species on to a stock of another, branches may be obtained which consist either of longitudinal sectors of the two different species (sectorial chimaeras) or of a core of

[1] McClintock 1932.   [2] Stern and Sekiguti 1931.
[3] *Revs.* Jones 1935, Weiss 1930, cf. Jørgensen and Crane 1927.

one species covered by a layer, which may be one, two, or more cells thick, of the other species (periclinal chimaeras).

Similar mosaic parts may be produced by a somatic elimination or a mutation (factorial or chromosomal) in a growing point, and consist in such cases not of mosaics between different species, but between different mutant types of the same species. Thus we may get (e.g. from callus tissue formed on the cut back stems of Solanum) periclinal chimaeras with a core of tetraploid tissue in a skin of normal diploid. The most famous of the inter-specific hybrid chimaeras are perhaps the Bronvaux Medlars, whose nature was for a long time in doubt but which are probably periclinal chimaeras with one or more layers of *Crataegus mespilus* over a core of *C. monogyna*; another is *Cytisus adami* which has a skin of *C. purpurus* and a core of *C. laburnum*. The morphological characters of such hybrids usually show the effects of both sorts of tissue, giving the impression that each specific type has retained its own characteristics in so far as the mechanical conditions imposed by the neighbouring tissue will permit. A very common type of chimaera is between normal tissues and tissues (of the same species) lacking chlorophyll; many green and white variegated plants are of this type.

CHAPTER 4

The Linear Differentiation of the Chromosomes

A. CROSSING-OVER[1]

1. *Cross-over Maps*

The fact on which the building up of a cross-over map is based has been mentioned in Chap. 1; the additive theorem of linkage states that if A, B, and C are three closely linked factors the linkage value between A and C is the sum of the values between A and B, and B and C. A, B and C can therefore be taken to lie at definite points along the length of a chromosome, the distances between any two being proportional to the linkage value between these two. A chromosome may therefore be represented by a simple straight line with the position of the genes marked on it according to a scale in which unit distance represents one crossing over per 100 gametes (the "Morgan," a name not in general use).

The additive theorem of linkage is only true when the linkage values are small. The number of recombinations between two genes which lie far apart on the chromosome is less than the sum of the linkage values for the separate intervals into which the intervening part of the chromosome can be divided. The explanation of this can be found in double crossing-over; two simultaneous cross-overs, one in the interval AB and the other in the interval BC, will remove B from between A and C but leave A and C in their original association. The total number of recombinations between A and C should be compensated for this according to the formula $AC = AB + BC - 2AB.BC$, since the chance of a cross-over simultaneously in AB and BC is clearly $AB.BC$, and each double cross-over must be counted twice.

This theoretical compensation is not exactly correct in practice, since it is found that the occurrence of a cross-over at one place lessens the probability that another will happen in the immediate neighbourhood (interference).[2] There are therefore less than $AB.BC$ double cross-overs and the formula becomes $AC = AB + BC - 2AB.BC.I$, where I (the coincidence) is an empirically determined constant, usually less than 1. The coincidence varies considerably from place to place within

---

[1] *General reference:* Mather 1938.   [2] Muller 1916.

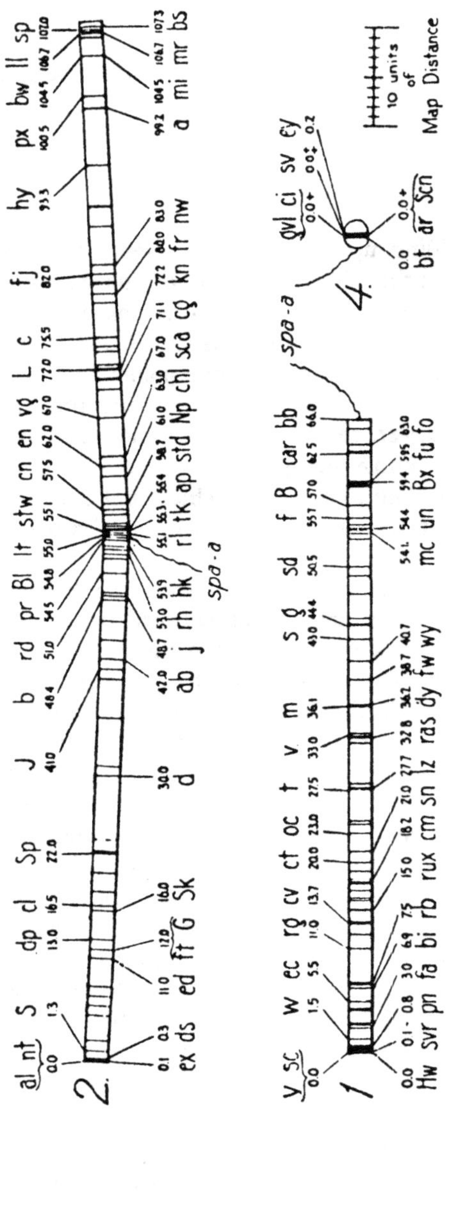

Fig. 39. **Cross-over Maps of *Drosophila melanogaster*.**—The maps are drawn so as to suggest the actual shapes of the chromosomes: V-shaped for the 2nd and 3rd, rod-shaped for the X and round for the small 4th. But note that the distances are calculated from cross-over data, not from the actual sizes of the chromosomes. *spa-a* = spindle attachment or centromere.

(From Bridges.)

a chromosome and in different chromosomes (and according to the methods adopted by different people for calculating it!).[1] There is no interference between cross-overs on different sides of the centromere in the V-shaped autosomes of Drosophila. For a cytological discussion of interference, see p. 125. By adding to the linkage or recombination values the number of breakages which, because of double crossing-over, do not result in recombination, we can obtain the total frequency of cross-over between two factors. These values, which can only be arrived at by calculation from the crude recombination data, are known as cross-over values. A cross-over map represents the chromosome as a straight line on which the genes are placed at distances apart propor-

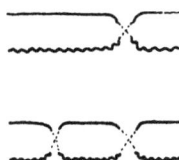

Fig. 40. **Diagram of Single** (above) and **Double** (below) **Cross-overs.**

tional to the cross-over values. For this reason the cross-over value between two genes is also known as the map-distance.

The most complete cross-over maps are those of Bridges for *D. melanogaster*, but considerable numbers of factors have been discovered and mapped in other Drosophila species.[2] The ten chromosomes of maize have also been fairly well filled with genes[3] and less complete maps can be prepared for the sweet-pea,[4] locusts,[5] etc.

The Drosophila maps show a moderately even spacing of genes throughout the chromosomes, but even here there are regions in which the genes appear to be unduly concentrated. These regions are the distal end of the *X*, that is the end farthest from the centromere, and the middle of the large autosomes immediately in the neighbourhood of the centromere, with less marked concentrations at the proximal end of the *X* and the ends of the arm of the autosomes. (In the small IVth chromosome, true cross-over is only doubtfully recorded, the few recombinations that occur may be due to mutation.) The points of crowding in the cross-over maps are much more marked in locusts.

The crowding of genes at particular parts of the map is the same as comparative lack of cross-over, and is a function of the position, since

[1] Cf. Anderson and Rhoades 1931, Stevens 1936.
[2] Cf. Morgan 1926, D. I. S. brochures *passim*.
[3] Emerson, Beadle and Frazer 1935.   [4] Punnett 1927.   [5] Nabours 1929.

parts of chromosome with an average frequency of cross-over have it reduced if translocated to a position with low cross-over.[1]

Maps Prepared on other Bases

The map based on the cross-over values in the triploid shows the genes arranged in the same order as in the diploid but with different intervals.[2] This modification of the ordinary cross-over values is probably due to varying amounts of pairing in prophase; in a triploid only two threads are paired at any one point, the third chromosome being left out of association at that part but pairing in some other region with one of the sister chromosomes (p. 69). Crossing-over is only possible in the paired regions, and the percentage of crossing-over between two chromosomes in a region is therefore dependent on the occurrence of pairing between these two out of the three chromosomes.

Muller[3] has prepared a map based on mutation frequency, arguing that the chance of a mutation is dependent only on the number of genes in a section of chromosome and that this number may be directly proportional to the length of chromosome; this involves the further assumption that the average mutation rate of the genes in a moderately long section of chromosome is constant (cf. p. 380). Maps prepared on this basis are slightly modified according as one considers mutations to different allelomorphs of the same locus as being different, or whether one simply measures the distance between two genes by the number of loci known between them. The fullest and most recent maps of this kind are based on the latter consideration and have been given by Schweitzer.[4]

3. Genetic and Environmental Effects on Crossing-over

The amount of crossing-over is under genetic control. This is shown most strikingly by the complete suppression of crossing-over in males of Diptera and its general lowering in the heterogametic sex. Its suppression in male Drosophila has been shown to depend on the development of an abnormal method of chromosome pairing in meiosis. Genes are also known which suppress or lower the frequency of crossing-over in all the chromosomes of the fly which contained them.[5] These factors seem also to perform the suppression by some action on chromosome

---

[1] Offermann, Stone and Muller 1931.   [2] Cf. Sansome and Philp 1932.
[3] Muller and Painter 1932.   [4] Schweitzer 1935.
[5] Gowen and Gowen 1922, 1933a; for a similar gene in maize, see Beadle 1933.

pairing, since they reduce the fertility of the flies; in some cases the effect is confined to females, in others the fertility of males is also affected.

Many factors are known which have a local effect in reducing crossing-over (so-called C factors).[1] These crossing-over suppressors are extremely useful in preparing flies of a required constitution. They are also used to carry on stocks of mutations which are lethal when homozygous by means of the balanced lethal technique; if one has a different recessive lethal in each of a pair of chromosomes, and suppresses all crossing-over, the only viable type is the heterozygote which therefore breeds true.[2]

The C factors are probably all inversions and probably obtain their effect by reducing zygotene pairing. The pairing in inverted loops

Fig. 41. A Balanced Lethal System.—Beaded *Bd* is a gene in the 3rd chromosome of *D. melanogaster*, which has a dominant effect on the wing margins and is lethal when homozygous. A stock was found in which one 3rd carried *Bd* and the other carried a recessive lethal *l3a* and a cross-over suppressor (? inversion) *C3a*. In this stock only the double heterozygote could survive and it therefore bred true, thus:

Bd/l3a, C3a   (selfed)

Offspring	1 BdBd	2 Bd/l3a C3a	1 l3a C3a/l3a C3a
	(dies)		(dies)

(p. 100) which inversions show in salivary gland chromosomes is probably very incompletely carried out in the short time available during prophase, though pairing may be more nearly complete in flies heterozygous for very long inversions and crossing-over may then be more nearly normal. Some of the cross-over gametes, however, will probably be inviable (p. 131).

Crossing-over is also affected by environmental conditions. It varies with the temperature[3] and in Drosophila with the age of the fly.[4] In neither case is the curve of variation simple, and the effect is also not uniform in different parts of the chromosome, the greatest effect always being found in the region of the centromere. The effect of temperature is paralleled by its effect on chiasma frequency.[5] X-irradiation also has an effect on crossing-over, again most markedly in the centromere region.[6] Crossing-over can indeed be induced in Drosophila males by irradiation, at least near the centromere in the autosomes.[7] It is not easy

[1] Sturtevant 1926a.   [2] Muller 1918.   [3] Plough 1917.
[4] Bridges 1927, 1929.   [5] White 1934.
[6] Mavor 1923, Muller 1925a.   [7] Friesen 1936a, b.

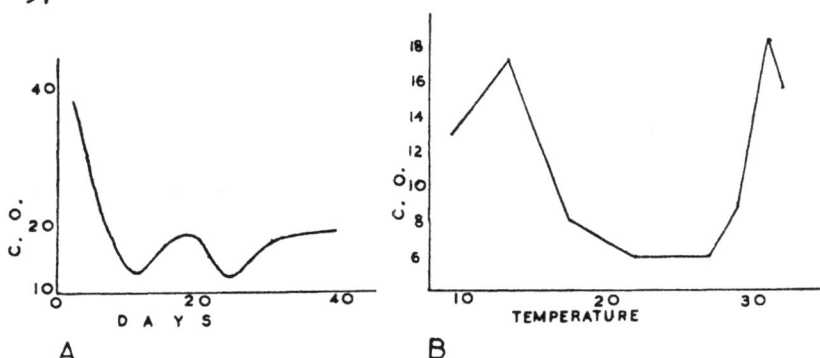

Fig. 42. Variation of Crossing-over.—A in chromosome III with age of female B in chromosome II with temperature,[1] in *Drosophila melanogaster*.
(After Bridges, and Plough.)

to reconcile this with the peculiar method of pairing of the autosomes in the male, and it is probable that the crossing-over is of an abnormal type and occurs at mitosis some time before meiosis (p. 373).[2]

4. *The Physical Reality of the Cross-over Map*

The demonstration of the physical reality of the cross-over map came from the discovery of cytologically observable alterations in the chromosomes which could be correlated with alterations in the position of definite sections of the cross-over map.

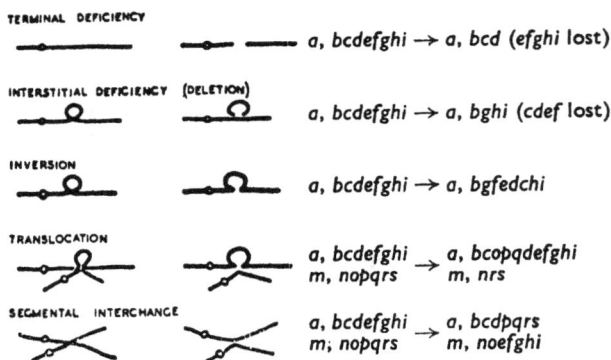

TERMINAL DEFICIENCY
$a, bcdefghi \rightarrow a, bcd$ (*efghi* lost)

INTERSTITIAL DEFICIENCY (DELETION)
$a, bcdefghi \rightarrow a, bghi$ (*cdef* lost)

INVERSION
$a, bcdefghi \rightarrow a, bgfedchi$

TRANSLOCATION
$a, bcdefghi \rightarrow a, bcopqdefghi$
$m, nopqrs \quad m, nrs$

SEGMENTAL INTERCHANGE
$a, bcdefghi \rightarrow a, bcdpqrs$
$m, nopqrs \quad m, noefghi$

Fig. 43. **Chromosome Rearrangements.**—On the left are shown the chromosomes before breakage, in the middle after breakage and rejoining. On the right the nature of the rearrangement is indicated by changes in the sequence of letters, which represent the genes (the centromere is symbolized by a comma). It is doubtful whether terminal deficiencies ever actually occur (cf. p. 368).

[1] This curve requires some correction, cf. Smith 1936.   [2] Whittinghill 1937.

The chromosome alterations are of the following types:

1. Deficiencies, in which a part of the chromosome is lost. In the homozygous condition they usually act as lethals and they may have a dominant phenotypic effect in the heterozygote. In the heterozygote recessive genes in the partner to the missing section may show pseudo-dominance and exaggeration similar to that mentioned in connection with Haplo-IV. No crossing-over takes place in the deficient region, which shows, genetically, that this part of the cross-over map is absent, not merely inactivated.

2. Deletions.—This name is often used for large deficiencies of the central parts of the chromosome, both ends of which are still present. They do not differ in essentials from deficiencies, in which the ends of the chromosome are probably also always retained even if this does not at first appear to be the case.

3. Translocations in which part of one chromosome is broken off from its normal place and joined on somewhere else, either to the same or another chromosome. They are probably always interstitial, that is, never at the very end of the chromosome. Branched chromosomes from lateral translocations have been observed cytologically but their permanence is doubtful and they have not been analysed genetically.[1]

4. Duplications are products of breeding from translocations. The organism contains its normal set of chromosomes, attached to one of which is a translocated fragment which is therefore present in excess and is said to be duplicated. The presence of a dominant gene in a duplicated fragment frequently overcomes the presence of two recessive allelomorphs in the normal position, whence duplications may appear as gene-suppressors.[2]

5. Inversions are cases in which a section of the chromosome becomes reversed in order. They can be regarded as a particular type of translocation. They are detected chiefly by their effect on crossing-over, which they suppress.

6. Simple fragmentation probably never occurs; a breakage of a chromosome seems in fact always to be followed by a rejoining of broken ends. If it does occur it would not be permanent as one chromosome would lack a centromere and would therefore be lost at division.

7. Simple fusion of whole chromosomes does seem to be a possibility if they become united in the centromere region[3]—if they are

[1] Cf. Darlington 1937, Offermann 1936.
[2] Schultz and Bridges 1932.   [3] Painter and Stone 1935.

united elsewhere they may be pulled apart in division by the separation of the two centromeres to opposite poles.

The origin of these chromosome abnormalities is probably by breakage while the chromosomes are lying in random loops in contact with one another. Their production is greatly increased by treatment

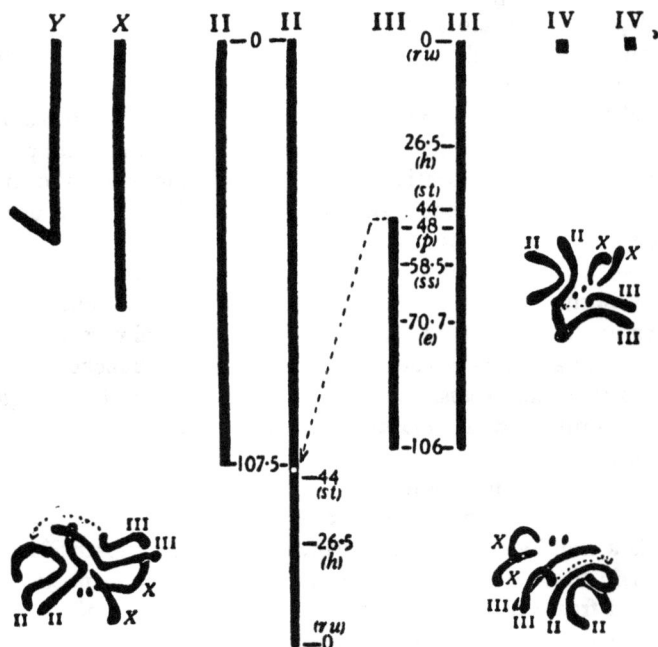

Fig. 44. Cytological and Genetical Evidence of a Translocation.—A translocation, known as the "Star Curly" translocation, appeared in X-rayed *D. melanogaster*. It was found by linkage studies that the "left" end of one 3rd chromosome had been broken between the loci of scarlet (*st*) and pink (*p*) and the fragment attached to the "right" end of a 2nd chromosome. The cytological findings were in accordance with this, as may be seen from the mitotic metaphases shown.

(From Painter and Muller.)

of the parent organisms with high frequency radiations (X rays, γ rays) or electrons (β rays).

Genetic analysis of a chromosome abnormality enables one to find what section of the chromosome map has been shifted, e.g. by determining which genes are suppressed by a duplication.[1] The first fact to notice is that a translocation or deletion affects a set of genes contained

[1] Bridges 1917, 1919a.

in a definite piece of the map, not a collection which is scattered haphazard, as might be expected if the map did not correspond with material reality. Further, cytological investigation often reveals an actual chromatin fragment in an unusual position, and this fragment can be correlated with that part of the map which genetic analysis shows to be affected. A chromosome map can thus be prepared, in which the position of the genes on the actual chromosome is marked.

The maps prepared by study of translocations in metaphase chromosomes in Drosophila show considerable differences from the cross-over maps.[1] In particular, regions where the genes are crowded together in

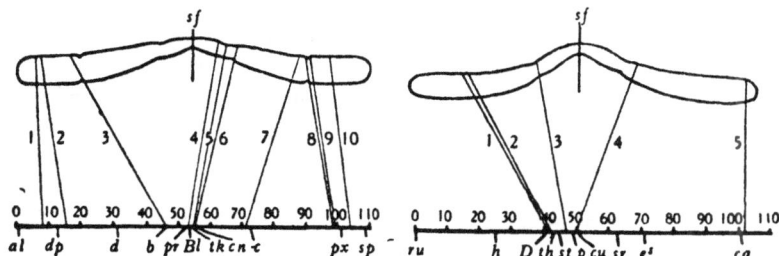

Fig. 45. Cross Over and Mitotic Metaphase Maps of the 2nd and 3rd chromosomes of *D. melanogaster*.
(From Timofeef Ressovsky, after Muller, Painter, and Dobzhansky.)

the cross-over map are found to occupy a considerably greater proportion of the metaphase chromosome map. In *Drosophila melanogaster* the crowded regions are mostly near the centromeres (middle of II and III, left end of X). The crowding here is partly explained by the fact that these regions of the chromosome are occupied by "inert" or heterochromatic material, in which no genes, or very few, are known and which are therefore not represented on the cross-over map. But even if we make allowance for this, and add to the cross-over map a section corresponding to the inert region of the metaphase chromosomes, we still find that there is some crowding of genes in the regions just outside the heterochromatic parts. This represents a lower frequency of cross-over per unit of material length than in other parts. The scale on which a cross-over unit represents material length is thus not constant. The greatest amount of crossing-over per unit of material length is in the middle of the *X* chromosome and in the middle of the autosomes, that is to say, it is as far as possible away from the

[1] Dobzhansky 1929, 1930*b*, Painter and Muller 1929.

attachment constriction and from the free end of the chromosome (cf. p. 125).

The maps based on mutation frequency are not affected by regional differences in crossing-over and agree very much better with maps based on the metaphase lengths of translocations, etc.

5. Salivary Gland Chromosomes

The metaphase chromosomes of Drosophila possess few external marks which can serve as fixed points for mapping. More detailed maps can be made of organisms in which the chromosomes show morphological differentiation along their length, for instance by possessing visible chromomeres. Further, at metaphase the gene thread or chromonema is coiled in a spiral and the length of chromosome occupied by a section of chromosomes will depend on the tightness of the spiral, while in the prophase stages where chromomeres are visible the spiral is practically uncoiled and this source of error does not occur. Until recently no organism was known in which both detailed genetical analysis and well defined chromomere structures were available. But in 1933 attention was called to the chromosomes in the larval salivary glands of Diptera[1] and particularly of Drosophila,[2] which have a very peculiar structure in which the linear differentiation is clearly visible.

In the very large nuclei of the salivary gland the chromosomes are as though in permanent meiotic prophase; they are paired and are thus present in the haploid number. Each chromosome, after the pairing has taken place, seems to have divided many times, so that the paired chromosomes consist of a series of some hundreds of threads lying parallel, forming the walls of a hollow cylinder.[3] The threads bear chromomeres, which show as rows of dots around the cylinder, or may apparently fuse into discs. The heterochromatic (inert) parts of the chromosomes may in some species, of which *Drosophila melanogaster* is one, fuse into a mass (the chromocentre) to which the euchromatic parts are attached. The length of the chromosomes may be as much as 100–150 times that of mitotic metaphase chromosomes; one can probably regard the parallel threads of which they consist as the completely unwound chromonemata which make up the short coil of the metaphase chromosomes.

The complicated chromomere structure and large size of the salivary

---

[1] Heitz and Baur 1933. [2] Painter 1934, 1935.
[3] Some authors consider that the cylinder is solid or has a honeycomb structure, cf. Metz 1935, *Rev.* of structure and origin, Cooper 1938, Painter 1939.

gland chromosomes allows an exact allocation of definite genes to definite positions in the chromosome;[1] and the fact that the chromosomes
are paired, enables one to draw detailed conclusions as to the sorts of
pairing to be expected in organisms with chromosomal abnormalities.
In salivary glands many chromonemata appear to be able to pair at a
point, while in zygotene pairing is only in twos (p. 69). Pairing of

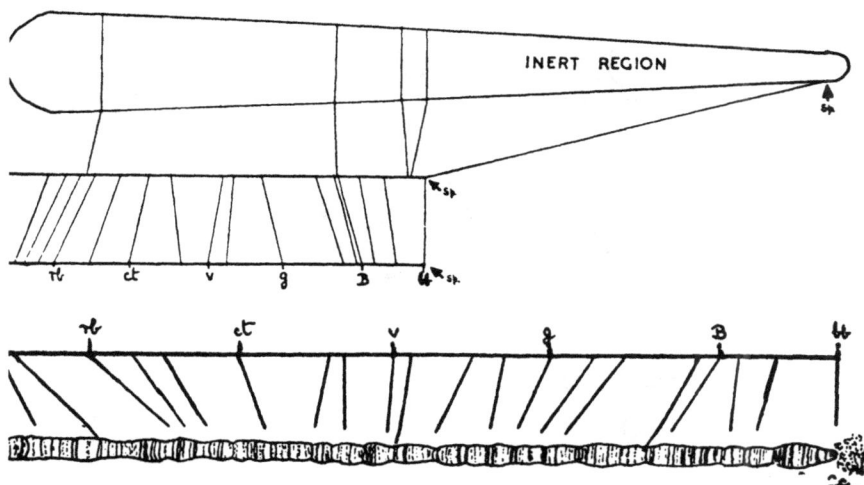

Fig. 46. **Maps of the X Chromosome of** D. melanogaster.—Above is the mitotic
metaphase chromosome with the position of some breaks occurring in translocations, etc., and below it the mutation frequency map (both after Muller and
Painter) and the normal cross-over map. The cross-over map is also shown on a
larger scale and compared with the salivary gland chromosome map (after Muller,
Painter, and Mackensen). Notice how the genes appear concentrated at the
ends of the cross-over map. Sp spindle fibre attachment or centromere, Cc
chromocentre.

salivary chromosomes must therefore be brought about by some forces
different to those concerned in normal meiotic pairing; they may be
related to the forces which produce somatic pairing (p. 74), which is
strong in Diptera.

The application of observations of salivary gland pairing to zygotene
pairing is probably justified; so far it has not led to any contradictions
but to extensions of principles which had already been deduced from
direct observation. The most important of these principles is that pairing is strictly between similar points of the chromosome. The pointwise
pairing may be complete even between a deficient chromosome and a

[1] Bridges 1935, 1938, Painter 1934, 1935.

translocated fragment, the limitation on completeness being apparently only the mechanical hindrance offered by the other chromosomes in the closely packed nucleus.

In the salivary chromosome map the genes are more or less evenly spaced along the chromosome, although even here there are regions in

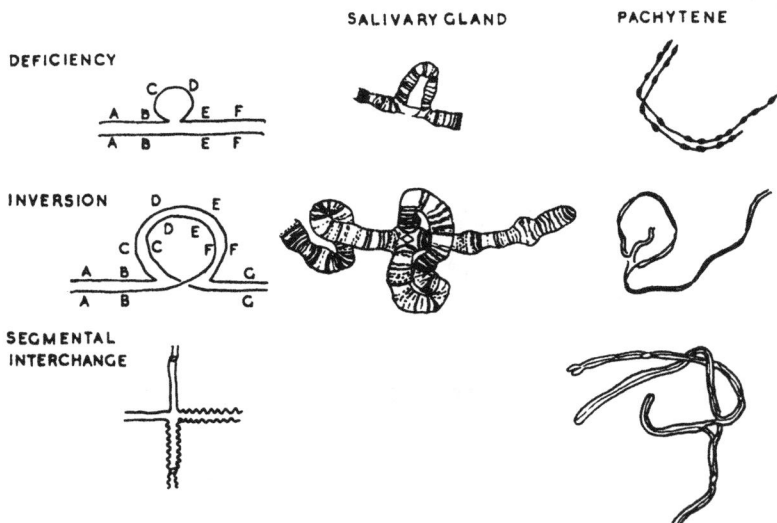

Fig. 47. **Pairing in Chromosomal Rearrangements.**—In inversions, deficiencies, translocations, etc., the pairing is between homologous genes, and not between the chromosomes as wholes. The figure shows diagrams of the pairing on the left, examples from the salivary glands of Drosophila (after Painter) in the middle and from pachytene figures on the right. The figure of a terminal deficiency (top right) and segmental interchange (bottom right) were found in maize by McClintock;[1] the latter has been slightly simplified. The pachytene inversion (middle right) is from Chorthippus (Darlington) and has also been slightly simplified.

which the known genes are more concentrated. The chromomeres (bands) however are quite evenly spaced and as each band probably represents one or a few genes, it is probable that if all genes are known, no regions would be more crowded than any other. The apparent crowding of the known genes indicates regions in which the genes are more mutable than normal, and the salivary map therefore shows the same sort of crowding as Muller's map in which the distance apart of the genes represents the mutation frequency.

The salivary chromosomes reveal the presence of small inverted duplications in the normal *Drosophila melanogaster*; for instance, in

[1] Cf. Rhoades and McClintock 1935.

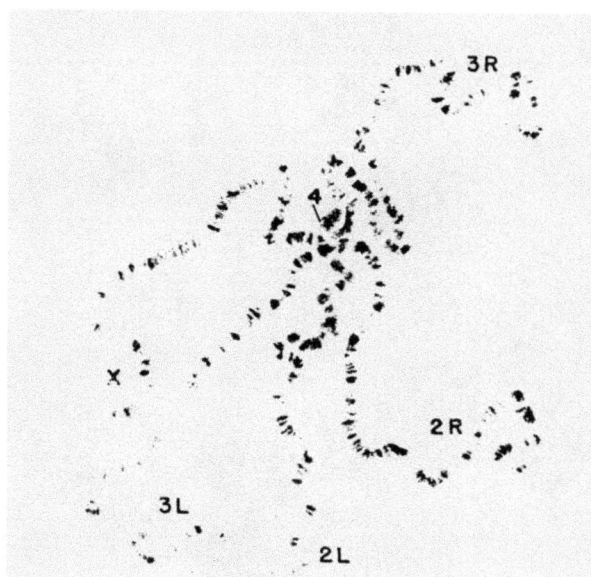

A

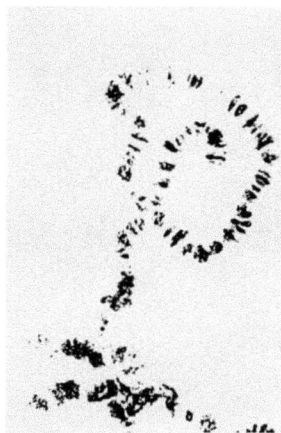

B

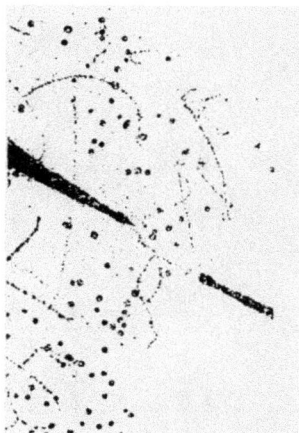

C

Plate 3. **A** shows the chromosome complex from a salivary gland nucleus in *Drosophila melanogaster*. Notice that similar chromosomes are paired, and the inert regions united into a single chromocentre (upper middle, just below the small IVth chromosome), so that the V-shaped IInd and IIIrd chromosomes each have the appearance of two long arms (2 right and 2 left, 3 right and 3 left).

(Courtesy of B. Kaufmann.)

B shows loop pairing in an inversion (*ClB* inversion) in the X-chromosome of *D. melanogaster*, as it appears in salivary gland nuclei. The chromocentre is below.

(Courtesy of M. E. Hoover.)

C shows a "lamp-brush" chromosome from an oocyte of the newt *Triturus*. The chromosome, which is unstained, is being stretched between two micro-needles. Notice the chiasmata. The preparation is from a fairly late oocyte stage, in which the chromosomes have already lost much of their fuzzy appearance.

(× 100. Courtesy of W. E. Duryee.)

the left arm of the IInd chromosome. These pair in characteristic inverted loops; they have not yet been detected genetically, presumably because no mutant allelomorphs of the genes concerned have been discovered. There are also several small bilaterally symmetrical elements which may represent similar inverted duplications in which the two elements are actually contiguous. These phenomena provide evidence that translocation and duplication has been involved in the evolution of *Drosophila melanogaster*; we shall see later that a comparison of the different species of Drosophila demonstrates the same thing.

6. "*Lampbrush*" *Chromosomes*[1]

During ovogenesis in some yolky eggs (sharks, amphibia, birds, etc.) the diplotene chromosomes become enormously enlarged, being comparable in length to salivary gland chromosomes. The pachytene stage may last a very long time, up to two years in Amphibia, and during this period the chromosomes undergo a peculiar change whose exact nature is not yet fully understood. In the early stages of the process, the paired chromatids lie closely side by side and show a series of swellings, which look much like other chromomeres. As the increase in length continues, the chromomeres seem to give off filaments, consisting of rows of granules, which lie transversely to the length of the chromosome, and bend round to form loops. The sequences of these loops seem to be characteristic of particular regions of the chromosome, just as are the sequences of bands in the salivary gland chromosomes, and presumably the growth in length of the whole chromosome is a result of the uncoiling of the chromonemata to their full extent. Eventually the loops and threads disappear from the sides of the chromosomes, which contract again to a more usual size, and the division proceeds. The formation and shedding of the side branches releases into the nuclear sap, and thence probably into the cytoplasm, substances formed under direct control of the chromosomes, and these substances may perhaps play an important part in endowing the egg with its developmental capacities.

7. *Genetic Analysis of the Process of Crossing-over*

The effect of temperature on crossing-over is shown only by gametes which were in the prophase of meiosis when the abnormal temperature was applied.[2] This is evidence that crossing-over occurs during meiotic prophase;[3] but the validity of this hypothesis rests mainly on the demon-

[1] Koltzoff 1938. Duryee, 1938.   [2] Plough 1917   [3] Cf. Gowen 1929.

stration, which we must now examine, that crossing-over takes place during the stage when the chromosomes are divided into chromatids.

The consequences of crossing-over (in the'sense of the breakage hypothesis) would be different if it occurred between two whole chromosomes or between two out of four chromatids. In the first case the four gametes formed from a germ-mother cell would consist of two similar pairs, in the second case all four would be dissimilar. In an ordinary diploid organism, the gametes resulting from one mother cell cannot be separately identified, and the condition of affairs cannot be directly determined. It is otherwise, however, in organisms with a well-developed haploid phase: the four haploid spores are often aggregated

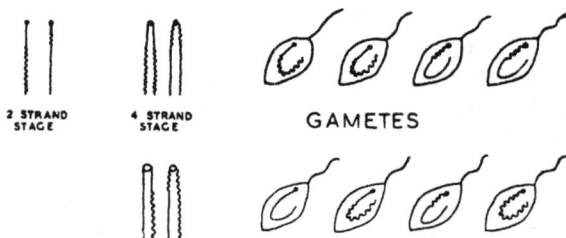

Fig. 48. Two- and Four-Strand Crossing-Over.—With two-strand crossing-over, shown above, the process of crossing-over is complete before the chromosomes have split into daughter chromatids, which will therefore form two pairs of similars; and from them there must form two pairs of similar gametes. If crossing-over takes place after the chromatids are formed, i.e. in the four-strand stage, all four chromatids and therefore all four gametes, will be different.

together in a cluster and can be isolated and their gametic constitution separately determined. Until recently not very many cases of the segregation of linked factors have been analysed in diplo-haplonts, since not many certain cases of linkage are known. Wettstein[1] found only 2-typè tetrads in Funaria (i.e. crossing-over apparently between chromosomes), but Allen[2] (in Sphaerocarpus) has shown that 4-type tetrads occur. In the last few years several examples of the inheritance of linked factors in diplo-haplonts have been studied. The most complete study is that of Lindegren on Neurospora, an Ascomycete.[3] The essential feature of the chromosome cycle is the formation, from the fertilized zygotes, of an ascus in which the two reduction divisions occur, giving four haploid cells which are arranged in a row, and which then undergo a further, mitotic, division to give eight haploid ascospores, arranged in a row as four pairs; from the ascospores the

[1] Wettstein 1924, 1928b.   [2] Allen 1926, 1935.   [3] Lindegren 1933, 1936a, b.

haploid mycelia are developed. By a micromanipulation technique the ascospores can be isolated and grown individually and their genetic constitution determined. We shall consider the data on crossing-over relating to the factor P (pale ascus as against normal p) and the alternative sex factors known as $+$ and $-$. Four-type "tetrads" occur; owing to the extra mitotic division, the tetrads here consist of groups of four pairs of ascospores instead of four single spores. The cross-over percentage was determined by the number of recombinations in the usual way and was $22 \cdot 5$ per cent.

Other evidence that crossing-over takes place between chromatids and not between chromosomes can be produced if we make the assumption, which will be shown to be true in higher forms, that the centromeres of the two chromosomes always disjoin from one another in the first division. If this is the case, at the centromere the two sister chromatids from one chromosome are always separated from the two sisters from the other chromosome (a reductional separation). The only way in which two non-sister chromatids can get to the same pole after the first division is by the occurrence of a cross-over between chromatids, when a separation of this kind (equational separation) is possible for the regions distal to the crossing-over. Now it can easily be shown that equational separation does occur. If we have two factors which are not linked and if they are both separated reductionally at the first division, they must give tetrads consisting of two pairs of similar spores (2-type tetrads), while if either or both are separated equationally at the first division and reductionally at the second, 4-type tetrads will result (see Fig. 49).

Very many investigations on non-linked factors have revealed the presence of such 4-type tetrads and therefore the occurrence of a cross-over, in the chromatid stage, between the segregating factor and the attachment constriction. Segregation of this kind is known as chromatid segregation, since the mechanism segregates the two allelomorphs from one another when they are in a body consisting of four chromatids instead of segregating them from a body consisting of two chromosomes.

The considerations we have just discussed enables one to obtain an estimate of the distance, in cross-over units, between a certain factor and the centromere. We have seen that if in a heterozygote a cross-over occurs (in the chromatid stage) between the factor and the centromere, one of the two chromatids going to each pole of the 1st division will contain one allelomorph and the other the other, and the segregation of the factor will not occur until the second division. The fact that segregation occurs in the second division can be recognized, as we have seen, by the occurrence of 4-type tetrads from a double heterozygote; but clearly the occurrence of 4-type tetrads in itself does not indicate which of the two pairs of factors is the one having second division reduction. In Neurospora, however, the arrangement of the ascospores

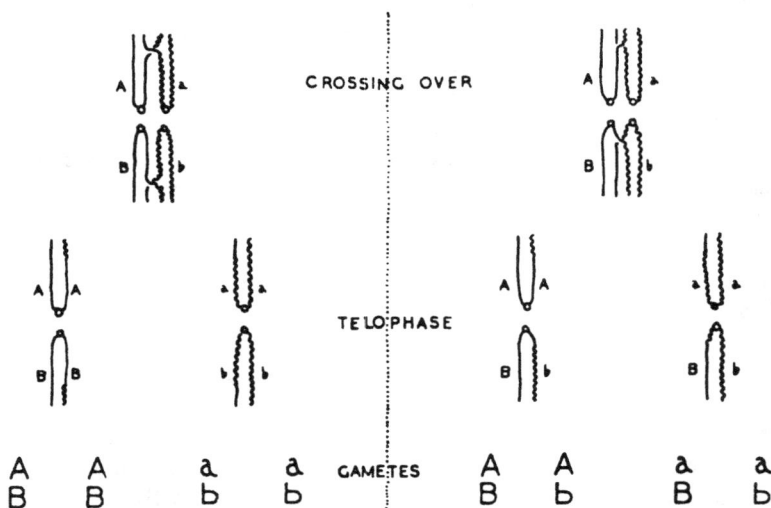

Fig. 49. **Diagram of Crossing-Over and First and Second Division Segregation.**—On the left, above, are two pairs of chromosomes, both heterozygous, the upper with *Aa* and the lower with *Bb*. The chiasmata leave these factors attached to their original centromeres. In the first division telophase A is therefore separated from *a* and B from *b*; the gametes then form a 2-type tetrad, with two *AB*'s and two *ab*'s. On the right, crossing-over takes place between B and the centromere; at first division telophase A is separated from *a*, but both daughter nuclei contain B and *b*; the gametes form a 4-type tetrad, with *AB*, *Ab*, *aB*, and *ab*.

in a row makes it possible to decide this; if a pair of factors is reduced in the first division one allelomorph must be entirely at one end of the row and the other at the other end.

The proportion of second division reductions is equal to the proportion of cases in which crossing-over has occurred. Now if we consider ordinary crossing-over between linked factors, in each case in which one cross-over occurs, two of the daughter chromatids will show new combinations and the other two the original combinations, so that the proportions of cases of crossing-over is twice the proportion of new combinations; we therefore have only to divide the proportion of second division reductions by 2 to obtain the cross-over distance between the gene and the centromere. In Neurospora, Lindegren obtained values of $16 \cdot 5$ for the cross-over distances between pale and the centromere and $6 \cdot 5$ between the sex factor and the centromere. Hence, disregarding double cross-overs, the distance between pale and the sex factor should be either $16 \cdot 5 - 6 \cdot 5 = 10$ or $16 \cdot 5 + 6 \cdot 5 = 23$. Actually we saw that direct measurement gave $22 \cdot 5$ in extremely good agreement with the latter hypothesis; the two estimates confirm one another and the general line of argument, and we can draw a map of the sex chromosome containing P and $\pm$ on different sides of the centromere.

8. *Crossing-Over in Polyploids*

Chromatid segregation can be discovered in higher organisms if some way can be found of getting two, instead of one, of the segregating chromatids into a single gamete. This happens in a race such as the attached-X race in Drosophila, where the two X chromosomes of the female are permanently attached together and form XX eggs.[1] It also occurs in polyploids, where there are more than four homologous chromatids to be distributed to the four gametes. In both these cases the occurrence of chromatid segregation can be proved. Thus in a tetraploid *Datura* of the constitution $AAAa$, gametes aa could be produced (giving $aaaa$ on selfing).[2] This can only occur if in the chromatid stage $AAAAAAAa$ crossing-over occurs so that the two a's become associated with two different centromeres which are passing to the same pole so that the factors have a chance to pass to the same pole in the second division.

The best analysed cases of chromatid segregations in polyploids are in triploid *Drosophila melanogaster*.[3] One can label all the chromosomes of one kind (e.g. the X) by different recessive factors distributed along the whole length of the chromosome. Crossing-over between whole chromosomes (in the so-called 2-strand stage) cannot give more than three sorts of chromatids (2 of each kind) to be distributed to the four gametes, and each sort must be completely unlike any other. At the second division the two sister members of each pair must be separated from each other by the separation of the attachment constriction, so that the diploid gametes can only contain two chromatids which are wholly unlike each other. Crossing-over between chromatids, in the 4-strand stage, can on the other hand give rise to six different sorts of

[1] Anderson 1925. The same phenomenon occurs in non-disjunctional eggs, cf. Bridges 1916.　　　　　　　　　[2] Blakeslee, Belling and Farnham 1923.
[3] Bridges and Anderson 1925, Redfield 1930, 1932, Rhoades 1933.

chromatids which need not be completely different, and the diploid gametes can contain two chromatids which are partly alike and partly different. The parts for which the two chromatids are alike have shown equational reduction; that is the reduction has failed to separate the two alike sister chromatids. The flies which develop from these eggs are therefore known as equational exceptions and may show, in a homozygous condition, a factor contained in a single dose in their mother.

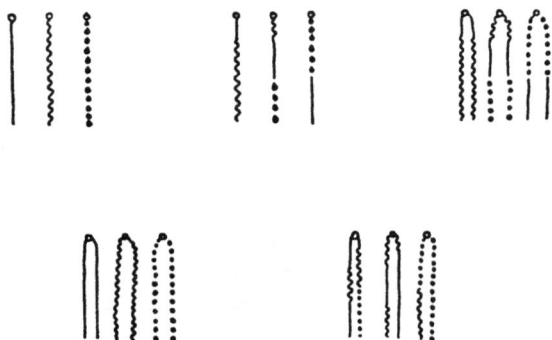

Fig. 50. **Crossing-Over in Triploids.**—The upper row shows crossing-over in the "2-strand" stage, i.e. before the chromosomes have split. The six chromatids form three pairs of similars. Below is crossing-over in the "4-strand" stage, i.e. after the split; the six chromatids are all different, and if two of them are chosen, they may be alike in some parts but not in others. The first and third chromatids, counting from the left, are recurrent double cross-overs, while the second is a progressive.

We have explained the occurrence of equational exceptions or chromatid segregations, which is another name for the same phenomenon, as a consequence of crossing-over among the chromatids between the factor in question and the centromere. If this is true the frequency of equationals should be high for factors lying far away from the centromere, for which there will be a high chance of a cross-over between the factor and the centromere, and low for factors lying nearer to it. The highest frequency which can be expected is that due to pure chance; if one diploid gamete contains a certain factor, the probability that it will also contain the same allelomorph from the sister chromatid cannot be greater than the chance of choosing the sister from among the five that are left, namely 20 per cent. The actual frequencies of equationals accordingly vary from 0 near the centromere towards 20 per cent at the distal end of the X (actually the highest value obtained is about 12 per cent, so even at the distal end of the X there has not been enough crossing-over for the chromatid segregation to be on a purely chance basis). The fact that no equationals occur near the centromere is a proof that we are correct in assuming that the two centromeres of sister chromosomes must separate from one another in the first division.

The analysis of triploid crossing-over also shows that any of the six chromatids may cross-over with any other different chromatid; but one cannot determine in this way whether crossing-over may happen between two sister chromatids, since it would have no observable consequences. At any one locus, however, crossing-over only occurs between two chromatids; one never finds crossing-over between four chromatids since that would involve two sister chromatids crossing-over in the same place with chromatids belonging to two other different chromosomes, and this could only happen if all three chromosomes were paired at one point in zygotene. Zygotene pairing is, however, strictly in twos in those organisms in which it can be cytologicaly observed. Some evidence as to whether it is in twos in Drosophila can be derived from a consideration of double cross-overs. The second cross-over may be between the same two chromosomes as the first (recurrent double cross-over), or between one of the two involved in the first cross-over and the third chromosome (progressive double cross-over). There should be a high proportion of progressive double cross-overs if all the chromosomes were paired throughout their length, but more recurrent double cross-overs if they are paired in twos with only infrequent changes of partner. The evidence is not yet perfectly clear but seems to be in favour of the second alternative, which fits in with the cytological observations.

The calculation of gamete ratios in experiments involving crossing-over becomes progressively more complicated in higher multiple polyploids,[1] and becomes theoretically impossible when the frequency of chiasma formation, i.e. degree of metaphase association is not known. In tetraploid *Primula sinensis*,[2] the chromosomes are nearly always associated in pairs at metaphase and the results obtained for the segregation of various linked factors agree quite well with a calculation based on random association of the chromosomes in pairs and an absence of chromatid segregation; the latter of these hypotheses can hardly be more than approximately correct.

---

[1] Cf. Mather 1936b. [2] de Winton and Haldane 1931.

B. TRANSLOCATION

1. *Datura Secondaries and Tertiaries*[1]

In Datura, the tomato, and some other plants, translocation and duplication is often found combined with trisomy. The trisomic chromosome is in these plants not simply an extra member of the

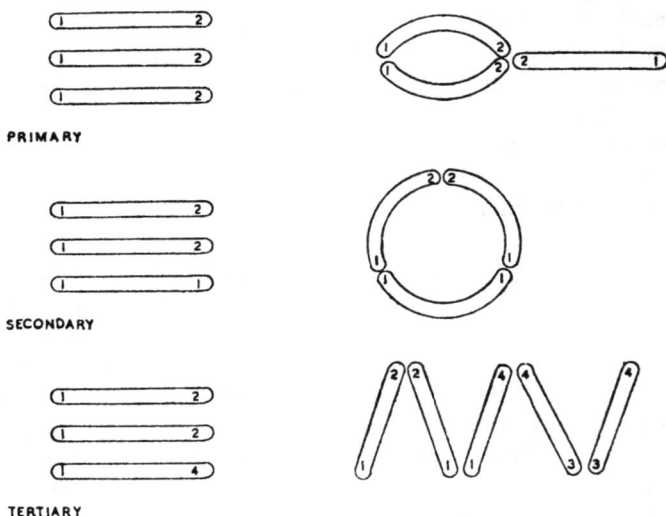

PRIMARY

SECONDARY

TERTIARY

Fig. 51. Chromosome Constitution and Association in Trisomic Types in Datura.—The column on the left shows the chromosome constitution; in primary trisomics the reduplicated chromosome is one of the normal set, in the secondaries it has two similar ends, in the tertiaries it has an end of one chromosome united with part of a different chromosome. The ends are distinguished by numbers. The main data about the behaviour of the chromosomes relate to their associations at metaphase of meiosis, when the chiasmata are completely terminalized (right-hand column). In the primaries, the characteristic configuration is a ring of two with a third attached; in the secondaries, rings of three may be formed, and in the tertiaries the reduplicated chromosome may unite two other pairs, giving chains of five.   (Based on Belling.)

normal set, but involves a translocation. In one type, called secondary trisomics, the trisomic chromosome consists of a reduplicated half: from a chromosome pair *abcd*, *abcd*, the translocations *abba*, *dccd* are formed, and one of these compounds is present in the trisomic, giving, for instance, a form with the normal *abcd*, *abcd*, plus *abba*. In the other type (tertiaries) the trisomic body is a translocation between two non-homologous chromosomes, e.g. *abpq*.

[1] *Revs.* Blakeslee 1930, 1934.

The nature of these chromosomes can be recognized by their effect on the balance of the organism (p. 172), but even more definitely by their association in meiosis.[1] In secondaries associations in rings of three may be found, and, even more strikingly, the two *abcd*, *abcd* chromosomes may associate, while the *abba* chromosome associates with itself to form a ring. This association must be due to the formation of chiasmata, which are subsequently completely terminalized, and the fact that here the two ends of a single chromosome pair with one another is a remarkable demonstration that zygotene pairing is a relation between individual loci and not between two chromosomes as wholes (p. 99). In tertiary trisomics associations of five chromosomes are possible and are sometimes found.

2. *Segmental Interchange*

The name segmental interchange has been given to the phenomenon of mutual translocation by which chromosomes *abcd*, *efgh* become *ebcd*, *afgh*. The formation of such mutual interchanges may be supposed to be by crossing-over of non-homologous chromosomes or by breakage of two chromosomes which happen to be lying against each other in a mitosis. The importance of the phenomenon arises from the peculiar results to be expected in the hybrid between races with the original and with the interchanged segments. Such a hybrid will contain chromosomes *abcd*, *efgh*, *ebcd*, *afgh*, which can pair and be associated in a ring of four, *abcd*, *dcbe*, *efgh*, *hgfa*. Examples of this have been found in several plants (Zea, Datura, Pisum) and in Drosophila and probably Orthoptera.[2] Such hybrids are spoken of as interchange heterozygotes.

The segregation from such rings of four or more chromosomes is in most organisms fairly irregular, since the rings are usually orientated at random on the metaphase plate. The irregular gametes with reduplicated or deficient segments often fail to function. Regular arrangement is found in Oenothera where the chiasmata are completely terminalized and the centromeres median.

Chromosome complements of this type have been developed in nature as a method of evolution of true-breeding organisms, which are nevertheless fundamentally hybrid. The most remarkable example of this rather exceptional method of evolutionary divergence is in the genus Oenothera. Here we find, not only separate races which give interchange heterozygotes when crossed (as in Pisum and Datura) but

[1] Cf. Belling 1927.   [2] *Rev.* Darlington 1937.

actual species which are permanent interchange heterozygotes. The genus Oenothera was one of the earliest to be investigated genetically on a large scale and its peculiar behaviour led de Vries[1] to generalizations which do not apply fully to other organisms. In particular, at a time when Mendel's results were only just being rediscovered, de Vries attached the word mutation to a phenomenon which, as we shall see, is quite different in nature to gene-mutation, although in some ways resembling it in phenotypic effect.

The main facts of inheritance in Oenothera are that the varieties breed more or less true when self-fertilized, but produce a fairly large proportion of bad seed and also throw a certain proportion, up to about 2 per cent, of unlike types which are de Vries's mutants; while on

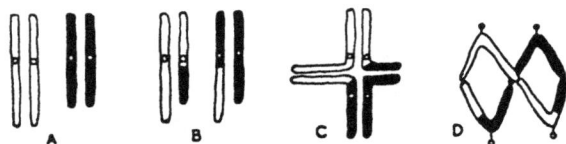

Fig. 52. **The Formation of Metaphase Rings as a consequence of Segmental Interchange.**—A shows the chromosomes before the change; B after it; C shows the pairing. If chiasmata are formed near the ends of the chromosomes, and become terminalized, the four chromosomes will be associated in a ring at metaphase. The centromeres are represented by circles.

crossing, the resulting hybrids often differ according to which species is the female in the cross. The first real elucidation of this behaviour came from the suggestions of Renner and Muller.[2] Renner suggested that the species of Oenothera are really heterozygotes, consisting of two associated complexes of genes, and forming, in the main, two classes of gametes each containing one of the complexes. Muller suggested a comparison with the balanced lethal system in Drosophila, and advanced the hypothesis that neither of the complexes could survive in the homozygous condition. A heterozygote AB would therefore breed true, since the AA and BB zygotes would die. Renner then proceeded to show that not only were lethal factors involved which kill the zygote in the homozygous condition, but that gametic lethal factors are also found, so that some of the species transmit only one of their complexes through the pollen or through the ovules.

The chromosomal basis for this mechanism has only recently been discovered. It has been known for a long time[3] that the chromosomes in

[1] de Vries 1901.
[2] Cf. Muller 1918, Renner 1925, 1929, Sturtevant 1926b.   [3] Gates 1908.

Oenothera are often associated in a ring at metaphase, but this was originally interpreted on the basis of the theory of telosynapsis (p. 113), the assumption being that the postulated continuous spireme of prophase was for some unknown reason not broken up into separate chromosomes in these forms. Cleland[1] was the first to show that the phenomenon is regular and that rings (sometimes replaced by chains) of definite even numbers of chromosomes are characteristic of the different races and hybrids. The ring formation can be explained as a consequence of segmental interchange similar to that described by Belling in Datura.[2] As an example, *Oe. Lamarckiana* forms a ring of

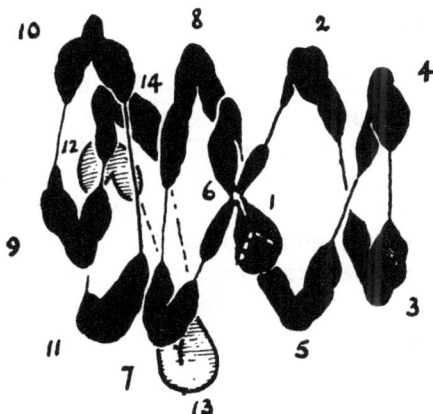

Fig. 53. **Chromosome Rings in Oenothera.**—The figure shows the side view of meiotic metaphase in Oenothera. The chromosomes are held together in two rings by terminal chiasmata, alternate chromosomes going to the same pole (1, 3, 5, 7 downwards and 2, 4, 6, 8 upwards in the ring on the right; 9, 11, 13 downwards and 10, 12, 14 upwards in the ring on the left). In this figure the regularity of the arrangement is disturbed by one of the rare cross-overs between chromosomes belonging to different complexes (1 and 6). This will give rise to a mutant. (From Darlington.)

twelve and one pair at meiosis, and the chromosomes in the ring of twelve become associated terminally and are arranged so that alternate chromosomes regularly pass to the different poles. The species contains two complexes, *velans* and *gaudens*, and we may suppose that the genes composing these complexes lie in the middle regions of the chromosomes, while the end segments are the mutually translocated parts which pair. The constancy of the complexes depends on the fact that the genes of one complex lie in alternate chromosomes, and on the regularity of the arrangement of the ring on the spindle, while the true-breeding of the heterozygote depends on recessive or gametic lethal effects of the complexes.

The formation of de Vriesian mutants can occur in two ways.

(1) by a segmental interchange between the two complexes which

[1] Cleland 1922, 1931.   [2] Darlington 1929*b*, 1931*c*.

will give a new ring formation and a new association of parts of one complex with parts of the other;

(2) by segmental interchange between two chromosomes within one complex.

Darlington has shown how a process analogous to the second of these would account for the formation of the rather complicated system of complexes, lethals and ring-mechanism which has to be postulated. If we start with an organism with the two pairs of chromosomes AxB, AxB, and CD, CD, let one of the x segments be translocated to the middle of the CD chromosome. We shall have AxB, AB and CxD, CD, both pairs being terminally associated. Then a crossing-over in the two x segments (near one end of the segment) may occasionally occur and give AxD, CxB, and it will be possible to obtain an organism $AxD.DC.CxB.BA.$ in which there are two complexes, one with and one without the x segment, and also a balanced lethal mechanism since the AxD, AxD, CxB, CxB type with reduplicated x and the AB, AB, CD, CD type with deficient x might be expected to be inviable. Definite evidence that the Oenothera complexes have originated in some way similar to this has been found. Chiasma formation can be occasionally observed between the middle sections of the chromosomes in one complex, which shows that these middle sections are homologous, corresponding to the x segment in the example above.

The detailed working out of the hypothesis of segmental interchange has been undertaken by various workers, and the hypothesis has survived the important test of prediction. Cleland[1] was able to deduce, from the pairing relations in parent species and some of their hybrids, what should be the pairing in other hybrids which had not yet been examined, and subsequent observation showed the correctness of his deductions.

[1] Cleland 1931.

CHAPTER 5

The Mechanics of the Chromosomes[1]

---

1. The Relation of Meiosis and Mitosis

We must now expand the summary description of mitosis and meiosis given in Chap. 1.

The most important difference between the two processes is that in mitosis the prophase chromosomes appear in the diploid number of paired threads, while in meiosis they appear in the diploid number of single threads, which then come together in pairs to form the haploid number of double bodies. This pairing side by side of the meiotic chromosomes is known as parasynapsis and is generally accepted at the present day. Until a few years ago it was disputed by some cytologists, who maintained that the early prophase chromosomes, both in mitosis and meiosis, were joined by their ends into a continuous thread or spireme, which was double in mitosis, but single in meiosis. This thread was supposed to segment into the diploid number of paired chromosomes in the mitotic prophase, but in meiosis into the haploid number of bodies each of which consisted of two homologous chromosomes paired end to end (telosynapsis); these two chromosomes then became bent at their point of junction so that they lay side by side in zygotene. In another variation of the theory it was supposed that the thread was at first single both in mitosis and meiosis, and that in mitosis also the side-by-side arrangement of the daughter threads, which can be seen in later stages, was arrived at by the doubling back on themselves of two chromatids which originally were paired end to end. The hypothesis of telosynapsis, then, supposed that both in meiosis and mitosis the chromosomes first appeared as a continuous spireme which became segmented between each chromosome in mitosis and between alternate chromosomes in meiosis. No explanation was offered for this difference in the method of segmentation, but it was the lack of a satisfactory observational basis which led to the abandonment of the hypothesis. The existence of a continuous spireme could only be plausibly suggested in those organisms where the large number of chromosomes made observations of early prophase very difficult; in

---

[1] *General references:* Belar 1928, Darlington 1937, Geitler 1934, Sharp 1934, White 1937, Wilson 1928.

the simpler cases the fact that the chromosomes were separate at their first appearance could be conclusively demonstrated. The only other

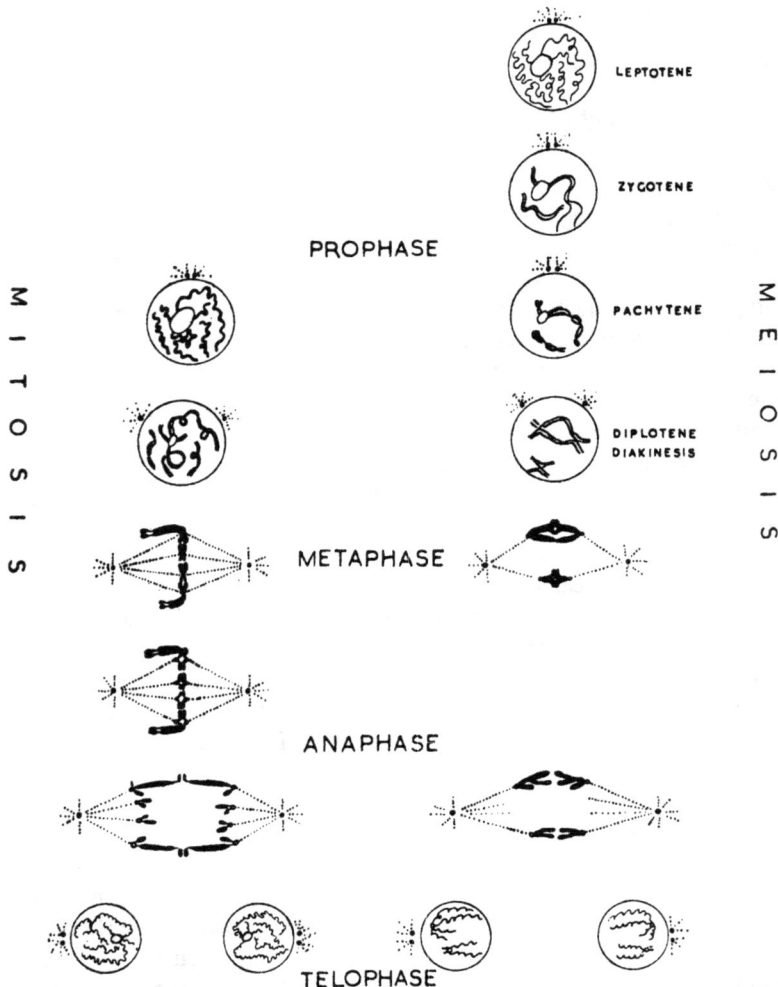

Fig. 54. **The Relation of Mitosis and Meiosis** (according to Darlington).—The stages of mitosis are shown on the left, and the corresponding stages of meiosis on the right. Note that the chromosomes are double at their first appearance in mitosis, but not in the earliest (leptotene) stage of meiotic prophase. They pair in zygotene, beginning to become double (split) in pachytene, and are double in the last stages of meiotic prophase (diplotene and diakinesis). In mitosis anaphase begins by the division of the centromeres, but in meiosis this does not occur till the anaphase of the second division (not shown).

evidence in support of telosynapsis was adduced from the occurrence of continuous rings of chromosomes at meiotic metaphase in some organisms (chiefly Oenothera), but another more satisfactory explanation can now be given for this (p. 111).

Darlington was the first to suggest that if the parasynaptic interpretation is adopted, all the differences between meiosis and mitosis can, with the help of very few hypotheses, be deduced from the original difference in the singleness or doubleness of the prophase threads. He suggested that homologous chromosomes, or rather homologous chromonemata, attract one another in pairs. This attraction is satisfied in mitotic prophase, but not in early meiotic prophase, where the chromosomes are at first single, and can only satisfy the attraction by coming together in zygotene pairing. In meiosis, the splitting of the chromosomes, which occurs in the interphase before a mitotic division, does not happen till pachytene when the chromosomes are already associated in pairs. It therefore produces a set of four associated threads and these fall apart into the diplotene loops, being held together by the changes of partner at chiasmata. The falling apart of the four threads is evidence that although two homologous chromosomes attract one another, one pair of homologous chromonemata repels another similar pair.

Thus Darlington supposed that the splitting of a chromosome in preparation for the next mitotic division takes place in the interphase before that division, and that the singleness of the meiotic chromosomes is due simply to the fact that they begin condensation and contraction for the division before the splitting has occurred. This hypothesis is known as the Precocity Theory.

A modification of the simple precocity theory has been proposed by Huskins,[1] who accepts Darlington's hypotheses that the differences between the two sorts of division are due to the repulsion between pairs of chromonemata and attraction between single chromonemata, but rejects Darlington's account of the origin of the singleness of the meiotic chromosomes in prophase. According to Huskins, the splitting of the chromosomes for one division takes place during the previous division, forming a so-called tertiary split which causes the chromosome pairs at mitotic metaphase to be actually quadripartite, those at meiotic metaphase octopartite bodies. During the interphase before meiosis, this tertiary split must be supposed to be destroyed in some way, only to be restored at pachytene; in about the diakinesis stage it

---

[1] Cf. Huskins and Smith 1934.

is succeeded by another split for the division which follows the second meiotic division.

The two theories give quite different accounts of anaphase separation. According to Darlington, the initial separation of the chromosomes in anaphase is in mitosis due to the division of the centromeres at this stage and in meiosis to an increased repulsion between the centromeres which now begins to overcome the forces holding the chromosomes together at the chiasmata. According to Huskins, on the other hand, the separation of the mitotic chromosomes is analogous to the formation of diplotene loops; it is a consequence of the formation of the tertiary split which converts the paired mitotic chromosomes into a pair of mutually repelling paired threads. Huskins claims that the centromeres are double throughout the mitotic prophase, not remaining single until metaphase as Darlington supposes. For the meiotic metaphase separation Huskins invokes the same mechanism as does Darlington; the tertiary split occurring in late diakinesis has no consequences until two divisions later.

It may appear that a decision between the two theories could easily be taken by a simple observation of the single or double nature of the chromosome thread at relevant stages. Indeed, such observation is the main basis on which Huskins advances his theory. Darlington points out, however, that the observations which have been published of the tertiary split are highly contradictory among themselves, and suggests that the diameter of the chromosome spiral is sufficiently near the wave-length of ordinary light for optical illusion to play a large part in determining the microscopical appearance of the thread. Huskins retorts that, while it might be possible to mistake a single cylinder for a double thread when observing it from the side, an end-on view which shows a double thread can hardly be in error.[1]

The question has, however, recently been investigated by rays of wave-length very short compared with the structures involved, namely X-rays. Nuclei in the resting stage are X-rayed and examined for breakages in the next mitotic metaphase.[2] If at the time of radiation the chromosomes are single, breakages (with rejoining of the ends) will give dicentric and acentric fragments (i.e. with two or with no centromeres) each consisting of two chromatids formed by splitting after the operation. If on the other hand the chromosomes are already double, pseudo-chiasmata will arise. It has been found that the critical period

---

[1] *Rev.* Kaufman 1936.
[2] Mather and Stone 1933, Riley 1936, Mather 1937a.

is different in different organisms but that splitting takes place some time after the telophase and before the prophase, e.g. in locusts about 24 hours before. This seems a complete vindication of Darlington's original form of the hypothesis.[1]

The critical evidence in this line of argument is provided by chromo-

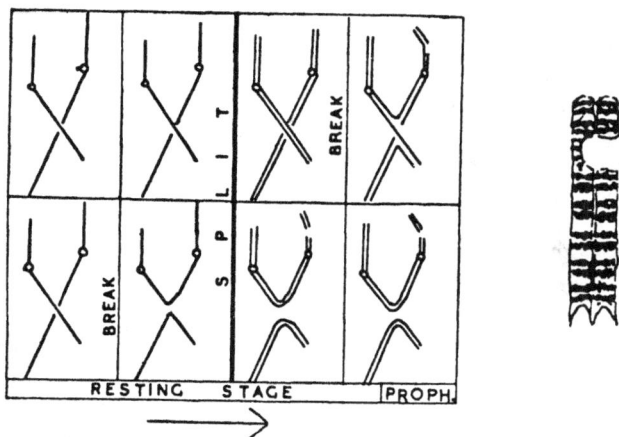

Fig. 55. **The Time of Division of the Chromosomes.**—Chromosomes are broken by X-rays at various intervals before entering mitosis and the metaphase figures examined. If the break occurs before the chromosome splits into two chromatids, one may find figures where two chromosomes have been broken and rejoined with each other (lower row). But if the break occurs after splitting into chromatids, pseudo-chiasmata can occur between the four chromatids from non-homologous chromosomes which happen to be lying near one another (upper row). Similarly, early irradiation can break a chromosome in two, later irradiation will only affect one of the two chromatids (top right of each drawing).

(After Mather.)

On the right is a drawing of a chromosome at mitotic metaphase, showing its alleged quadripartite structure. In the injured region three of the four half-chromatids of which the chromosome is supposed to consist have been destroyed by X-rays.

(After Nebel.)

some figures which are large enough to be incontrovertibly established. It has been stated, however, that changes on a smaller scale occur which lead to the hypothesis that the fundamental structure of the chromosome shortly before mitotic prophase is not that of a pair of chromatids but of four half-chromatids, and figures have been published showing three of these wiped out by X-rays and only one left.[2] However, one is here again attempting to discriminate between two

[1] But cf. Huskins 1937, where he gives up his own theory but does not accept Darlington's.   [2] Nebel 1936.

appearances which are both on the limit of visibility, and it is hardly possible to accept the evidence for the existence of half-chromatids as firmly established. As yet investigators who criticize Darlington's hypothesis of precocity on these lines neither agree among themselves nor suggest a coherent alternative theory of the relation between the two types of division. Even if it should eventually be established that there are units smaller than the half-chromosomes which are usually known as chromatids, it must be remembered that the precocity theory is concerned only with the behaviour of these units in attracting and repelling one another, and is not finally invalidated if it turns out that quarter-chromosomes behave as units in other respects, for instance, in relation to X-ray breakage.

2. *Chiasmata*

Chiasmata appear first in diplotene as the "nodes" between the loops into which the quadripartite chromosome pairs are opening out. They are at first always interstitial, that is to say, located somewhere along the length of the chromosomes, and not directly at the ends. They may also be either distributed more or less at random along the lengths of the chromosomes, the randomness being modified by interference (p. 125), or localized in particular regions. Their position, however, does not remain constant: as diplotene and diakinesis progress, the chiasmata move towards one or both ends of the chromosomes, usually only towards the non-attachment end in chromosomes which have a terminal attachment and towards both ends in chromosomes with an interstitial or median attachment. This movement is spoken of as terminalization and occurs to different degrees in different organisms. It is weakest in organisms with the longest chromosomes. The terminalization leads not only to a concentration of chiasmata at the ends of the chromosomes, which can be expressed as a terminalization coefficient (i.e. ratio of terminal chiasmata to total number of chiasmata) but also to an actual reduction in the total number of chiasmata. This reduction might be due to the breakage of interstitial chiasmata, but the evidence rather suggests that it is due to the pushing off of the first chiasma to arrive at the end of a chromosome by the second chiasma which becomes terminalized to that end later (see p. 127).

The terminalization of a chiasma can be arrested if there is a change of homology along the length of a chromosome, and the unusual failure of terminalization may thus give evidence of a dissimilarity between chromosomes, other parts of which may be able to pair.

The number of chiasmata found in any one bivalent in different nuclei is not constant but gives a frequency distribution with the mode favoured more than would be expected on a basis of pure chance. This can be explained by supposing that chiasmata show the phenomenon of

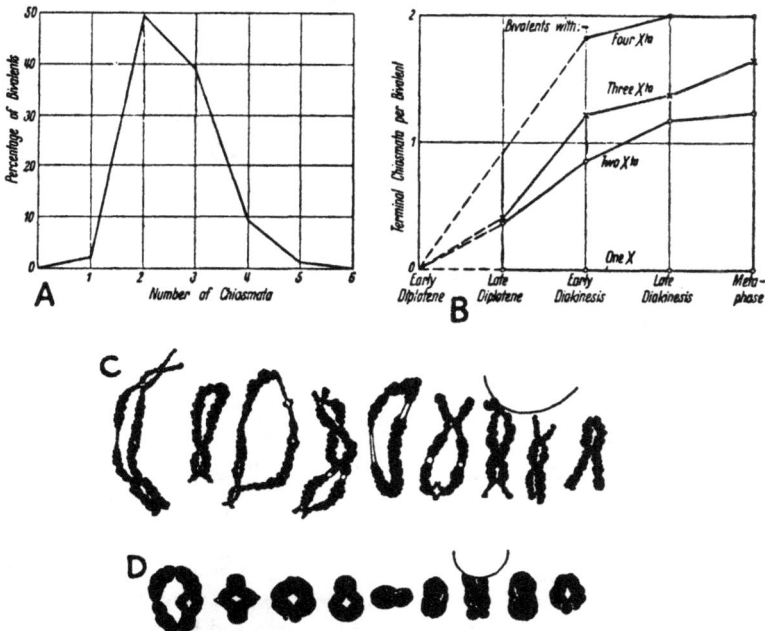

Fig. 56. **The Cytological Behaviour of Chiasmata.**—The graph to the left above gives the percentage of bivalents having different numbers of chiasmata. The graph to the right shows the increase of terminal chiasmata in bivalents with one, two, three, or four chiasmata. Below are the chromosomes from a diplotene stage and, in the lowest row, from a late diakinesis, to show the increase in terminal chiasmata. (The association on the left is a ring of four, consequent on a segmental interchange for which the plant was heterozygous.) All figures relate to maize (*Xta* = chiasmata).

(From Darlington.)

interference, the formation of one chiasma lowering the probability that another will be found in its immediate neighbourhood.[1] This is one of the parallels between chiasma formation and crossing-over which will be discussed in more detail later.

The average number of chiasmata formed in different bivalents often show a correlation with the length of the chromosome.[2] This is only

[1] Haldane 1931.   [2] *Rev.* Hearne and Huskins 1935.

clearly found in organisms with unlocalized chiasmata, but where the
chiasmata are localized there might be a similar relation between
frequency and the length of the chiasma forming segment, but this cannot easily be determined since there is no way of telling how long the
chiasma forming segment is. In organisms with random chiasmata the

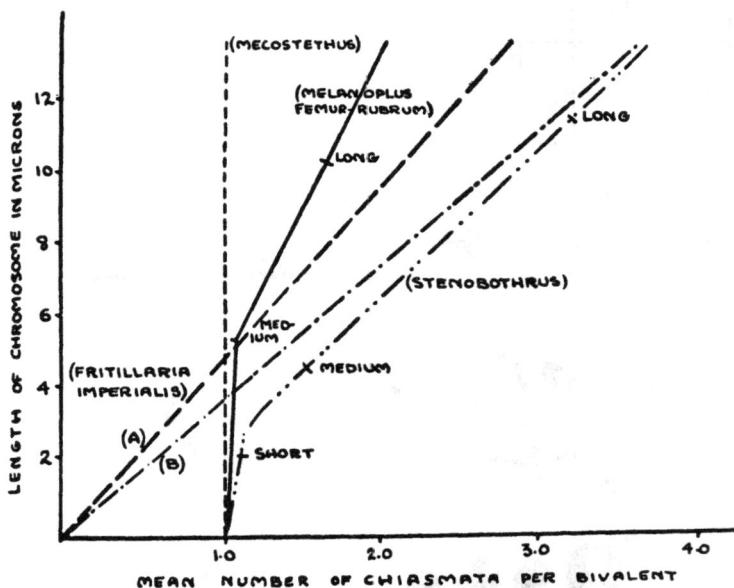

Fig. 57. **Relation between Length of Chromosome and Number of Chiasmata.**—In Mecostethus all chromosomes have one chiasma, whatever their
length. In other organisms the short chromosomes usually have one chiasma and
the longer ones more. In *Fritillaria imperialis* there are two clones, *a* and *b*, with
slightly different chiasma frequencies; in both, the short chromosomes, which are
probably fragments broken off normal chromosomes, may have less than one
chiasma per bivalent.

(From Hearne and Huskins.)

relation with length is not always of the same form. In Fritillaria there
is simple proportionality, but in most organisms there is a tendency for
one chiasma at least to be formed even in the shortest chromosomes. In
Mecostethus with localized pairing there is very rarely more than one
chiasma found whatever the length of the chromosome, though occasional chiasmata are found at the distal ends of some chromosomes.[1]

The relation between chiasma formation and length led Darlington
to formulate the theory that the chromosomes forming a meiotic
bivalent are only held together by the occurrence of chiasmata between

[1] White 1936.

them. The general necessity of some such hypothesis is shown by the fact that the chromosomes are not always associated in threes in triploids, fours in tetraploids, etc., as they would be if the association was due to a general attraction between homologues. Darlington[1] points out that the attraction between homologues (in pairs) is responsible for zygotene pairing, and that this pairing must be distinguished from metaphase association (which he unfortunately also refers to as "pairing"; in this work the word pairing is kept for zygotene pairing and pairing in salivary glands, etc., while the formation of metaphase bivalents is spoken of as association). The suggestion that association is a result of chiasma formation allowed Darlington to use the chiasma-length relation to predict the frequency of association in short chromosome fragments in Fritillaria; the long chromosomes form 2·58 chiasmata per bivalent, fragments one-ninth as long should form 0·29 chiasmata and should be associated in 29 per cent of nuclei; the actual association found was 22 per cent, which is as good an agreement as can be expected.

3. The Theory of Crossing-Over

The earliest speculations on the breakage theory of crossing-over related the phenomenon to chiasma formation: Morgan[2] pointed to the observation of Janssens[3] as providing a material basis for the theory, but did not attempt to distinguish between "total" chiasmatypy, in which breakage and rejoining was supposed to occur between whole chromosomes, and "partial" chiasmatypy, in which it occurs between chromatids. Recent genetic (p. 103) and cytological evidence shows that only the second of these mechanisms need be considered: crossing-over takes place in the four-strand stage. The partial chiasmatype hypothesis, as it is at present put forward (Belling, Darlington),[4] states that the occurrence of a cross-over is the cause of the change of partner among four chromatids associated in pairs which is made apparent as a chiasma. Its only rival for serious consideration is the hypothesis, at one time suggested by Darlington but more recently upheld mainly by Sax,[5] that crossing-over takes place by the breakage of interstitial chiasmata during the reduction in number between diplotene and metaphase.

The partial chiasmatype hypothesis supposes that all chiasmata are the results of previous crossing-over. This cannot be proved directly,

---

[1] Darlington 1930b, 1937.   [2] Cf. Morgan 1926.   [3] Janssens 1909, 1924.
[4] Cf. Belling 1931, 1933, Darlington 1937.   [5] Sax 1930, 1931.

but it can be shown with a high degree of certainty that in some cases chiasma formation has involved a crossing-over. The simplest case is that of a heteromorphic bivalent (e.g. one of the sister chromosomes has a deficiency) which divides into two equal halves (equationally) at the first division: either the two chromatids at one attachment constriction were not sisters or there has been a crossing-over, and the first of these possibilities can be ruled out as inconsistent with all genetic data.

If a chiasma involves a crossing-over, the sister chromatids remain paired together through their length. This must be so, and thus the chiasma must involve a cross-over, in all cases in which it can be

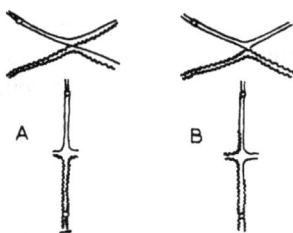

Fig. 58. **Chiasmata and Crossing-Over.**—The "classical" interpretation of a chiasma is shown in A; no crossing-over is involved, as may be seen from the anaphase figure below. The "chiasmatype" interpretation is shown in B; note that sister chromatids remain together in all parts of the bivalent. The lower part of B shows the anaphase configuration, with partially terminalized chiasma; note that crossing-over has occurred.

shown that the paired threads on each side of a chiasma have acted together as a unit. If we find the two threads of one pair coiled as a unit round the two threads of the other pair, this must mean that each pair represents a single chromosome, and that the members of the pairs are sisters. The other special cases in which a cross-over can be shown to be associated with a chiasma are more complicated.[1] (Cf. Fig. 62.)

The generalization of these deductions to the theory that all chiasmata involve crossing-over is prompted partly by a desire to simplify the working hypotheses. It has to its credit two main predictions.

In the first place, it enables one, by ascertaining the average chiasma frequencies of different bivalents, to predict the length in cross-over units of the genetic maps of these chromosomes, since the average occurrence of one chiasma between two points is equivalent to 50 per cent crossing-over. This has been done for maize;[2] and the predicted

[1] Cf Darlington 1930a, 1937.   [2] Darlington 1934b.

lengths are reasonably larger than the maps at present known for the chromosomes, which is an error on the right side, particularly in view of the comparatively small number of genes mapped. In a hybrid between Zea and Euchlaena, Beadle[1] could measure both the chiasma frequency and the frequency of crossing-over in a certain section of chromosome, which was relatively translocated in the two species. He found a crossing-over value of 12 per cent, which should correspond to a chiasma frequency of 24 per cent; the actual number of chiasmata found were 20 per cent, which is a satisfactory agreement.

Secondly, Darlington was led to predict that there must be something very odd about chiasma formation in male Drosophila, in which

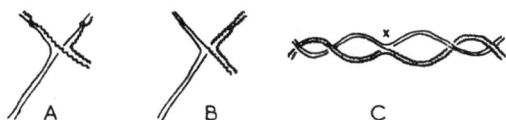

Fig. 59. **Proof that some Chiasmata involve Crossing-Over.**—A shows the "classical" interpretation of a heteromorphic bivalent; the two chromatids paired at a centromere must be different, which is not the case. The "Chiasmatype" interpretation is shown at B. In C two chromosomes have a chiasma at X, and on each side of it the two pairs of chromatids are coiled round one another; each pair therefore behaves as a unit and consists of two sisters, so that a crossing-over must have occurred at the chiasma.

there is no crossing-over but in which the reduction takes place with such regularity that some sort of metaphase association must be supposed to occur. This prediction has been fulfilled: in male Drosophila the chromosomes associate at metaphase by a process other than chiasma formation, being held together in groups of 4 chromatids by some generalized attraction, perhaps analogous to that which causes somatic pairing. The only chiasma formation is between the X and Y (in the inert region) and these two chromosomes show very rare crossing-over.[2]

There are further evidences of the general parallelism between crossing-over and chiasma formation. (1) It has been shown that the variation of chiasma frequency with temperature in mice[3] and Orthoptera[4] is similar to the rather complicated variation in crossing-over value with temperature in Drosophila. This argues at least a strict proportionality between chiasma frequency and cross-over frequency. (2) The total amount of crossing-over is the same in triploid Drosophila as

[1] Beadle 1932.   [2] Darlington 1934a.   [3] Bryden 1935.   [4] White 1934.

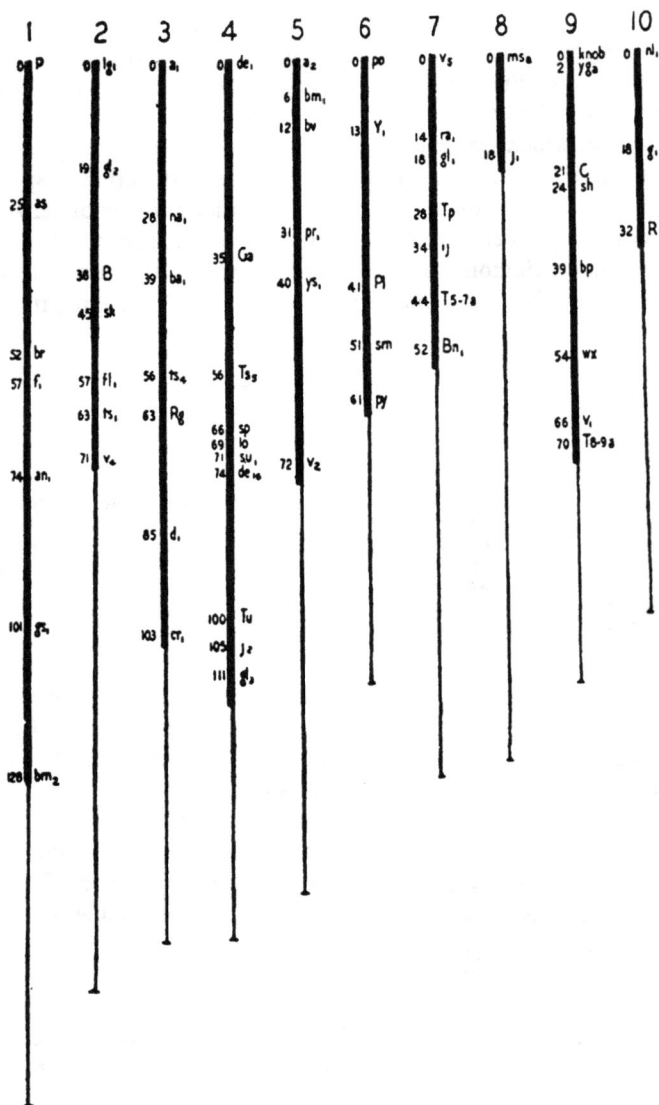

Fig. 60. **Cross-over Maps in Maize.**—The thick lines indicate parts of the map already known, the thin lines the lengths predicted from the chiasma frequencies.
(From Darlington.)

in diploid,[1] that is the cross-over maps are of the same total length, although the distribution of genes along the chromosomes is different in the two cases, probably owing to competition at zygotene pairing. If each cross-over determines a chiasma, there must be the same number of chiasmata per chromosome in diploid and triploid organisms, that is, the total number of chiasmata in triploids must be $\frac{3}{2}$ times the number in a diploid. This prediction has been verified for related diploid and triploid plants.[2]

4. Chiasma Interference

Mather[3] has pointed out that the normal linear relationship between chiasma frequency and chromosome length could be explained if we supposed that in every chromosome pair one chiasma is first formed

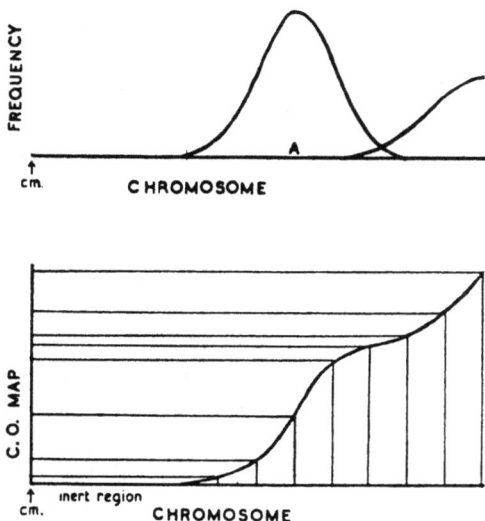

Fig. 61. **The Distribution of Chiasmata and Crossing-Over in the Chromosome.**—The upper diagram expresses the hypothesis that the first chiasma is formed at a certain average distance from the centromere (cm), occurring with a normal distribution round the point A; subsequently a second chiasma may be formed further distally. In the lower diagram the curve gives the total frequency of chiasmata between any point and the centromere, and thus the cross-over distance between that point and the centromere. Note how points equidistant along the chromosome become crowded together in regions of the map where chiasmata are rare.

(After Mather.)

[1] Cf. Redfield 1930, 1932.
[2] Darlington and Mather 1932.
[3] Mather 1936c, 1937.

somewhere near the centromere, and that subsequent chiasmata are formed at distances which vary around a certain interval determined by interference. If we plotted a curve giving the frequency of occurrence of a chiasma at different places on the chromosome, we should obtain a series of overlapping normal frequency curves. The peak of a curve would correspond to a place of high chiasma frequency, i.e. of high frequency of crossing-over. Thus two genes separated by a certain length of chromosome would show crossing-over more frequently if they lay in such a region than if they lay in the trough between the peaks, where chiasmata are rarer. The peaks should thus correspond to lengths of the cross-over map in which genes appear unduly separated, while in the troughs we should expect to find them crowded together. A preliminary analysis of Drosophila chromosomes can be made on these lines.

Chiasma interference and crossing-over interference has been found between different chromosomes in a nucleus. If crossing-over is reduced in one chromosome by the presence of an inversion, there is abnormally frequent crossing-over in the other chromosomes.[1] Similarly, there is a negative correlation between chiasma frequencies in different chromosomes.[2] The mechanism of this is quite obscure.

5. Other Theories of Crossing-Over and Chiasma Formation

The alternative (so-called classical) theory[3] of chiasma formation explains a chiasma as resulting from the opening out of the bundle of four pachytene chromatids alternately along the plane separating the two chromosomes (reductional plane) and the plane separating the two daughter chromatids derived from each chromosome (equational plane). Sax's[4] theory of crossing-over assumes that chiasma formation is of the classical type, and that crossing-over takes place by a breakage and rejoining of the two chromatids which cross one another at a chiasma. This breakage at the same time destroys the chiasma. Two lines of argument have been advanced against this theory:[5] (a) that chiasmata are not actually of the classical type. We have seen above that at least some chiasmata must be of the chiasmatype kind. Darlington also points out that various configurations which would be expected in the classical hypothesis actually do not occur. For instance, different chromosomes sometimes become entangled during zygotene pairing, so

---

[1] Schultz and Redfield 1932, 1933.
[2] Mather and Lamm 1935, Mather 1936a.
[3] Rev. McClung 1927.   [4] Sax 1930, 1931.   [5] Darlington 1932a.

that a chromosome may be pinched in between the two pairing threads of another chromosome pair. If, in the formation of the diplotene loops, sister chromosomes become separated by a split along the reductional plane, this entangled chromosome should be pinched between the two paired chromatids on each side of the loop, but such configurations are never found. Entangled chromosomes always lie in the space between the two sides of the loop. It might, however, be held that the presence of the entangled chromosome determines the plane of separation in this region. If entangled chromosomes are found on both sides of the middle chiasma in a bivalent with at least three

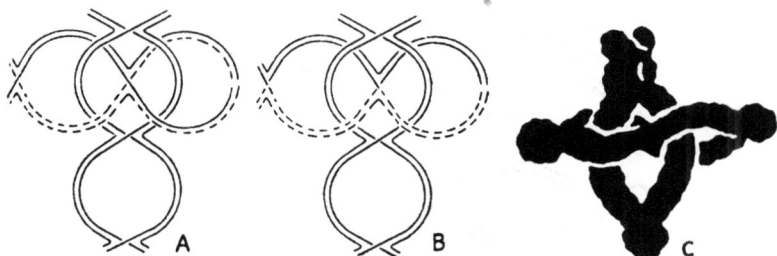

Fig. 62. **Chromosome Interlocking as a Demonstration of Crossing-Over.—** The diagram shows the "classical" interpretation on the left; note that in one region a chromosome would have to lie between sister chromatids, which is impossible. The "chiasmatype" interpretation is in the middle, and an actual example (in Eremurus) on the right.

(From Mather, and Upcott.)

chiasmata, it is clear from the above that the chiasma must involve a cross-over.[1]

(b) It is argued that breakage of chiasmata does not occur. In some organisms with little terminalization (e.g. Lathyrus, Pisum) very little or no reduction of chiasmata number occurs before metaphase, but cross-over frequencies seem to be much the same as in organisms such as Primula with complete terminalization and a great reduction in chiasma number. Further, the statistical evidence suggests that only sufficient chiasmata are formed to account for the observed cross-over frequencies, and as not all these chiasmata become broken the number of chiasma-breakages could not be sufficient to account for the amount of crossing-over. (Cf. prediction of lengths of cross-over maps in maize.) Moreover, the decrease in numbers of chiasmata between diplotene and metaphase, when it occurs at all, is exactly paralleled by the progress of terminalization and can be explained by the pushing off

[1] Mather 1933a.

E

of the first terminal chiasmata without the hypothesis of breakage having to be invoked.

Crossing-over by chiasma breakage is, therefore, a conceivable mechanism, but one whose occurrence has not been demonstrated and is for the above reasons not likely to be general. In his most recent paper Sax[1] seems to have himself abandoned the theory.

6. *Compound Crossing-Over and Chromatid Interference*

A bivalent very often contains more than one chiasma, and there are several ways in which the successive exchanges of partner may be

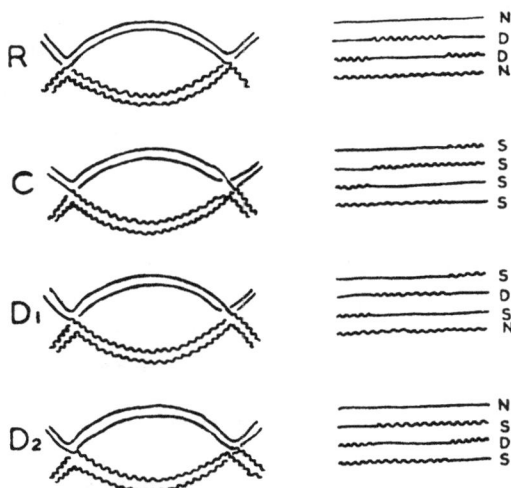

Fig. 63. **Compound Crossing-Over and Sequences of Chiasmata.**—The diagram shows, on the left, the four possible sequences of two chiasmata in a bivalent consisting of four chromatids. R reciprocal, C complementary, D1 and D2 the two types of disparate. At the right are given the corresponding chromatids after crossing over; the non-cross-over chromatids are labelled N, the single cross-overs S and the doubles D.

related to one another. The sequences of chiasmata are said to give rise to compound crossing-over; the term double cross-over is primarily derived from breeding, and implies that two cross-overs have occurred in the same sequence of genes, and we shall see that only some of the chromatids resulting from compound cross-overs are actually double cross-overs in this sense.

Compound crossing-over can be observed cytologically as sequences

[1] Sax 1936.

of chiasmata in the meiotic bivalent. The chromatid configurations which occur can be classified as follows. If A, A' are two sister chromatids and aa' the other two sisters, suppose A and a cross-over at the first chiasma. Then at the second chiasma there are four possibilities. If A, a cross-over again the chiasmata are spoken of as reciprocal (recurrent compound cross-overs), if $A'a'$ cross-over, they are complementary (independent compound cross-overs). In both cases the original arrangement of the partners is restored and the two chiasmata are said to be compensated or comparate. The other two possibilities are that Aa' or $A'a$ cross-over at the second chiasma and these give non-compensating or disparate (diagonal and progressive) chiasmata.

The types of gametes produced are different in the different cases. They can be identified in diplo-haplonts such as Neurospora.[1] It will be noticed that if all types occur with equal frequency, it is only in half the gametes that the original association of the genes at the ends of the chromosome are still associated, i.e. that a double cross-over has occurred. The variation in the intensity of interference along the length of a chromosome might therefore be a variation, not in the frequency with which successive chiasmata occur, but in the type which occur. Actually Hearne and Huskins[2] have described a case in which the cytological types do not apparently occur at random, there being double as many non-compensating as compensating arrangements. In Neurospora the genetic evidence indicates a considerable excess of recurrents, but in other organisms the types seem to occur as would be expected on a basis of chance, so that there is no "chromatid interference."[3]

Note the distinction between these types of compound crossing-over between the chromatids derived from one pair of chromosomes and the recurrent and progressive types of double crossing-over between different chromosomes in a polyploid (p. 107).

7. Crossing-Over in Structural Hybrids

Organisms which are heterozygous for some structural change, such as a translocation or inversion, may be called structural hybrids.[4] Chiasma formation is dependent on zygotene pairing and in such organisms this is often interfered with by the hybridity. If, for instance, part of a chromosome A is translocated on to the end of chromosome B,

[1] Lindegren and Lindegren 1937.   [2] Hearne and Huskins 1935.
[3] Beadle and Emerson 1935, Mather 1933, cf. Darlington 1937.
[4] Darlington 1929.

the two *B* chromosomes when they attempt to pair will be pulled apart by the fragment of *A* attempting to get into contact with its homologue, and unless the chromosomes are perfectly flexible this conflict of forces will prevent some of the normal pairing. In salivary gland nuclei, where pairing can be observed in much more detail than in zygotene nuclei, the chromosomes are usually flexible enough for full pairing of all sections, even when translocated. But this seems not to be the case in meiotic prophase, since it is found that a translocated segment lowers the crossing-over frequency in its neighbourhood and this can be most easily explained as the result of an inhibition of pairing

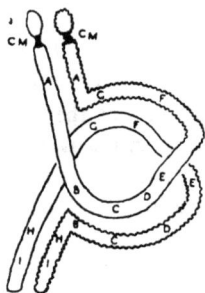

Fig. 64. **Crossing-Over in Inversions.**—The figure shows a cross-over in an organism heterozygous for an inversion, which has paired in the usual loop (see Fig. 47). Only one chromatid of each chromosome is shown, for clearness. Through the cross-over, the two centromeres (*cm*) become attached to the same chromatid, which is therefore pulled in two opposite directions at anaphase, and forms a "bridge." The other cross-over chromatid has no centromere and remains passive. The two non-cross-over chromatids (not shown) are normal.

caused by the relative inflexibility of the chromosome threads. (Other explanations, have, however, been advanced.)[1]

In heterozygous inversions, pairing is possible in two ways. If we have chromosomes *abcdefg* and *abedcfg*, the inverted segments *cde* may pair, leaving the sections *ab* and *fg* together and unpaired at each end; or there may be complete pairing by the formation of a loop. The first type of pairing reduces the total paired length, since only *cde* is paired instead of the whole *abcdefg*; it therefore reduces the number of chiasmata formed and thus reduces crossing-over. (The non-homologous ends *ab* and *fg* may become associated by "snarl" pairing (p. 367), but this probably does not lead to chiasma formation.) The second method of pairing, in loops, is probably the commoner. It is usual in salivary gland chromosomes and has been observed in pachytene in maize. It allows all sections of the chromosome to be paired and crossing-over may occur at any point.

The results of crossing-over in inversions depend on the relation between the point of crossing-over and the centromere. Suppose we have two chromosomes *abc,defg* and *abc,dfeg*, where the comma repre-

[1] Dobzhansky 1931.

sents the centromere. Clearly crossing-over between e and f would give chromatids $abc,defd,cba$ and $gefg$. That is to say, one of the chromatids would have two centromeres (dicentric) the other none (acentric). Similar types of chromosomes appear if there are two cross-overs, and a new type, a loop with one centromere, is also possible. The different types of compound cross-overs give rise to characteristic sets of chromatids; for instance, two complementary compound cross-overs in the inverted segment give rise to two dicentric and two acentric chromatids. The various types of compound cross-overs can be recognized, since the different chromatids have characteristic behaviours.[1] The dicentric chromatids are pulled towards both poles at the first division and thus remain as a bridge between the anaphase chromosomes; the acentric chromatids are not moved from the metaphase plate and are lost, while the loop chromosomes form a bridge in the second division, when their centromere has divided into two daughter centromeres which pull in opposite directions. Most of the sorts of cross-over chromatids which arise are therefore eliminated in one way or another. Inversions, in fact, are often first discovered as cross-over suppressors or C factors.[2] The situation is peculiar in oogenesis in Drosophila, since the formation of chromosome bridges orientates the spindles in such a way that the abnormal chromosomes tend to be eliminated in the polar bodies. The reduction in recombination is therefore not accompanied by an equivalent reduction in fertility.[3]

8. *The Cause of Crossing-Over*

A complete elucidation of the mechanical forces which cause the breakage and rejoining of chromatids and thus give rise to crossing-over and chiasma formation cannot be provided until our knowledge of the nature and physical properties of the chromosomes is in a much more advanced state. One of the most plausible suggestions which has been made[4] is that the breakage might be caused by a torsion applied to the four threads at the moment when the chromosomes divide at pachytene. This hypothesis enables one to understand how chiasma interference may come about: the relief of the torsion by one breakage and rejoining will reduce the force acting on the chromatids in that immediate neighbourhood. The torsional force involved here may possibly be related to the forces producing the spiralization of the chromosomes. Regional differences in crossing-over, and therefore the

---

[1] Cf. Upcott 1937.
[2] Sturtevant 1926a.
[3] Sturtevant and Beadle 1936.
[4] Darlington 1935a, b, 1937.

regional differences in chiasma frequency, are determined by position, and the frequency characteristic of a given region is shown by any section of chromosome occurring there (in translocations, etc.), so they must be due to general mechanical conditions, not to local differences in elasticity, etc.[1]

Darlington has pointed out that when the chromosomes start to contract (by spiralization) while they are paired in zygotene, they begin to coil round one another (p. 366). This relational coiling just balances

Fig. 65. **Relational Coiling and Crossing-Over.**—The left diagram shows two chromosomes before crossing-over. Each chromosome consists of two chromatids, or is just going to split into two chromatids; these are coiled relationally (in a right-handed spiral in this diagram) and the whole chromosomes are relationally coiled (left-handedly). At the point where the chromosomes cross, one chromatid of each chromosome breaks, the broken ends curl away and join up in a new formation after relieving some of the torsion; the final state is shown on the right.

(From Darlington.)

the spiralization which has brought it into being, until the division of the chromosomes in zygotene turns each chromosome into a pair of weaker chromatids, one of which may break under the strain; if one breaks, all the stress will fall on the other chromosome and one of its chromatids will also break. This double break will allow the release of the stress by rotation, after which the broken ends may join up again, when it will be found that a crossing-over has resulted. Accordingly it is found that the relational coiling observable in zygotene is replaced by chiasmata in diplotene. A test of the hypothesis is provided by the behaviour of the unpaired central regions which are characteristic of the chromosomes of some species. The homologues here lie too far apart for chiasmata to be formed, and the relational coiling is therefore

[1] Offermann, Stone, and Muller 1931.

not replaced but persists more or less unaltered. (In an unpaired region at the end of a chromosome, the two homologues can spiralize independently and no relational coiling develops and hence no chiasmata are formed.)

An alternative method of crossing-over, which has been put forward by several authors,[1] is based on the supposition that the division of the chromosome into chromatids takes place in two stages. It is supposed that first the chromomeres (presumably the genes) divide, and that the newly formed chromomeres then form a new connecting thread which becomes the chromonema. Crossing-over would occur if the newly formed thread sometimes joined up with a chromomere derived from the other chromosome. This mechanism is a possible one, but has much less observational basis than the last; there is in fact no good evidence that the connecting thread does grow out from one chromomere till it meets the next. It is perhaps worth while pointing out that the usual diagrams of this process look much more convincing in two dimensions than when they are interpreted as representations of three-dimensional objects.

[1] Belling 1933, Lindegren and Lindegren 1937.

PART TWO

Genetics and Development

Genetic factors are identified in the first instance by the analysis of the inheritance of characters. Now that the mechanism of inheritance is known, in its main outlines at least, it is possible to tackle the next question, of how the genes affect the developmental processes which convert the fertilized egg into the adult organism. Genetics here links up with experimental embryology as another aspect of the general biological problem of reproduction. The first chapter of this part gives a summary of modern views on the causal processes of development in animals, with especial reference to what is known, from the experimental embryoiogical side, of the role of the nucleus. The next two chapters discuss the actions and interactions of genes which control the kinds and quantities of substances produced during development. A separate chapter is devoted to the effect of genes on animal patterns, which is connected with the fundamental problem of the development of biological form. The last chapter deals with the special problem of sex; there are few topics of general biological importance to which genetics has made more significant contributions.

CHAPTER 6

Genes and Development[1]

The mode of action of genes during development can be investigated in two ways: (1) by experiments on the mechanism of development, (2) by examination of the changes produced in developing organisms by gene-changes. The first field of inquiry is usually considered to constitute the separate science of experimental embryology, while the second is a part of genetics, and is often referred to as phaenogenetics. Such a separation is, however, very artificial; both subjects fall into the general field of investigation of how an adult organism arises from the individuals of the previous generation. Thus some at least of the results of experimental embryology are essential for a full understanding of genetics, and in this chapter we shall consider the general mechanism of the development of animals and the ways in which genes may act to control the course of the reactions involved. Unfortunately there are no adequate data for a similar survey of genetic control of plant development.

1. *The Equal Division of the Nucleus*

The simplest hypothesis to account for the genetic control of development might seem to be that the genes were gradually sorted out by unequal divisions of the zygote nucleus, so that each part of the embryo received a different collection of genes, which could then determine the particular way in which that part should develop. Actually the hypothesis is not so simple as it seems; it would be difficult to imagine a mechanism for the differential divisions without assuming that the kind of genes which went into a particular nucleus was determined by the kind of cytoplasm surrounding the nucleus; without, in fact, assuming exactly that differentiation of the embryonic regions which the hypothesis was invented to explain. However, the facts spare us the necessity of devoting further thought to the hypothesis. It is untrue. There is evidence from many different groups of animals that all the first cleavage nuclei, at any rate, are equivalent and can be substituted for one another.[2] The first steps in differentiation therefore

---

[1] *General references:* Daleq 1935, 1938, Huxley and de Beer 1934, Morgan 1927, 1934, Schleip 1929, Spemann 1938, Weiss 1930.　　 [2] *Rev.* Morgan 1927.

do not depend on differential divisions; in some organisms, in fact, such as Protista and a few Metazoa of which Chaetopterus[1] is the best

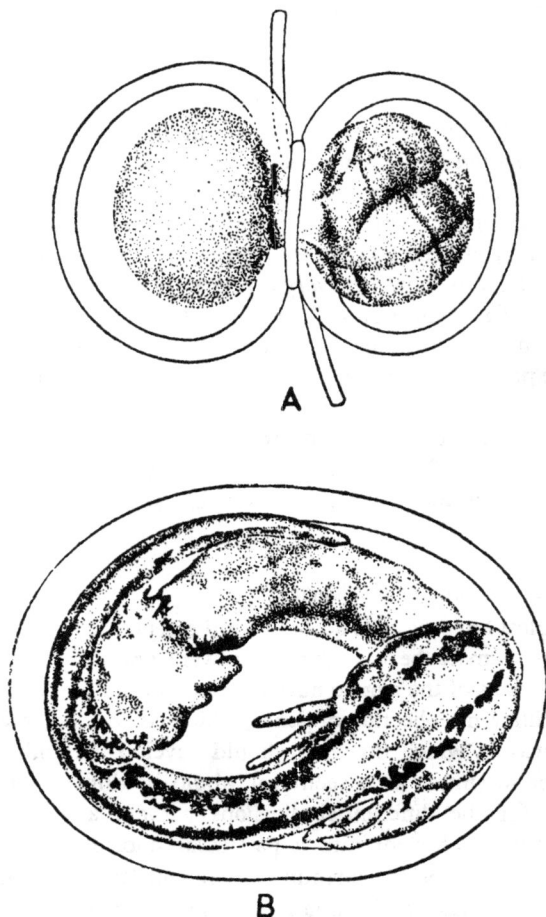

Fig. 66. **The Equality of Nuclear Division during Cleavage.**—*A* a fertilized newt's egg, lying in its gelatinous capsule, was constricted by a hair loop before the first division. The zygote nucleus, in the right half, cleaved four times, and in the 16-cell stage one cleavage nucleus passed over into the left half of the egg. *B* both halves developed into normal larvae, showing that any one of the first 16 nuclei is capable of taking part in the formation of any organ. This result is only obtained if the constriction divides the egg in such a way that both halves possess part of the cytoplasmic organizer; a half lacking the organizer develops no embryonic organs whatever nuclei it may possess.

(After Spemann, from Huxley and de Beer.)

[1] Lillie 1902, Brachet 1937.

known, the first steps of development can proceed even if nuclear division is entirely suppressed. The only example of differential nuclear division in the early stages of development is in Ascaris,[1] where it is probably caused by, and not the cause of, cytoplasmic differences.

The occurrence of differential divisions in later stages cannot be so confidently denied. In plants and lower animals the possibility of complete regeneration from small parts shows that they must contain the whole set of hereditary factors. In higher animals there are a few cases in which a region regenerates an organ other than the one removed (homoesis). But usually the only evidence is the occurrence of small mosaic patches due to somatic mutation in late developmental stages, and these can in the nature of the case only reveal the presence of those genes which affect the mosaic organ, and which would have to be assumed to be there on either hypothesis.

2. General Concepts of Developmental Mechanics

Development is a historical process and we can only understand it if we begin at the beginning. The first-formed organs of an animal may be developed in either of two ways: (1) a particular part of the egg is, from the earliest stage in which we can investigate it, determined to become the organ and does become it under any conditions in which it can develop at all, or (2) this irrevocable determination takes place later and can often be shown to depend on the interaction between the part in question and some other nearby part of the egg. Development of the former kind occurs in the so-called mosaic eggs, and the parts which are determined to become particular organs are said to contain organ-forming substances. A deeper insight into the processes involved can be obtained from eggs which follow the second scheme. Here the various regions of the egg are at first capable of being altered from their normal course of development; they are "indifferent" and may become something other than their "presumptive fate." During a certain period, a process takes place which fixes the future course of development of the various regions; this process is "determination." The sort of organ which any region is determined to become depends on its position within the whole egg. When determination is complete, the egg is in the same state as was the mosaic egg at the beginning of its development; it consists of a set of regions, each determined to develop into an organ of a certain kind, arranged in a definite pattern. Eggs of this

[1] Boveri, Rev. Schleip 1929.

second kind, in which alterations of development are possible in the early stages, are known as "regulation eggs."

In both regulation and mosaic eggs the most fundamental factor in early development is the arrangement of the different regions into a definite pattern. In addition, in regulation eggs the determined regions are themselves formed by a process of interaction. In vertebrates this interaction can be further analysed. The determination is effected in accordance with the position of the region relative to a controlling region (the organization centre) situated near the blastopore, and the whole pattern of the egg is derived from a simpler pattern which is already present in the organization centre. Patterns of this kind are referred to as "individuation fields." The elements which are arranged in the patterns are the determined regions, or originally, the processes by which the regions become determined. These processes are interactions between the indifferent parts of the egg and substances (evocators) given out by the organization centre. The indifferent regions are only capable of reacting with the evocators during a comparatively short period when they are said to be "competent."

In a vertebrate embryo we have therefore a system of three parts. Competent parts of the egg react with evocators to give regions which are determined to develop in certain ways; and the organization centre, which gives out the evocators, also possesses a pattern (individuation field) which controls these reactions so that the determined regions are arranged in a pattern to give normal development. A further insight into the nature of the individuation field can be gained from the fact that a part of an organization centre, when isolated, develops not only into that part of the embryo which it would normally form, but into more; in favourable circumstances it may develop into a whole embryo. Similarly, two organization centres, placed together during the time when they are active in inducing development, may amalgamate and induce only a single complete embryo. The pattern of a whole embryo is therefore an equilibrium to which parts of the organization centre tend to return. The organization centre must consist of different regions which arrange themselves in a pattern which is the equilibrium resulting from their mutual interactions.

An essentially similar scheme may be applied to other regulation eggs, but in most cases we cannot yet analyse the reactions into competent tissues and evocators; we only know that processes of determination go on in definite patterns (e.g. echinoderms, some insects). In mosaic eggs, our knowledge is even less complete. Determined regions appear in a pattern, but we cannot even demonstrate any process which has been involved in the determination.

The concepts which have been summarized above have been worked out mainly on the first formed organs, but they also apply to those which appear later in development. Complications enter, however, with the functioning of the blood stream, nerves, etc., which allow reactions to occur between parts of the body which are far removed from each other. In some cases, e.g. hormonal control, it might also be possible, though perhaps unnecessary, to use the analysis into evocators and competent tissues. If the tissue responds to a hormonal stimulus by any particular pattern of reaction, this pattern is presumably a property of the tissue itself, i.e. of the competence rather than of the stimulus. We shall see that even in early development the competent tissues may not be completely without pattern properties.

After this general account, which has attempted to show the fundamental similarities of developmental processes throughout the animal kingdom, it will be as well to exhibit the variations found in different groups and their dependence on genetic factors.

3. Mosaic Eggs

The eggs of Ascidians[1] may be considered as the best-known example of mosaic eggs, though even in these recent work has shown that some regulation is possible in very early stages. In Styela, Conklin found that at least three organ-forming substances are present before fertilization; a clear cytoplasm surrounding the egg nucleus, a yellow peripheral cytoplasm and a grey inner mass of cytoplasm. These regions cannot of course be formed under the influence of the zygote nucleus, since they are present before it is constituted. Before fertilization they are already arranged in a pattern which is related to the primary egg axis defined by the position of the egg nucleus and the polar bodies. The fertilizing sperm enters in a restricted region near the vegetative pole, and the cytoplasmic regions then move, in a definite way related to the movements of the nuclei, and also break up into further demarcated regions, which, by the time the nuclei fuse, lie in the fundamental pattern of the vertebrate egg which we shall meet again in the Amphibian blastula.

Although these movements are related to the two parental nuclei, there is no evidence that the relation is specific for the particular genetic constitution of the nuclei, which act as triggers releasing a reaction rather than as determinants of the course of events. In any

[1] Conklin 1905, 1924, Dalcq 1932, 1935a.

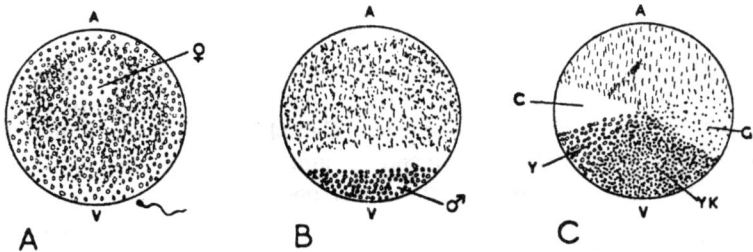

Fig. 67. Organ-Forming Substances in the egg of the Ascidian Styela.—The animal pole A is uppermost, the vegetative pole v down. In A, just before fertilization, there is a layer of yellow cytoplasm on the surface, and a clear area round the female pronucleus in the upper part of the egg. The sperm enters near the vegetative pole, and after fertilization (B) the yellow cytoplasm collects round the sperm nucleus, with the clear cytoplasm from the female nucleus just above it. In C the sperm nucleus moves up and meets the egg nucleus just below the equator on one side, and the clear and yellow cytoplasms move up with it, making two crescents (C and Y) on this side, which will be ventral; the yellow cytoplasm eventually forms muscles and mesenchyme, the clear forms ectoderm. Meanwhile, the yolk tends to collect at the vegetative pole (YK), and above it on the dorsal side appears a light grey area (G) which will form notochord and neural plate.
(After Conklin.)

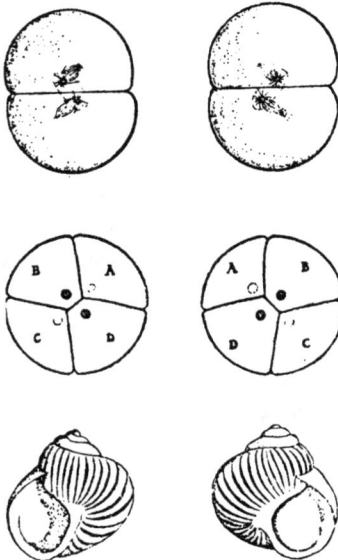

Fig. 68. Cleavage and Shell Form in Left- and Right-handed Gastropods. The left-handed (sinistral) form is on the left, the right-handed (dextral) on the right. The species shown is not *Limnea*. (After Morgan, from Huxley and de Beer.)

case, the whole of the formation of the substances and their arrangement into the definitive pattern occurs before the zygote nucleus is formed. It is clear, therefore, that as regards one individual life-history, the fundamental features of the organism which depend on these cytoplasmic arrangements are transmitted cytoplasmically and not through the nucleus.

Work on another organism with mosaic eggs, the snail Limnea,[1] has raised the possibility that although the fundamental plan of the body is dependent on the cytoplasmic arrangement of the egg, that arrangement itself may be controlled by the genes of the mother. Two forms of the snail are known, one of which is coiled in a right-handed, the other in a left-handed spiral. The way in which any given snail coils is determined by the cytoplasm of the egg from which it developed; but the cytoplasm of the egg is dependent on the genetic constitution of the

Fig. 69. The Inheritance of Coiling in *Limnea.*—The gene D for right-handed (dextral) coiling is dominant to d for left-handed (sinistral) coiling, but the direction in which a snail coils is determined not by the genes it contains but by those its mother contained, e.g.

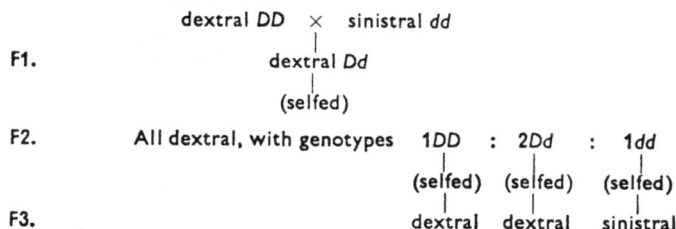

	dextral *DD*	×	sinistral *dd*

F1. dextral *Dd*

(selfed)

F2. All dextral, with genotypes   1*DD* : 2*Dd* : 1*dd*

(selfed) (selfed) (selfed)

F3. dextral   dextral   sinistral

mother in which the egg was formed, right-handedness being dominant. Thus from the evolutionary point of view, where we are interested not in single ontogenies but in whole series of them, we can say that the direction of coiling is inherited by a nuclear, genetic mechanism. The same may perhaps be true for the organ-forming substances of, for instance, the Ascidian egg.

4. *Insects*

The embryology of insects[2] is of particular interest to geneticists because that group has provided, in Drosophila, the best material yet discovered for genetical investigation. Unfortunately it is not equally good as a subject of experimental embryological research.

The early investigations of insect eggs showed them to be highly

[1] Boycott and Diver 1923, 1930, Sturtevant 1923.
[2] *Revs.* Bodenstein 1936, Richards and Miller 1937, Seidel 1936.

mosaic; localized injuries to the eggs of Musca result in corresponding local injuries in the larvae. Musca belongs to the group of insects whose cleavage follows a fairly definite pattern. Recently work has been done on other insects whose early development is more indeterminate. Seidel, working on the dragonfly Platycnemis, has elucidated a series of phenomena which may occur quite widely in insect development. The first event of importance, after fertilization, is the arrival of one of

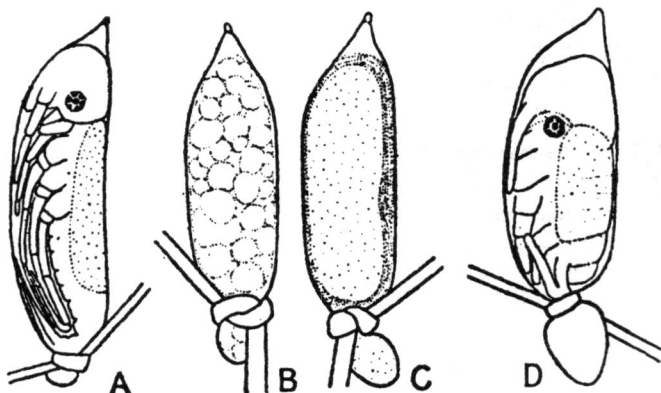

Fig. 70. **The Activation and Differentiation Centres in Insects.**—A. A constriction at the most posterior end of the egg, behind the activation centre, allows a normal embryo to develop. B. If a constriction is made at an early stage just in front of the activation centre, no embryo develops, the anterior part of the egg forming extra-embryonic blastoderm as in C. D. If the same constriction is made at a later stage, after the activation centre has finished its action, an embryo develops in that part which is still in cellular contact with the differentiation centre.

(From Seidel.)

the cleavage nuclei, accompanied by a small quantity of cytoplasm, in the posterior end of the yolk-filled egg. The movement towards this region takes place under the influence of a general repulsion between the cleavage nuclei, and by various means it is possible to make different members of the group of early nuclei arrive first at the posterior end. Any nucleus, no matter which, when it arrives at the posterior end, reacts with the particular cytoplasm localized there and sets free a substance which diffuses forward throughout the entire egg. In any region which this substance is prevented from reaching, no development occurs; but the substance merely activates development and does not determine the nature of the development of the different regions. Again there is no evidence that the particular genetic nature of the nucleus affects the activating substance, but no particular investigations of this point have been made.

The most important of the regions activated by the substance lies on the dorsal surface of the egg and is known as the differentiation centre. It is also presumably dependent on a cytoplasmic localization. At this centre the formation of the embryonic rudiments begins, and around it the egg is organized as a unitary organism. It is not clear how far one can say that the differentiation centre impresses a pattern on the developing egg or egg-fragment, but it is clear that the centre is as it were the focus round which the pattern is formed. Parts of the egg which contain the activator substance but which have been deprived of cellular contact with the differentiation centre can only develop into incomplete structures in which no regulation occurs.

The occurrence of regulation in fairly late stages of Platycnemis led to a re-examination of the possibility of regulation in other insects. In certain other forms with indeterminate cleavage, regulation by means of a differentiation centre can be shown to occur in the blastoderm stage, as in Platycnemis, or even later. Regulation is certainly much less in determinate types, which include Drosophila. Geigy,[1] and also Howland,[2] have, however, shown that complete larvae may be obtained from eggs injured immediately after fertilization. Most of the genetic characters studied in Drosophila are characters of the adult, and very little is known about their determination. Geigy has reported that after making injuries in young egg stages, he obtained perfect larvae but defective adults, so that it is possible that the adult characters were already determined and were affected by the injury: on the other hand, injuries to the larvae may have been overlooked.

5. Echinoderms

The echinoderm egg is often taken as a nearly perfect example of a regulation egg. But actually regulation of a fragment to give a complete larva is only possible if the egg is cut along a plane parallel to its main animal-vegetative axis. This axis expresses a fundamental cytoplasmic pattern which is present even in the unfertilized egg. The pattern can be considered as two opposed gradients;[3] one, having its high end at the animal pole, tends to produce ectoderm and its derivatives, the other with its high end at the vegetative pole, tends to produce an invagina-

[1] Geigy 1931a, b. For Drosophila development see Poulson 1937, Robertson 1936.
[2] Howland and Child 1935, Howland and Sonnenblick 1936.
[3] Hörstadius 1928, 1935, 1936.

tion and the development of endoderm. Embryonic determination is a result of the interaction of these two gradients. Whereas in the vertebrates the pattern of the determined elements is apparently impressed upon them by an active organization centre working on relatively passive competent material, here the pattern is a balance between two

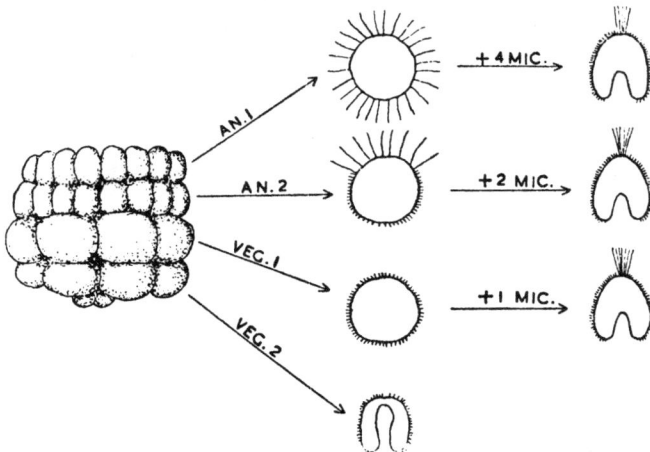

Fig. 71. Diagram illustrating the Gradients in Echinoderm Eggs.—A cleavage stage is shown on the left with the animal pole upwards and vegetative pole downwards. The four main circlets of cells can be separated and allowed to develop in isolation. The middle column shows that the uppermost (animal—1) forms long sensory cilia over the whole surface, and fails to gastrulate. In animal—2 the sensory cilia are confined to the upper half. Vegetative—1 usually gives no sensory cilia or gut, but sometimes one or other of these is present in a reduced form. Vegetative—2 gastrulates and develops an abnormally large gut. If the small cells at the vegetative pole (micromeres) are isolated they fall apart and do not develop. The column on the right shows the normal gastrulae formed from combinations of an.—1 + 4 micromeres, or an.—2 + 2 micromeres or veg.—1 + 1 micromere.

(After Hörstadius.)

active regions. Possibly such a balance, which involves the whole extent of the echinoderm egg, will be found within the organization centre of the vertebrates.

The two opposing gradients are already present in the unfertilized egg, and are clearly cytoplasmic in nature. Something is already known about their metabolic basis;[1] it is suggested that the animal gradient involves a carbohydrate oxidation which can be inhibited by lithium salts, while the vegetative gradient depends on an anabolic process,

[1] Lindahl 1936.

which is probably proteolytic, and which gives rise to poisonous phenolic and indolic end-products which are normally neutralized by combination with sulphate ions.

The egg cytoplasm not only contains the fundamental plan of the developing gastrula, but also seems to control the speed of cleavage. Thus in several hybrids between species in which the eggs cleave at different rates, it has been stated that the rate of cleavage is exactly that of the maternal species and that no influence of the sperm can be found.[1] An example[2] is a cross between Dendraster ♀ and Strongylocentrotus ♂. In the original species the times from fertilization to the first and second cleavages are (at 20° C.) Dendraster 57 and 28 minutes, Strongylocentrotus 95 and 47 minutes. Not only did the hybrids cleave at the faster rate, which excludes the possibility that the effect might be due to any sort of injury consequent on cross-fertilization, but even enucleated fragments, fertilized by Strongylocentrotus sperm, cleaved at the faster rate characteristic of their cytoplasm. This behaviour is not always found, since in crosses between large and small varieties of rabbits, the influence of the sperm on the cleavage-rate can be detected as early as the four-cell stage.[3]

In most echinoderm hybrids, the paternal characters begin to be shown at about the gastrula stage. The position in which the mesoderm cells are given off from the endoderm, and the type of skeleton formed, are usually intermediate between the two parental types, though in some cases (e.g. Echinid × Antedon) even these comparatively late characters are purely maternal. von Ubisch[4] has recently re-investigated several cases in connection with his studies on "germ-layer chimaeras," which are obtained by adding the presumptive mesoderm cells of one species to a blastula of another. The foreign mesoderm can be added in various proportions, and if the host mesoderm is removed, forms can be produced in which all the mesoderm belongs to a species other than that of the ectoderm and endoderm; these correspond among animals to periclinal chimaeras among plants. On comparing the skeletons formed in hybrids between species *A* and *B* with those formed in chimaeras containing mixtures of *A* and *B* mesoderm, the shape was found to be much the same in both cases, and intermediate between the two parental types. Thus for this late character the chromosomes alone in the hybrids had almost the same effect as the chromosomes and cytoplasm of the added mesoderm in the chimaeras.

[1] *Rev.* Morgan 1934, Schleip 1929.   [2] Moore 1933.
[3] Castle and Gregory 1929.   [4] v. Ubisch 1937.

6. *Amphibia*

Amphibian eggs are the most fully studied of the vertebrate types, and provide the classic example of the organization centre. Spemann[1] showed that the region round the blastopore can, when transplanted into new surroundings in a gastrula, cause those surroundings to develop into part of the embryonic body including neural tube, notochord, somites, etc. This active region round the blastopore, which is

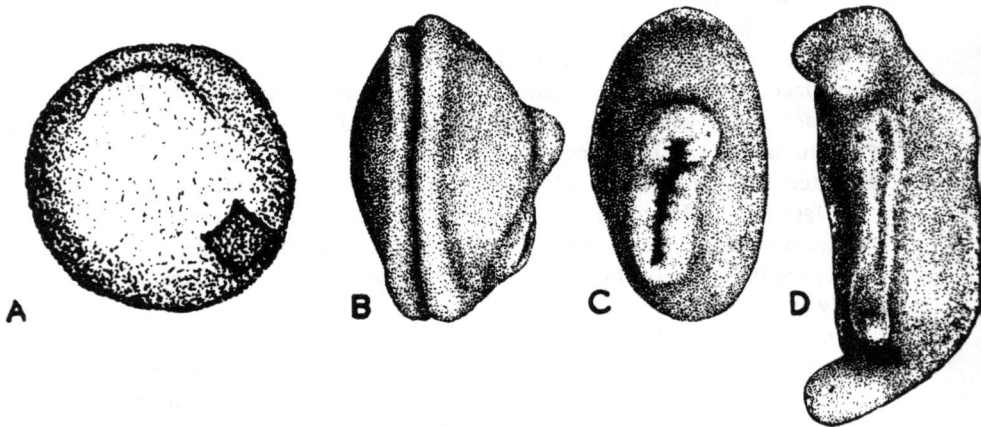

Fig. 72. **Organizers in Amphibia.**—A. Ventral view of a young gastrula of a newt, showing the blastopore at the top; the dorsal lip of the blastopore from another egg has been grafted to the bottom right. B. Dorsal view of neurula stage, showing the neural folds of the host embryo, with induced embryo on right. C. Neural folds of induced embryo. D. The host and induced embryos at a later stage (sides reversed).

(After Spemann and H. Mangold.)

known as the organization centre, is formed in a region of cytoplasm which becomes localized at fertilization or just afterwards; there is no evidence that the zygotic nucleus has anything to do with this localization which seems to be dependent on the interaction between the gradient of yolk-content from the animal to the vegetative pole and another dorso-ventral gradient.[2]

One part of the process by which the organization centre induces the formation of a new embryonic rudiment is concerned with the production of particular kinds of tissues, such as neural tissue, etc. This process consists of a reaction between the tissue and a substance or

[1] Spemann and H. Mangold 1924.   [2] Dalcq and Pasteels 1937.

substances produced by the organizer. The substance concerned with the production of neural tissue (known as the neural evocator) can be isolated in an impure form; its nature is still in doubt, but there are some grounds for suggesting that it may be sterol-like.[1] It is extremely unspecific, in the sense that it is quite independent, both in origin and in action, of the specific constitution of the tissues involved. Thus frog evocator will work on newt tissue, and chick evocator on mammals. It is even found in groups widely separated from the vertebrates; most remarkably of all in hydroids, which do not even possess a nerve cord.

This suggests that the genetic constitution of an egg acts by modifying the reaction to evocators, and not by determining the character of the evocators themselves. Thus when frog evocator acts on newt tissues, it induces newt neural tissues. It seems fairly certain, in fact, that genetic factors control histological type by controlling the reactivity or "competence" and not by controlling the evocators.

The situation is more complicated when we consider not only the histological type of the induced tissues but also the pattern in which they are arranged. Again most of the experiments have been made between different species or genera, and the genetic analysis is still lacking. It is certain that the pattern in which the induced tissues are arranged is partly dependent on the organizer. The process by which the organizer influences the pattern is called "individuation," and it is probably performed by some method other than the diffusion of a single homogeneous chemical substance; evocator extracts do not transmit any pattern to the reacting tissues. The organizer-pattern is, as might be expected, closely related to the specific nature of the organizer. A large Axolotl organizer induces an Axolotl-sized neural plate in frog ectoderm,[2] and similarly for other combinations.

It is remarkable that organizers from different groups share pattern-properties which may not be at all apparent in normal development. A newt has nothing to correspond to the sucker of a frog tadpole. But if frog ectoderm is grafted into a newt embryo in a position similar to that in which the frog develops a sucker, there is some influence there which induces a sucker to form.[3] In this position there must be some similarity between the two organization centres. It is difficult to believe, however, that the similarity could be sufficiently detailed to determine the actual structure and arrangement of the sucker, and it is probable that the

---

[1] Waddington, Needham, and Brachet 1936, Needham 1936.
[2] Holtfreter 1935.
[3] Spemann and Schotte 1932, cf. Rotman 1935a, b, Twitty 1936.

reacting tissue is itself able to produce some pattern properties. But we are still largely ignorant of how extensive the pattern-forming properties of the competent tissues may be. As regards neural tissue, it seems that the normal pattern is almost entirely dependent on the action of the organizer, but in other organs the competent tissue may be more important.

There are already some indications of the relative importance of the nucleus and cytoplasm in determining the pattern properties of organizers and competent tissues. The formation of the pattern within the

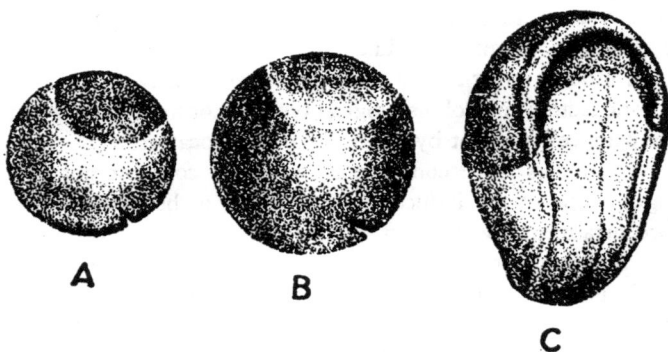

A   **B**   **C**

Fig. 73. **Control of Pattern by the Organizer.**—The two figures on the left show an exchange-graft of ectoderm between a small light newt gastrula and a large dark Axolotl gastrula. The figure on the right shows the Axolotl material taking part harmoniously in the formation of the newt neural plate.
(From Holtfreter.)

organization centre seems to be connected with the original localization of the organizer within the egg, and this process, as we have seen, is mainly cytoplasmic or at least independent of the zygote nucleus. In the unfolding of the pattern, both components appear to play a part. Dalcq[1] showed that eggs fertilized by sperm which had been injured by X-rays or trypaflavine fail to gastrulate properly, probably because the injury to the genetic constitution upsets the pattern of the organizer. Similarly, Hamburger[2] has described characteristic abnormalities of development which must be due to incompatibility between the chromosomes and cytoplasm. The same conclusion can be more certainly drawn from experiments on bastard merogony, that is, experiments in which the egg nucleus was removed and the enucleated

[1] Dalcq and Simon 1932.   [2] Hamburger 1936.

fragment fertilized by foreign sperm. Baltzer and Hadorn[1] showed that the bastard merogons die by reason of a necrosis of certain particular regions of tissue. In the *Triton palmatus* ♀ × *Triton cristatus* ♂ hybrid, the limiting tissue is the head mesenchyme, which dies at the neurula stage. This is part of the organization centre, so here the incompatibility of nucleus and cytoplasm affects the organizer pattern. The competence of tissues is also affected. Pieces of the merogon, other than

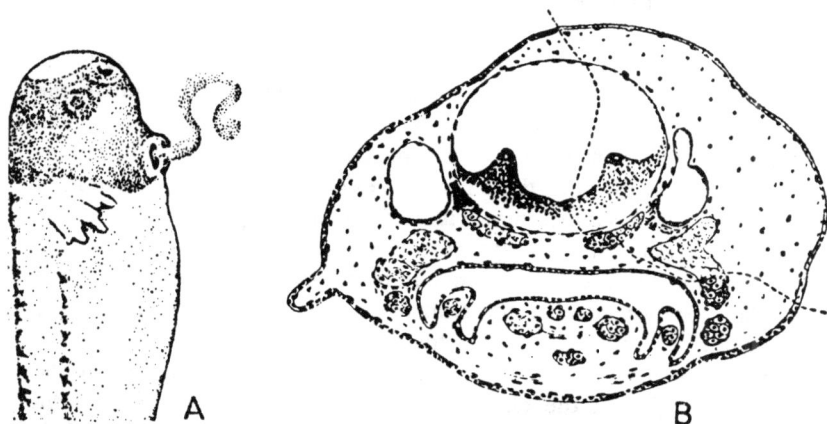

Fig. 74. **Specific Differences expressed through Competence.**—A newt larva on to which Anuran ectoderm had been transplanted in the gastrula stage. A shows that the ectoderm forms suckers, although no such organs occur in this position in normal newt larvae. B is a section showing the Anuran material above and to the right of the dotted line; it forms a harmonious part of the embryo, the pattern of its development being largely controlled by the host organizer; but it retains its own histological type.

(From Holtfreter.)

the head mesenchyme, may survive for a considerable time if transplanted into other surroundings in a normal embryo and one, at least, of the tissues formed (epidermis), shows the specific characteristics of the species from which its cytoplasm was derived. It is not known whether the influence of the nuclear factors would be shown in organs formed still later in development, as would be suggested by comparison with the echinoderm experiments mentioned above.

We may sum up by saying (1) that an organization centre induces the formation of new sorts of tissues by means of stimulating substances which are apparently the same in many different species; (2) the type of tissue produced in response to the organizer is dependent on the

[1] Cf. Baltzer 1933, Hadorn 1935, 1936, 1937.

specific nature of the reacting tissue; (3) the specific nature of an organizer affects the pattern in which the induced tissues are arranged, but this pattern is also influenced by the specific nature of the reacting tissues. In all these processes, the specific constitution of the tissues seems to be dependent on cytoplasmic as well as nuclear factors.

7. Other Vertebrates

Organizers have been shown to exist in all other classes of vertebrates except the reptiles, which present technical difficulties which have not yet been overcome. The mechanism of their action is probably the same in principle as that of the amphibian organizers, namely the production of new types of tissue by the action of evocator substances acting on competent tissues, with some degree of control by the organizer of the way in which these new tissues are arranged. The position of the main organization centre within the embryo differs, of course, in different classes. It is always related to the focus of the gastrulation movements. Thus in fish[1] the endoderm and mesoderm are both formed at the posterior edge of the embryonic disc, and this region is the organizer. In birds,[2] the gastrulation is split up into two phases, and so is the organizer. In the first phase, the endoderm is formed in the posterior region of the blastoderm and the newly formed endoderm is an organizer which induces the next stage of gastrulation, which is marked by the appearance of the primitive streak from which the mesoderm is given off. The region in which the mesoderm is being most actively invaginated is then in its turn a second organization centre and induces the formation of the neural plate. In mammals,[3] it has been shown that the primitive streak can induce neural tissue (when transplanted into a chick blastoderm) but nothing is known of the organizing powers of the endoderm; the technical difficulties of testing it are very great.

8. Acetabularia

A unique opportunity for investigating the role of the nucleus in development occurs in the rather special case of the unicellular green alga *Acetabularia mediterranea*. In spite of the fact that this plant consists of only a single cell, with a single nucleus, it attains a considerable size, up to about 5 cm. In shape it is rather like a toadstool, with a long narrow stalk bearing at one end an umbrella-like "hat" and at

[1] Luther 1935, Oppenheimer 1936.
[2] Waddington 1932, 1933.   [3] Waddington 1937.

the other a cluster of rhizoids. The nucleus always lies near the rhizoid end, and can easily be removed by cutting off this end. Hämmerling[1] found that enucleate fragments can live for a considerable time, and can also regenerate missing parts. The regeneration of a hat occurs more frequently in fragments originating near the hat end, while

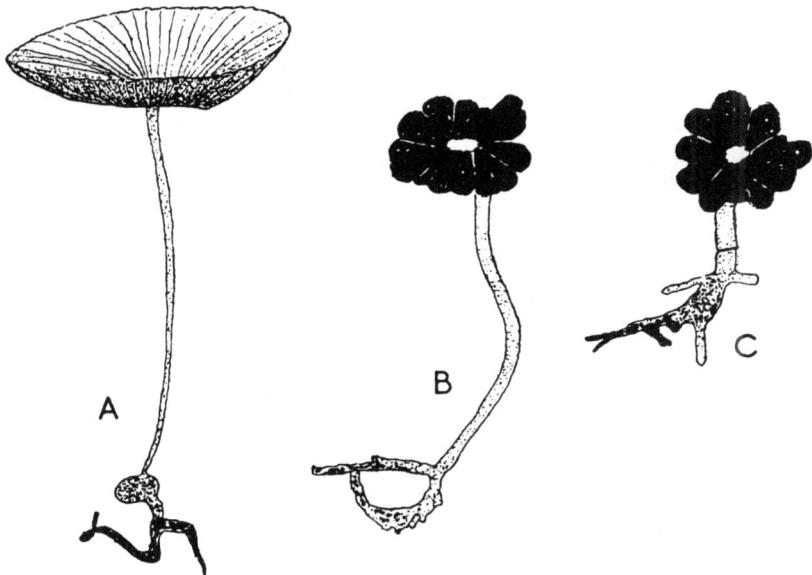

Fig. 75. **Acetabularia.**—A. *Acetabularia mediterranea*, the stem has been some-what shortened in the drawing. B. A. *Wettsteinii*. C. Hat regenerated from a piece of *mediterranea* stem grafted on to *Wettsteinii* rhizoid containing the nucleus.
(After Hämmerling.)

rhizoids tend to be regenerated more easily from the other end. By a series of regeneration and grafting experiments, Hämmerling demonstrated the existence of two gradients of morphogenetic substances, a "hat substance" concentrated at the hat end, and a "rhizoid substance" concentrated at the other end. These substances must exist independently of the nucleus in the enucleate fragments, but there is good evidence that the nucleus is ultimately responsible for their production. If, for instance, a piece is isolated from the middle of the stalk, it contains very little of either substance, and regeneration only rarely occurs. If, however, the rhizoid end containing the nucleus is left

[1] Hämmerling 1932, 1934.

attached to such a piece, hats are much more frequently regenerated, though this process does not begin until a certain latent period has elapsed. The natural conclusion is that new hat substance has been produced under the influence of the nucleus

The nuclear determination of the morphogenetic substances is most conclusively demonstrated by transplantations between different species. A piece of the stem of *A. mediterranea*, without a nucleus, may be transplanted on to the nucleus-containing rhizoid region of *A. Wettsteinii*, and then amputated and allowed to regenerate. It forms a *Wettsteinii* "hat." Thus the nucleus imposes its specific characteristics, even its pattern characters, on to the cytoplasm. Hämmerling concludes that the cytoplasm in this species has no determinative influence on the type of development, but he points out that no regeneration at all occurs unless the cytoplasm is in a reactive state; it must, in our language, be competent to react to the organizing action of the nuclear stuffs.

This is perhaps the only case in which the production of morphogenetic substances has been shown to be under the direct control of the nucleus. Unfortunately, we cannot take the conditions in a unicellular organism such as Acetabularia as typical of those in Metazoa. It is, however, very important to find a case in which pattern can be determined by a single substance. This is very difficult to understand unless we suppose that the pattern is formed as a result of the interaction of the substance with a spatially heterogeneous substrate. The importance of the Acetabularia case is that it suggests that the factors which interact to give rise to a pattern are chemical substances, just as are the factors which merely stimulate histological changes and which we have called evocators.

9. *Development as an Epigenetic Process*

One of the classical controversies in embryology was that between the preformationists and the epigenisists. The former supposed that all the characters of the adult organism were present in the fertilized egg and only needed to be "unfolded" in some way during ontogenesis, while the latter maintained that development involved the production of something absolutely new, which arose from the interaction of the original constituents of the zygote. After the account which has just been given of the modern view of development in actual concrete cases it is rather difficult to attach any very definite meaning to these old theories. It is, surely, obvious that the fertilized egg contains constituents which have definite properties which allow only a certain limited number of reactions to occur; in so far as this is true, one may say that development proceeds on a basis of the "preformed" qualities of the fertilized egg. But equally it is clear that the interaction of these constituents gives rise to new types of tissue and organ which were not present originally, and in so far development must be considered as "epigenetic." In particular, the discovery of the organizer, and of the way in which the primary organizer induces tissues or organs which later act as secondar·· organizers, has led to great emphasis being placed in recent years on the epigenetic character of developmental processes.

These considerations have a bearing on the common genetical terms genotype and phenotype. Neither of these is very easy to define precisely, partly because their usage is even now not strictly standardized. The genotype was originally defined[1] as the sum total of the genes contained in the fertilized egg, but the word is usually used to refer comprehensively to the whole genetic system of the zygote considered both as a set of potentialities for developmental reactions and as a set of hereditable units; that is to say, it includes not only the mere sum of the genes, but also their arrangement, as expressed in position effects, trans ocations, inversions, etc. The question arises as to whether the cytoplasmic characteristics of the zygote are to be included in the genotype, but although they are obviously a very important part of the developmental potentialities of the zygote, it seems advisable not to include them in the genotype: probably it is better to consider them as part of the phenotype determined by the genes of the mother.

The main difficulty in defining the concept of phenotype is caused by the fact that animals change in time. In its original sense, the word referred to the characters, both anatomical and physiological, of the adult. But clearly if we have an animal whose eye colour darkens with age there is no essential difference between the light eye of the young animal and the dark eye of the older one; both must be included in the phenotype. But if this is allowed, there is no reason to exclude from the phenotype the processes (about which we usually know very little) by which the eye pigments are synthesized during development. The phenotype in fact must be used as a name for the whole set of characters of an organism, considered as a developing entity. Phenotypic differences between two organisms may be caused by genotypic differences or may be produced by different environments acting on the same genotype.

[1] Johannsen 1911.

The concepts of genotype and phenotype are defined in the first place in relation to differences between whole organisms. They are not adequate or appropriate for the consideration of the development of differences within a single organism. In this connection we do, indeed, require the concept of the hereditary (chromosomal) constitution of the zygote, and we can without danger extend the meaning of the word genotype to cover this. But the difference between an eye and a nose, for instance, is clearly neither genotypic nor phenotypic. It is due, as we have seen, to the different sets of developmental processes which have occurred in the two masses of tissue; and these again can be traced back to local interactions between the various genes of the genotype and the already differentiated regions of the cytoplasm in the egg. One might say that the set of organizers and organizing relations to which a certain piece of tissue will be subject during development make up its "epigenetic constitution" or "epigenotype"; then the appearance of a particular organ is the product of the genotype and the epigenotype, reacting with the external environment. In transplantation experiments, such as those described above, it is the epigenotype which is altered.

10. Substance Genes and Pattern Genes

In discussing the effects of genes during development we must attempt to make hypotheses as to their actions which fit in with the general scheme of developmental mechanics which has just been given. We can start by making a distinction between genes which apparently affect the kinds or amounts of substances produced, and genes which affect the patterns in which the substances are arranged. This distinction, like the similar distinction between evocation and individuation in embryology, is not an absolute one. The "substances" with which we are concerned may not be well-defined chemical units but may in some cases be tissues, which of course themselves have a pattern on a small scale. One of the main problems of developmental physiology, however, is to discover the nature and origin of the comparatively large-scale patterns found in organisms, and to try to understand, for instance, how the characteristic arrangement of tissues in a limb can be explained in terms of physico-chemical entities which have a pattern on an altogether smaller scale. The category of pattern genes is introduced because it makes it impossible to forget this most fundamental problem. We shall class genes as pattern genes when they affect some pattern of a supra-molecular order, and usually quite large in relation to the whole organism. They are singled out from other genes, not because they are necessarily fundamentally different but because they are particularly interesting, just as vitamins and hormones are singled out from the group of food substances and chemical constituents of the plasma.

We shall see that every gene probably affects several different processes, and in discussing the action of a gene we shall usually be concentrating on one only of the whole set of reactions in which it is concerned. We should therefore, to be strictly accurate, speak not of substance and pattern genes but of substance and pattern aspects of genic action, but the gain in accuracy is not worth the clumsiness which such locutions would introduce. The context will make clear which aspect of the gene's total activity is being referred to.

CHAPTER 7

The Interaction of Genes: The Effects

1. *Allelomorphs and Multiple Factors*

In Mendel's original experiments, the two allelomorphic forms of the genes used were related to one another as dominant and recessive. That is, one of the pair, the dominant, came to complete expression in the heterozygote, suppressing all signs of the other recessive gene. This relationship is not a necessary one; all gradations are known between cases in which the heterozygote shows the full effect of one gene (complete dominance) and cases in which the heterozygote is strictly intermediate and it is impossible to speak of either gene as dominant. A considerable degree of dominance is, however, much more usual than would be expected on a basis of pure chance, and several attempts have been made to explain why this should be; they are considered later (p. 297).

The relation of dominance-recessiveness depends on the interaction between two allelomorphic genes, but interactions of a similar kind exist between genes belonging to different loci; in fact, all the genes in the genotype probably react with one another during development. We shall discuss the nature of the reactions in the next chapter: probably they are always reactions between gene-products rather than between genes themselves. Here we shall examine the evidence that such interactions do occur.

One form of this interaction is the dependence of a phenotypic character on the presence of two different genes. A well-known case is the walnut comb of fowls, which is the form assumed in the presence of both of two factors R and P, of which R alone gives the rose comb and P alone the pea comb, both of which are dominant over r or p which give the simple "single" comb. Another well-known example was found by Bateson and Punnett in sweet-peas, where the formation of coloured flowers is dependent on the presence of two dominant factors C and P.

The term complementary factors is usually reserved for cases in which a simple presence or absence of a character is determined by the presence or absence of two or more specific members of different loci. Many more cases are known in which the degree or kind of character is dependent on the interaction between different loci. Thus in *Drosophila melanogaster* there are at least 30 loci whose primary expression is an effect on the eye colour. Similarly in maize well over 50 genes have been described which affect the formation or distribution of

Generation	Parent Size	34	37	40	43	46	49	52	55	58	61	64	67	70	73	76	79	82	85	88	91	94	97	100
																								Mid-values of Classes in mms.
P1			1	21	140	49														13	45	91	19	1
F1									4	10	41	75	40	3										
F2	61							3	9	18	47	55	93	75	60	43	25	7	8	1				
F3	60				2	3		9	25	37	70	19	10											
	80											2	8	14	21	39	39	32	10	1				
F4	44			8	42	95	38	1																
	85													4	9	38	75	59	6	3	1			
F5	41	3	6	48	90	14																		
	90													2	3	8	14	20	25	25	20	8		

Fig. 76. **Inheritance by Polymeric Genes.**—The table shows the results of a cross between varieties of *Nicotiana longiflora* differing in corolla length. The corollas are grouped in classes whose mid-values are shown in the upper row of figures; the table gives the number of corollas falling into each class. The varieties are presumed to differ in several factors affecting corolla length. In the F1 these factors would be mainly heterozygous and the corolla length intermediate. In later generations plants are obtained which are homozygous for some of the factors and which therefore have more extreme corollas, either short or long. Owing to the number of factors involved, the pure parental types are not recovered till the F5.[1]

The increased variability of the F2 and later generations as compared with the F1 can only be explained as a result of the segregation of factors, and is usually accepted as sufficient evidence that the character in question is inherited through Mendelian factors even if these cannot be separately identified.

chlorophyll[2] and in *Primula sinensis* there are 11 genes for flower colour.[3] They are called multiple factors.

The detailed interaction between such genes has not always been worked out; the magnitude of such a task is obvious in cases like that mentioned in maize. In the simplest case the genes which affect the same character may interact in a purely quantitative way, either adding to or subtracting from the total effect. Genes of this kind are known as polymeric genes, and their presence may often be deduced from breeding results, since they give rise to an intermediate F1, but an F2 in

[1] Data from East 1910.
[2] Cf. Eyster 1934.
[3] de Winton and Haldane 1932.

F

which segregation can be detected by the increase in variability. One of the first cases to be interpreted in this way was that of the colour of the grain in wheat. Nilsson–Ehrle[1] showed that the red colour is produced under the influence of at least three factors, whose effects are cumulative. In a cross between a race with the three dominants and one with the three recessives the F1 will be intermediate, but the extreme grandparental forms will appear again in the F2, which will therefore be more variable than the F1. If the two original races each had some dominant extreme factors, these may be combined in some of the F2, with the production of forms more extreme than were the grandparents. Phenomena of this kind are common in investigations on the inheritance of what are frequently known as "quantitative" characters; naturally any character may be quantitative if we can find a way of measuring it; but the characters referred to here are those in which more or less continuous variation is observed, so that measurement has to be undertaken before the investigation can proceed at all. Standard examples are such things as height or weight in man, length of ear, height, in maize, etc. Similar phenomenon are also very common in crosses between natural species or geographic races (p. 273).

2. *Epistasis*

Another form of interaction between multiple factors is found when one factor is "dominant" over the others, i.e. suppresses their expression. We cannot use the terms dominance and recessiveness in such cases since the factors involved occupy different loci and are not allelomorphs. Bateson[2] proposed the words epistasis and hypostasis. Thus in mice the grey coat colour, produced by a factor spoken of by Bateson as *G*, is epistatic to the black coat colour produced by *B*, so that the *GGBB* mouse is grey.

3. *Suppressors*[3]

A gene *A* epistatic to *B* completely suppresses the expression of *B* in the compound *AB*: it also has an effect of its own. Suppressors are genes which fulfil the first of these conditions but not the second; in the compound with their "suppressee gene" they suppress its expression but have no other effect, and produce no change in the absence of the suppressee. Thus in *Drosophila melanogaster* there is a factor *pr* for

[1] Nilsson-Ehrle 1908.
[2] Cf. Bateson 1930.   [3] Bridges 1932*a*, *b*, Schultz and Bridges 1932.

purple eyed colour (IInd Chr) which is suppressed by a factor *su* lying in the IIIrd chr. The compound *pr pr su su* has normal eye colour. A suppressor is not the suppressor of a locus but of a gene. Thus the compound *pr pr su su* is not phenotypically the same as a deficiency of the *pr* locus: it is the same as $+^{pr} +^{pr}$. Similarly if *A* is epistatic to *B*, *AABB* is not the same as *AA* def.*B*, but the epistatic relation means that there is no difference between *AABB*, *AABb*, or *AAbb*.

Fig. 77. Modification of the Dihybrid Ratios.—On selfing an individual which is heterozygous for the two unlinked factors *A* and *B*, one obtains 9 zygotes containing both *A* and *B*, 3 with *A* and *b*, 3 with *B* and *a*, and 1 with neither *A* nor *B*. If both *A* and *B* are completely dominant, the phenotypic ratio will be modified from the fundamental genotypic ratio of 9 *AB* : 3 *Ab* : 3*a B* : 1 *ab* according as the classes are separately distinguishable. Some of the possible modifications are as follows:

9 : 3 : 3 : 1. Factors simple dominants, e.g. *A* = rose comb, *B* = pea comb, *AB* = walnut comb, *ab* = single comb in fowls.

12 : 3 : 1. *A* epistatic to *B*, so that *AB* looks the same as *A*; e.g. coat colours in mice *A* = *AB* = grey, *B* = black, *ab* = chocolate.

9 : 3 : 4. *B* ineffective in absence of *A* (or *a* epistatic to *B*); e.g. mice *A* = basic factor for colour formation, *a* = albino, *B* = black.

9 : 7. *A* and *B* complementary; e.g. *A* and *B* are the two factors (known as *C* and *R*) necessary for colour formation in sweet-peas.

13 : 3. *B* is a suppressor of *A*, having no effect itself; e.g. *A* is the factor for colour in fowls, *B* its suppressor (dominant white).

9 : 6 : 1. *A* and *B* are polymeric factors, each producing the same effect, the two effects being added in *AB*; e.g. polymeric factors for colour of the grain in wheat.

4. *Modifying factors*

Factors like the suppressors mentioned above, which have no effect unless they occur in compounds with some other definite genes, are known as modifiers or modifying genes for the factors with which they are effective.[1] Suppression is an extreme case of modification; usually the effect is less marked. Many modifying genes probably do not differ in any fundamental way from ordinary genes: they merely produce a slight effect which is concealed by the "factor of safety" (p. 185) characteristic of the dominant wild-type genes but which can be revealed in the less equilibrated mutant forms. Thus scute in *D. melanogaster* has a sub-threshold effect in certain bristles which is not revealed until the genotype is "sensitized" by the introduction of another bristle-gene, in this instance Hairless[2] (p. 371); scute might therefore be referred to as a modifier of Hairless. In other cases, where a particular allelomorph introduces a completely new reaction into the

[1] Bridges 1919.   [2] Sturtevant and Schultz 1931

developmental processes (is a neomorph, p. 166) the modifier may really have no action in the absence of the modified gene.

A particular example of the latter kind of factor interaction is found in "sex-limited" genes; the genes do not have any expression except in the presence of the (neomorphic) sex factors of one sex. Genes affecting primary and secondary sex characters but not essential to the determination of sex would of course fall into this category; so do factors controlling polymorphism of one sex as in Lepidoptera for example.[1]

5. *Multiple Effects of a Factor or Pleiotropy*

The example of scute and hairless mentioned above shows that easily recognizable genes may act as modifiers of one another. Such interactions are not limited to groups of genes all of which have as their main effect an alteration of the same organ of the phenotype. Thus the expression of an "eye colour gene" may be altered by the presence of a particular allelomorph at a locus whose main effect is on some other character such as bristle length. Any gene, in fact, has not only a main effect by which it is usually recognized but also a host of smaller effects which may be difficult to detect in the normal organism but may be apparent as modifying effects in mutant races. Dobzhansky[2] has described some of the subsidiary effects of the white series of allelomorphs in *D. melanogaster*; for instance, they affect the testis, colours and shape of the spermatheca as well as the colour of the eye. In some cases it is easy to see that the multiple effects of the gene are all of the same nature. This is so with the eye colour and testis colour in *Drosophila* mentioned above, and with the factors for flower colour in many plants (e.g. in Mendel's peas), which also cause colour to develop in various other parts of the plant. But sometimes the connection between the different effects is anything but obvious; for instance, the factor dumpy in *Drosophila melanogaster* which affects the wings, patterns of bristle, viability, etc. (p. 168). In some cases one set of phenotypic effects is produced under one kind of environmental condition while other conditions of temperature or humidity, etc., may produce quite different ones. The plurality of effects of one gene may be related to the cases of non-seriable multiple allelomorphs mentioned below, in which different allelomorphs belonging to the same locus affect different organ systems. The generality of phenomena of this kind is shown by the very usual effect of genes on viability. Practically any gene substitution in an organism alters the general capacity of the

[1] Gerould 1911, *Rev.* Ford 1937.   [2] Dobzhansky 1927, cf. 1930*a*.

organism to develop and maintain itself, and this alteration is probably an expression of the influence of the gene on the whole complex of reactions constituting the life of the organism. Similarly, cells usually die unless they have some representative of each locus; even very small deficiencies are nearly always lethal to cells in which they are homozygous.[1]

6. The Genotypic Milieu

We have seen that any given character may be affected by several different genes, and that any given gene may affect several of the trains of reactions going on in the development of an organism. There is, therefore, no simple one-one relation between a gene and a phenotypic character, but such a relation only exists between the phenotype and the genotype as a whole. This is sometimes referred to as the balance theory of genetic action. Thus it is strictly incorrect to say that w^+ corresponds to red eyes, and w to white eyes in *Drosophila melanogaster*: we should say that, in the usual genotypes met with in *Drosophila melanogaster* a substitution of w^+ for w will change the eyes from white to red. The whole of the genotype other than the particular gene in which we are interested can be referred to as the genotypic milieu or the genetic background.

Timofeeff-Ressovsky[2] has made a careful study of the effect of the genetic background on the expression of certain genes in *D. funebris* and has shown that the backgrounds of different local races may alter both the degree of expression (the expressivity), the penetrance (p. 190) and the actual mode of expression of certain genes (p. 191). Other studies of the same phenomenon have been made with particular reference to the effect of the genotypic milieu on relations of dominance and recessiveness: they are referred to on pp. 185, 297.

7. Dosage Relations of Genes

One of the most important variables in the genotypic milieu is the quantity of genes present. To investigate this aspect of the problem we require to know how the expression of a gene is altered firstly when more of it is added to one and the same background, and secondly when more of the background is added to the same quantity of the gene. Muller[3] has made an especial study of this question and has proposed dividing genes into five types. These types are founded on the relation between the gene considered and some other standard allelomorph

[1] Demerec 1934*a, b*.   [2] Timofeeff-Ressovsky 1934.   [3] Muller 1932*b*.

belonging to the same locus. The choice of a standard is obviously the wild-type, when such a thing exists, and it will be easiest to give the definitions in terms of such a scheme, which may be applied to an organism such as Drosophila. The classification is, however, always a classification of relations between genes, not of genes themselves, and this must be explicitly stated for organisms in which no standard or wild type is available.

a. Hypomorphs are genes which do the same thing as the standard gene, but do it less efficiently. An increase in the number of a hypo-

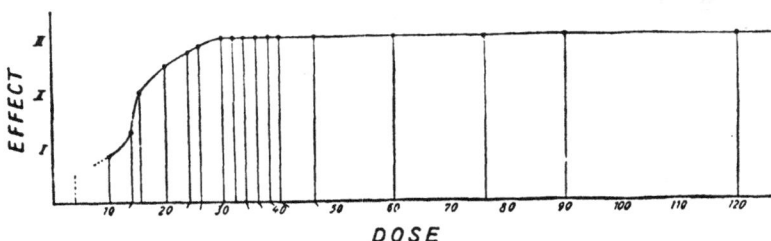

Fig. 78. The Relation between the Dose of a Hypomorph Gene and its Effect.—The locus concerned is that of bobbed in *Drosophila melanogaster*. Five allelomorphs were used, of which two, *bb'* and *bb''* lay in the Y chromosome and could, therefore, be added to flies without any difficulty. All the allelomorphs are concerned in the formation of bristles, but with different efficiencies. If one takes an arbitrary value of 30 for the minimum dose which produces full-sized bristles, the efficiencies of the allelomorphs can be estimated as: $+ = 30$, $bb' = 10$, $bb = 8$, $bb'' = 4$, $bb^l = 2$. The curve shows the effects of various combinations of genes in diploid females. Note that normal bristles can be obtained by adding sufficient mutant allelomorphs, e.g. *bb bb bb'' bb'* = 30, *bb bb bb' bb'* = 36, etc.

(After Stern.)

morphic gene in a constant genotypic milieu (e.g. by adding chromosome fragments containing it and only a few other genes[1] or by adding otherwise "empty" Y chromosomes)[2] increases the effect of the gene until eventually sufficient of the hypomorph may be present to give the same effect as the standard.

Schultz[3] worked with shaven, an apparently bristle-reducing (that is, actually an inefficiently bristle-producing) recessive gene in the IVth of *Drosophila melanogaster*. He could add IV's, which contain few genes and can be taken as nearly simple additions of shaven, until the diploid genotypic background with four doses of shaven showed extra bristles, that is, the four shavens gave more effect than the wild type gene. He also investigated the effect of increasing the dosage of the

[1] Muller 1933.   [2] Stern 1929.   [3] Schultz 1934.

genes other than shaven; in the triploid a given dose of shaven has less bristle-producing effect than does the same dose in the diploid. The rest of the genotype therefore works in a way opposed to shaven and the phenotype is the result of a balance between the two.

Hypomorphs can be detected by less comprehensive experiments than these by observing the effects of deficiencies for them. If a gene is working like the standard, but less efficiently, a deficiency should work less efficiently still and the heterozygote $\frac{\text{standard}}{\text{deficiency}}$ may show the recessive phenotype ("pseudo-dominance"); indeed, if the gene is nearly as efficient as the standard, the substitution for it of a deficiency may cause the heterozygote to show even greater difference from the

	Haplo-IV sv	Diplo-IV sv/sv	Diplo-IV sv/+	Triplo-IV sv/sv/sv	Tetra-IV sv/sv/sv/sv
Diploid ..	extreme shaven	shaven	wild type	slight shaven	extra bristles
Triploid ..	dies	extreme shaven	slight shaven	shaven	slight shaven

Fig. 79. **Different Doses** of the bristle-reducing gene *shaven* (IVth chromosome) in diploids and triploids of Drosophila.   (*After Schultz.*)

normal than does a homozygote hypomorph. This phenomenon is known as exaggeration.

b. Amorphs are an extreme case of hypomorphs. A gene may have very little of the same effect as the standard gene, and the limit of variation in this direction is for the gene to have no effect; these are amorphs. Additions of them would cause no alteration in the phenotype, and the heterozygote standard/deficiency would appear just like the normal heterozygote, and show no exaggeration. They are the contraries of neomorphs (see below).

c. Hypermorphs are genes which do the same thing as the standard but do it better. Since this classification is really a classification of relations between genes, the hypermorphic relations, being the converse of the hypomorphic one, must obviously be included. Cases such as Timofeeff-Ressovsky's[1] mutation (caused by X-rays) from the hypomorph forked to the wild (hypermorph) forked allelomorph demonstrate the actual possibility of such genes occurring. But in organisms such as Drosophila where the wild type is taken as the standard, hypermorphs are rarely found. Muller gives Abruptex as the only case for Drosophila. The reason for this is discussed below (p. 297).

d. Antimorphs are genes which have an effect opposite to that of the

[1] Timofeeff-Ressovsky 1932.

standard. An example in Drosophila is abnormal abdomen; the Ab gene has an effect more extensive than a deficiency for the wild-type allelomorph. One might say that the gene was doing the same thing as the wild-type but with negative efficiency. Another example is ebony: $\dfrac{e^+}{e}$ is lighter than $\dfrac{e^+}{e}$ plus e while $\dfrac{e}{e}$ is darker, i.e. adding e to $\dfrac{e^+}{e}$ darkens the fly, while adding e^+ to $\dfrac{e}{e}$ lightens it. Dubinin and Siderov[1] have described another antimorph in the IVth, *cubitus interruptus*.

e. *Neomorphs* are genes which are doing something quite different to anything done by the standard gene; in fact, the standard behaves towards them like an amorph. Hairy wing is an example, $\dfrac{Hw}{Hw^+}$ is less hairy than $\dfrac{Hw}{Hw}$ but $\dfrac{Hw}{Hw}$ plus Hw^+ is the same as $\dfrac{Hw}{Hw}$. Thus addition of Hw^+ has no effect. Bar behaves as a neomorph.

Mangelsdorf and Fraps[2] described a gene producing yellow pigment and vitamin A in the triploid endosperm of maize. The recessive allelomorph apparently produces no vitamin A, but each dose of the neomorphic dominant produces about two units of vitamin per gram of grain.

8. *Dosage compensation*

Sex-linked genes are present in only a single dose in the heterogametic sex, but in a double dose in the homogenetic sex. We should expect a female with two hypomorphic genes to show a different phenotype from a male with only one. But in Drosophila the males and females of mutant types are usually very alike; the differences in dosage have been compensated. This compensation can only be due to the action of modifiers also lying in the X and thus present in the same dosage as the gene whose effect they are modifying. The important thing for the expression of an X chromosome factor is therefore not its relation to the genotype as a whole but an intra-chromosomal balance between the gene and modifiers within the X chromosome itself.[3]

Muller[4] points out that such a system of dosage compensation could hardly be evolved if it had no function. Its function must be its effect on the wild type rather than on the mutant, since it can be of no particular evolutionary importance if the male and female mutant types,

---

[1] Dubinin and Siderov 1934. [2] Mangelsdorf and Fraps 1931.
[3] Stern 1929. [4] Muller 1932*b*.

both disadvantageous (otherwise they could not be mutants but would have become wild-type), are equally or unequally disadvantageous. The effect of the dosage compensation system must, he argues, be to make the effects of the wild-type genes the same in the two sexes. Now a change from one to two doses of a dominant gene appears to have no effect in the female. But, if Muller is correct, there must actually be a difference between the homozygous wild-type and the heterozygote. He suggests that the difference is one of variability. The homozygote is well on the horizontal part of the curve relating phenotypic effect to dosage (see Stern's curve for bobbed), while the heterozygote is near the inclined part of the curve, where the phenotypic effect probably varies with chance environmental influences as well as with dosages changes. This behaviour of the wild-type must have been evolved through the selection of the appropriate modifiers, which (1) give one dose of the gene an effect near the horizontal part of the curve, and (2) give two doses an effect well on the horizontal part. It should be noticed that (2) involves a restriction of the effect of higher doses; it is apparently not usually advantageous simply to allow the homozygote to produce double the minimum effect which can be allowed to the heterozygote, but is better to bring them both to the optimum.

The argument is therefore as follows: We can see that one dose of a hypomorphic gene in a male is brought to have the same effect as two doses in a female; therefore we must assume that the same thing applies to the wild-type allelomorphs. One dose of a wild-type gene in a female appears to have the same effect as two doses, but this cannot be true; two doses in a female or one compensated dose in a male must be less variable than one dose in the female.

Exactly the same considerations apply to autosomal wild-type genes, except that here the compensating mechanism need not lie in the same chromosome. But again it is presumably advantageous to get the homozygote well on to the horizontal part of the curve so as to limit its susceptibility to environmental variations, and this automatically brings the heterozygote nearly or quite to the same apparent effectiveness. The suggested mechanism therefore involves the selection of modifiers so as to make the wild-type gene dominant.[1] That selection of this kind may affect the wild-type gene as well as the modifiers is indicated by the fact that two known wild-type allelomorphs of white, found in different (American and Russian) races of

---

[1] For an account of dominance modifiers (acting on vestigial in *D. melanogaster*), see Goldschmidt 1937.

Drosophila melanogaster, have been shown by Muller[1] to have different efficiencies in producing red pigment, i.e. to be on different parts of the dose-effect curve. (Cf. pp. 185, 298.)

9. *Multiple Allelomorphs*

The simplest series of multiple allelomorphs are related to one another as hypo- and hypermorphs. That is, they all produce the same kind of effect but with different efficiencies. The genes at the locus of white eye in *Drosophila melanogaster* are a good example. These range down through coral, which is very little lighter than the normal red eye, through a whole series of members (eosin, cherry, apricot, buff, tinged, ivory, etc.) to white, which is practically an amorph, although not quite, as white-deficiency is even lighter in shade. Stern[2] in particular has stressed this quantitative relation between the effects of different members of an allelomorphic series (see p. 164). He has also drawn attention to the fact that simple relations of this kind, though common, are not by any means universal. Thus in the white series itself, there is the remarkable fact that the series of diminishing effectiveness is different for different phenotypic actions of the genes.[3]

Eye colour	..	..	$W > w^b > w^e > w$
Vitality	..	..	$W > w^e > w^b > w$
Fertility	..	..	$W > w > w^b > w^e$

In other cases it may be impossible to arrange the allelomorphs in series because their effects are non-commensurate, e.g. the locus of spineless has two allelomorphs, *ss* giving bristle reduction, and *ss^a* (aristopoedia) giving foot-like appendages in place of the arista in *Drosophila melanogaster*.

Another example in the same organism is the dumpy series.[4] This locus has three main effects: shortening the wing, causing the formation of vortices of the hairs on the thorax, and reducing the viability. Nearly all the possible combinations of these effects can be found in different allelomorphs, as is shown in the table, in which abnormal effects are indicated by minus signs, the number of which gives an indication of their strength.

---

[1] Muller 1933.   [2] Stern 1930.
[3] Timofeeff-Ressovsky 1933a.
[4] Muller (personal communication), cf. Stern 1930.

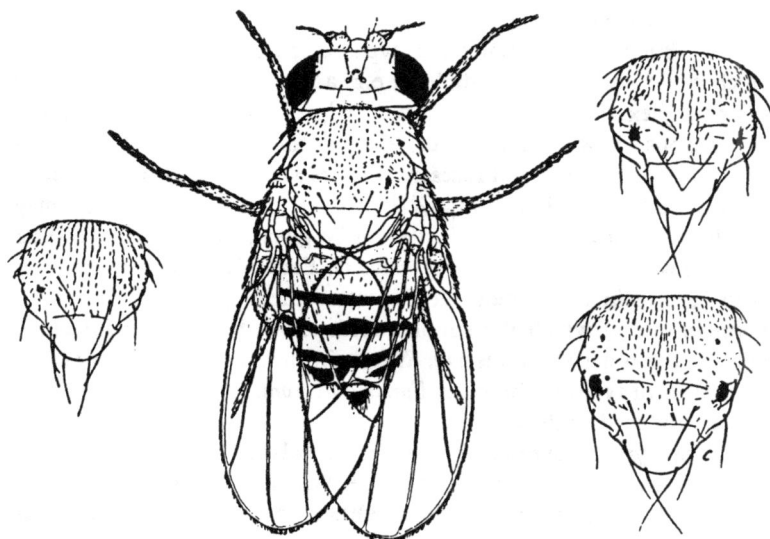

Fig. 80. **Vortex,** one of the allelomorphs of the dumpy locus, showing the whorls of bristles from which the gene takes its name.

(After Bridges and Mohr.)

			Shortening of wing	Vortices	Viability
Vortex ..	..	..	+	—	+
Oblique..	..	..	—	+	+
Dumpy..	..	..	— —	— —	+
Thoraxate	..	..	+	— — —	—
Lopped..	..	..	— — —	+	—
Truncate	..	..	— — —	— — —	—
Normal allele ..	..	+	+	+

The individual effects are usually recessive, so that for instance the compound oblique-thoraxate is phenotypically normal; their allelomorphism is shown by the fact that they both show some effect in combination with dumpy. In the presence of intensifiers, the more extreme effects become partially dominant.

10. *Dosage in Polyploids and Polysomics*

Phenotypic alterations due to dosage changes are also seen in polyploids and polysomics. The changes in homozygous autopolyploids are presumably mainly due to alteration in the balance between the chromosomes and cytoplasm and are not of a very far reaching nature (p. 67). But in an autopolyploid, types of heterozygosity are possible which could not occur in a diploid; e.g. a factor A may in a tetraploid occur simplex $Aaaa$, duplex $AAaa$, or triplex $AAAa$, as well as in the homozygous $aaaa$ (nulliplex) or $AAAA$ forms. The new balances between the dominant and recessive factors may give new intermediate phenotypes for the $Aaaa$ and $AAAa$ classes, or the simplex type may be fully dominant. No general rule can be given, the result depending on just how hypomorphic the recessive is. Examples have already been given (p. 57) in the section on characters affecting the triploid endosperm in one of which the simplex form showed the dominant character, in the other the recessive. Wettstein[1] has described a very complete series of types in the moss *Funaria hygrometrica* where polyploids can be produced artificially.

In allopolyploid hybrids clear cases of balance between the genes from the two parents may be found. A well known example is Karpechenko's[2] radish-cabbage hybrid. Other good cases are known in hybrid mosses (p. 67).

In polysomics the unbalance may give very, or only slightly, abnormal phenotypes. In such cases, the balance is between whole blocks of genes, and the units involved (whole chromosomes or large parts of chromosomes) have a general effect on nearly every part of the organism, usually with very little tendency for a main effect on one character: one is dealing as it were with a lump of the genetic background, not with a specific gene. It might have been thought that fairly large blocks of genes, such as whole chromosomes, would have nearly the same residual effect as the whole genotype and would therefore, when reduplicated, have less influence on the general balance of the genotypic milieu than would a small fragment containing just a few isolated genes. This is not so as a rule. The effect of a duplication is usually roughly proportional to the amount of reduplicated chromatin. The explanation is probably that most genes are working in the horizontal part of their dose-effect curve, and that an extra dose has little effect. The alteration caused by adding a new block of genes is dependent only on those comparatively few genes which are not at their maximum of effectiveness, and the number of these is roughly proportional to the total number of genes added.

The very large and varied series of polysomics synthesized by Blakeslee[3] and his collaborators in Datura provide perhaps the best

[1] Wettstein 1927.　　[2] Karpechenko 1924, 1928.　　[3] *Rev.* Blakeslee 1934.

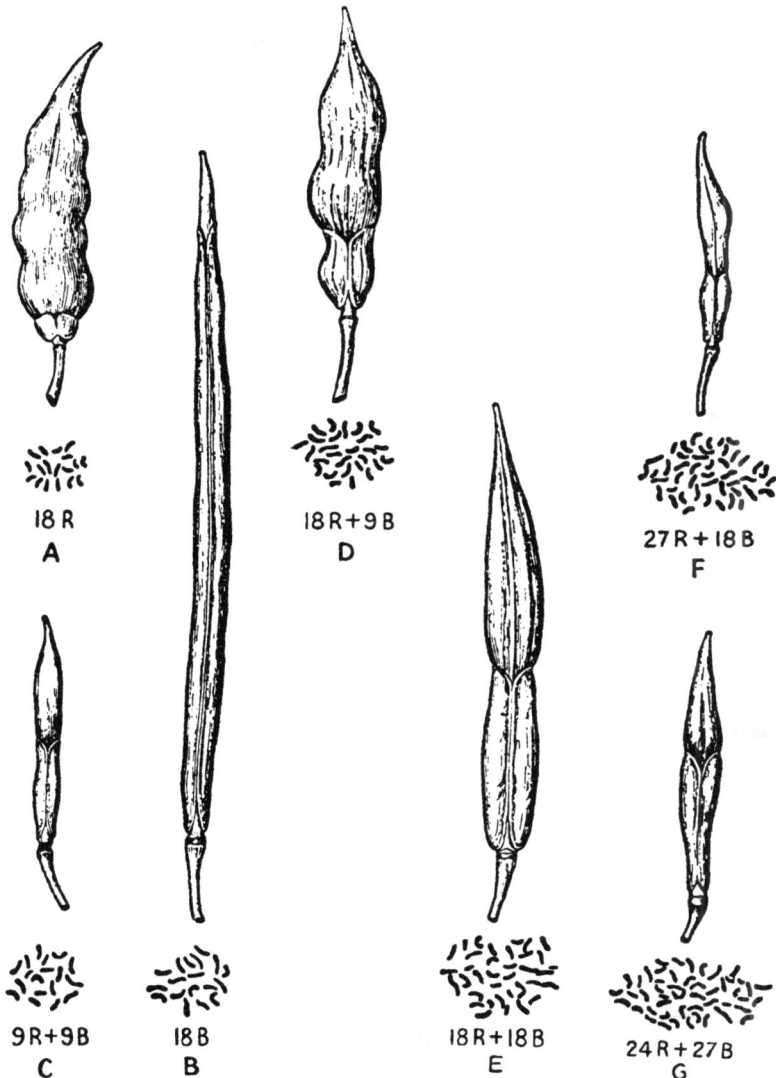

Fig. 81. **Polyploid Hybrids of Radish and Cabbage.**—A shows the seedpod of radish, B that of cabbage, C the diploid, sterile, hybrid. E is the tetraploid hybrid, D is a triploid with two sets of radish and one of cabbage chromosomes, F is a pentaploid (three radish, two cabbage) and G a hypoploid hexaploid (three cabbage and rather less than three radish). Note how the character of the pod depends on the proportion of radish and cabbage chromosomes.

(From Hurst, after Karpechenko.)

example of the effects of chromosome balance. The haploid chromosome number is twelve, and there are twelve primary trisomics, from which one may determine the developmental effect of each chromosome; as might be expected, the effects are usually not confined to a single part of the plant, but are rather general. The effects on the shape and spikiness of the seed-capsules have been very fully described and figured, and we shall confine our considerations to that organ.

The effect of a given chromosome, as expressed in the trisomic diploid, becomes exaggerated in the tetrasomic diploid, in which it is represented four times, but is reduced in the pentasomic tetraploid ($4n + 1$); the $4n + 2$ form is very like the $2n + 1$, and the $4n + 3$ is slightly less extreme than the $2n + 2$. Thus we have a clear case of the dependence of phenotypic expression on the balance between opposing tendencies; one can compare this with the balance theory of sex determination in Drosophila and the cases of balance in individual "quantity genes" discussed on p. 165.

In the secondary trisomics, the reduplicated chromosome consists of two similar ends; they can be compared, as regards dosage, with a $2n + 2$ form, but here it is only half a chromosome which is reduplicated, not a whole one. Typically the secondaries are less abnormal in appearance than $2n + 2$ plants, and, when compared with the primary from which they arose, show an exaggeration of some but not all of its features. Thus the primary Rolled has a reduplication of the chromosome whose ends are labelled as 1 : 2. From it two secondaries can be derived according to which end is doubled; in one secondary, Polycarpic, the trisome consists of two 1-ends, while in Sugarloaf it consists of two 2-ends. Each of these secondaries has an exaggeration of some of the abnormal features of Rolled, which is more or less intermediate between them.

A rather similar state of affairs is found in the doubly trisomic diploids, i.e. $2n + 2$ forms in which the two extra chromosomes are not alike. The double trisomic usually combines the characters of the two primaries. More complicated examples of balance occur in tertiary trisomics, which are $2n + 1$ forms in which the extra chromosome consists of two parts from two other chromosomes, e.g. $2n + 1 : 6$, where 1 is a chromosome end from the trisome in the primary Rolled and 6 is an end from the trisome in another primary Buckling. The tertiary trisomic is, as might be expected, more or less intermediate between the two secondaries to which it is related, e.g. $2n + 1 : 6$ is intermediate between $2n + 1 : 1$ and $2n + 6 : 6$.

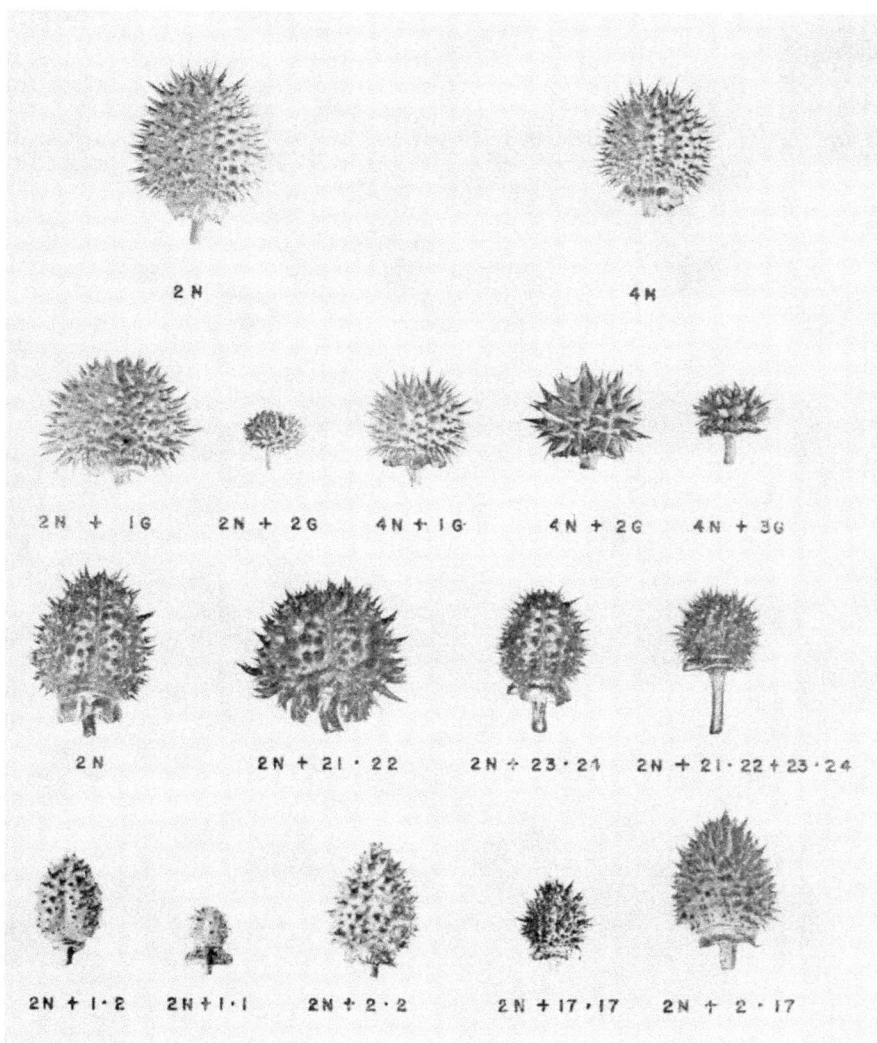

Plate 4. **Capsules of Datura,** to illustrate dosage.—Above are the diploid and tetraploid capsules. In the next row are a series illustrating the effects of different dosages of the so-called "Globe" chromosome (chromosome ends 21 and 22); notice the increase in effect with increasing dose relative to the rest of the genotype. The third row shows the normal capsule, with two primary trisomics (for chromosomes 21·22 (= Globe) and 23·24) for comparison with the double trisomic which contains three of both 21·22 and 23·24: note that this combines the globular shape and thin spines of the two primaries. In the lowest row, at the left is the primary for 1·2 for comparison with the two derived secondaries which have an extra chromosome consisting of two 1-ends or two 2-ends; note how each of the secondaries shows an exaggeration of some of the characteristics of the primary. At the right of the row are shown the secondary with extra 17·17, and the tertiary with extra 2·17, which is to some extent intermediate between the 2·2 and 17·17 secondaries.

(Courtesy of A. F. Blakeslee.)

CHAPTER 8

Gene Controlled Processes[1]

1. *Quantity of Effect and Rate of Production*

We have seen that genes which are related as hypo- and hypermorphs control the quantity of "effect" which occurs in the phenotype. In many cases, if not in all, these genes may be supposed to be responsible for the production of a greater or less quantity of some substance whose quantity controls the phenotypic effect. In most cases the nature of the substance in question is quite unknown; we have no idea, for instance, what substance may be involved in the control over bristle length shown by the gene bobbed. Indeed, this example raises the possibility that we may not be dealing with the quantity of a single chemical substance, but rather with the competence (the degree of reactivity to the bristle-forming stimulus) of a system consisting of many chemical parts.

We can analyse the phenomena further in simpler cases, where the substance whose quantity is controlled is more obvious and more susceptible of investigation. In such cases it is often clear that the genetic control over the quantity of substance is really a control over the rate of production of the substance: indeed it is difficult to see what else it could be. Goldschmidt[2] in particular has drawn attention to this genetic control of rate of reaction and has generalized it into a complete theory of gene action. Goldschmidt arrived at the conception in connection with his investigations on intersexuality in Lymantria (p. 216), but very many other almost diagrammatic examples of the same thing are known. Another example from Goldschmidt's work is the development of pigment in the skins of Lymantria caterpillars. Different races differ in the rate at which dark pigment is found; indeed, they differ not only in the overall rate, but in the detailed way in which the amount of pigment increases, so that the genetic control affects the whole form of the curve relating pigment to age, not only its end-point. Another well-known example of the same kind is the deposition of pigment in the eyes of the freshwater shrimp *Gammarus chevreuxi*.[3]

---

[1] *General references:* Goldschmidt 1938.
[2] Goldschmidt 1927, 1932.   [3] Ford and Huxley 1927, 1929.

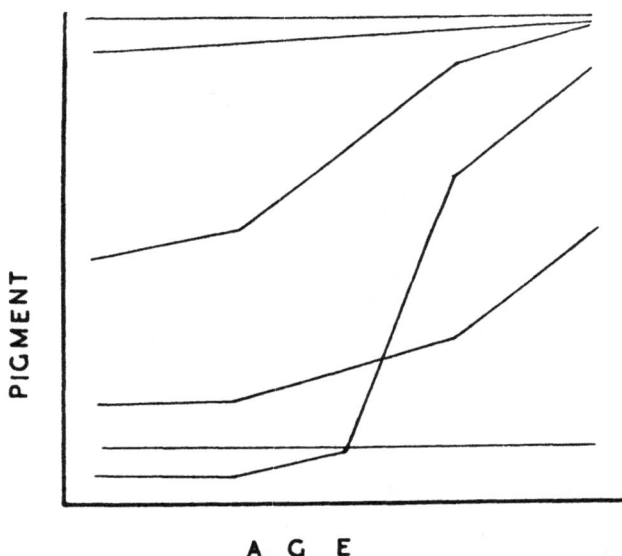

PIGMENT

A G E

A

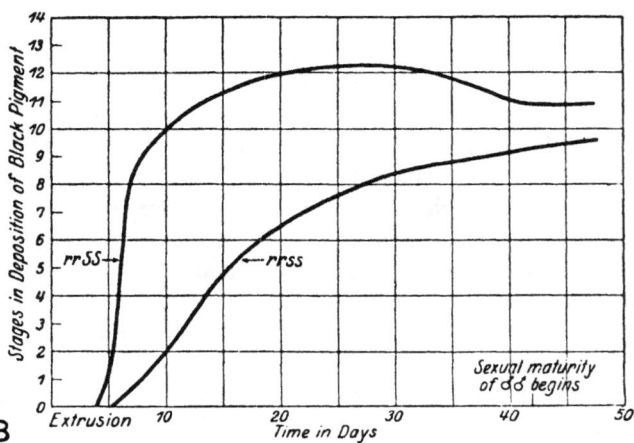

B

Fig. 82. **Genetic Control of the Rates of Developmental Processes.—**
A shows the deposition of pigment in the skins of caterpillars of different local races of Lymantria (after Goldschmidt); B shows the darkening of the eyes in Gammarus (after Ford and Huxley).

2. Genetic Control of the Kind of Substance Produced

Goldschmidt has argued that the quantitative control exerted by genes is a sign that different allelomorphs of the locus themselves differ only in quantity. They may represent different amounts of an enzyme catalyzing the reaction producing the substance, pigment or whatever it may be. This conclusion may be true of some hypo-hyper-morphs. But it is also possible to suppose that the genes differ in producing different enzymes which catalyze the same reaction to different degrees or even catalyze different reactions. In fact, in some cases there is no doubt that the two allelomorphs do cause the production of different enzymes or other chemical substances.[1] Perhaps the clearest cases are the blood-group genes of man. Two allelomorphs *A* and *B* each produce a specific isoagglutinogen quite independently of the presence of the other allelomorph or of any other genes in the nucleus. The third allelomorph at the locus produces no isoagglutinogen: it behaves as an amorph to the other two which are neomorphs to it and to each other. There is similar evidence of the production by a-, neo-morphs of substances which are actually enzymes. A well-known example is the recessive gene (amorph) which causes the loss of the enzyme enabling man to oxidize homogentisic acid, which is therefore excreted unchanged in the urine (a condition known as alkaptanuria). Similarly there is an a-, neo-morphic gene pair in rabbits, in which the dominant (normal, neomorphic) gene produces an enzyme which enables the animal to oxidize any xanthophyll in its diet, a process which is impossible to the mutant lacking this enzyme, so that its fat becomes coloured yellow.

The most fully studied example of genetic control of the kind of substance elaborated is that on the anthocyanin pigments of flowers, which we owe largely to Robinson and Scott-Moncrieff.[2] The nucleus of the anthocyanin molecule is affected by several kinds of substitutions, which are controlled by specific genes, which presumably give rise to substances, probably enzymes, which enable the various substitutions to occur. Further genetic control of colour is obtained by (1) variations of the P_H of the cell-sap, since many of the anthocyanins act as P_H indicators, and (2) the synthesis of co-pigments (substances, themselves colourless, e.g. flavones and tannins, which bring about a change of colour when added to solutions of anthocyanins).

Schultz[3] has studied the action of eye colour genes in Drosophila.

[1] *Rev.* Haldane 1935*a*.　　[2] Scott-Moncrieff 1936, 1937　　[3] Schultz 1935.

The chemical system is in this case probably very much simpler than in plants, although the actual chemical nature of the pigments (which are related to melanin) is still unknown. There seem to be only two pigments involved, a yellow and a red, which are related to one another as oxidation-reduction products.

Two processes occur in the development of the eye-colour: the synthesis of the pigment in the reduced (yellow) form, and its oxidation to the red form. Only in very few mutants is one of the pigments completely absent (no red in sepia, no yellow in vermilion), but all variations are found both in the total quantity of pigment, the time of its formation and the proportion of it which becomes oxidized. Schultz divides the genes into two groups: those in which pigment formation begins at the same time as in the wild-type, but follows a different

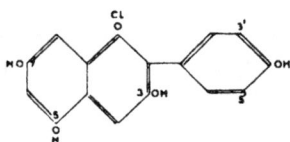

Fig. 83. The Anthocyanin Molecule.—There is always a sugar residue at 3. Genes are known with the following effects: (1) Oxidation at 3'; (2) Oxidation at 5' when 3' is already oxidized; (3) Oxidation at both 3' and 5'; (4) Methylation of the hydroxyl at 3'; (5) Methylation of hydroxyls at 3' and 5'; (6) Substitution of a sugar at 5; (7) Acylation with an organic acid residue (position uncertain).

course either in the amount of pigment formed or in the amount oxidized; and those in which the early stage of pigment formation is suppressed. Some of the genes concerned are, as we know, related as hyper-hypo-morphs; the double heterozygote shows an intermediate condition between the two homozygotes. These genes are mostly allelomorphic to one another. In combinations of non-allelomorphs, there is a difference according to whether or not the genes belong to the same group in Schultz's classification.[1] Those within one group seem to affect various elements in the complex chemical system leading to one of the two reactions, pigment-formation or pigment oxidation, and usually the gene with the most marked effect, i.e. the one giving the brighter coloured eye, is epistatic; presumably the part of the reaction it controls has become the limiting factor. In combinations between groups, the genes affect different systems and their effects are summated; early pigment formation is suppressed by the gene of one group, and when it does start the kind of pigment synthesized is governed by the gene of the other group.

The interaction of eye-colour genes has also been studied by bringing tissues containing different genes into physiological connection. It was

[1] Cf. Mainx 1937.

first found that in mosaics most genes appear to be quite "autarchic,"[1] that is to say, they are not affected by the genetic constitution of neighbouring tissue, but cause the small patch of tissue in which they are located to develop the characteristic colour. This reveals nothing about the nature of the gene reactions. Vermilion was an exception to the rule; it was "hyparchic" to wild, that is, a vermilion patch in a wild eye developed a wild eye colour. Ephrussi and Beadle[2] have carried the analysis much farther by transplanting imaginal discs of eyes of one genetic constitution into larvae of different constitutions and observing the colour of the eyes of the flies which develop. They have already studied about twenty-six different genes, and their results can be most easily summarized by relating them to the tentative hypothesis which they have put forward.

In this summary we shall deal only with the effects of the host on the colour of the implanted eyes and neglect the much feebler effects of the implant on the host. In a similar investigation on the moth Ephestia,[3] it has been shown that gonads containing the dominant factor may have a considerable effect on the development of colour when transplanted into a recessive host (p. 281).

The transplantation work cannot reveal anything about the majority of eye-colour genes, which are completely autarchic. In the non-autarchic gene-reactions three substances are concerned: *ca* (claret) substance, *v* (vermilion) substance, and *cn* (cinnabar) substance. There is a chain of reactions transforming *ca* into *v* and that into *cn* substance. This chain of reactions is brought about, not by the mutant genes as the nomenclature might suggest, but by the normal allelomorphs. Thus *cn* substance is lacking in homozygous *cn* eyes. It can be supplied by tissues of most other types, except those mentioned below, and if a *cn* eye disc is transplanted into a normal larva, it obtains *cn* substance from its host and develops wild pigmentation. Eyes homozygous for *v* vermilion, *cm* carmine, *p^p* peach or *rb* ruby have neither *cn* nor *v* substances, but these can be supplied if the eye discs are implanted into hosts of any other constitution except claret. If a *v* eye is grafted

---

[1] The word "autonomous" is usually used of the development of a gene which is not affected by neighbouring tissues. There is at present, however, no appropriate terminology for cases where interaction does occur between two genetically different tissues. It may be convenient to use the terms "hyparchic" and "eparchic," modelled on hypostatic and epistatic, and to use "autarchic" in place of "autonomous."

[2] *Revs.* Beadle and Ephrussi 1937, Becker 1938, Ephrussi 1938, Ephrussi and Beadle 1937.   [3] Kühn 1936, cf. Becker 1938.

into a *cn* host, the host supplies *v* substance from which the eye can make its own *cn* substance and develop wild-type pigment. Thus the lack of *cn* substance in *v* eyes is simply due to lack of *v* substance to

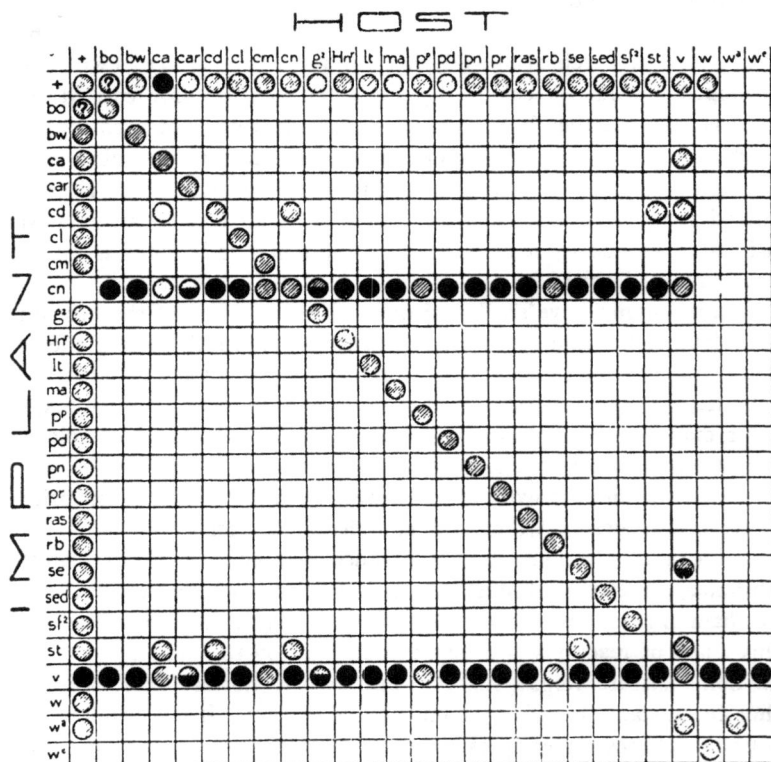

Fig. 84. The Results of Transplantation of Eye Discs in Drosophila.—Shaded circles indicate "autarchic" development, i.e. no influence of the host. Black circles indicate that the implant develops "hyparchically" in that combination, i.e. acquires the colour of the host eyes. If the influence of the host is incomplete the circle is shown partially black. The genes used are as follows: *bo* bordeaux, *bw* brown, *ca* claret, *car* carnation, *cd* cardinal, *cl* clot, *cm* carmine, *cn* cinnabar, *g²* garnet-2, *Hnr* Henna recessive, *lt* light, *ma* maroon, *pᵖ* peach, *pd* purploid, *pn* prune, *pr* purple, *ras* raspberry, *rb* ruby, *se* sepia, *sed* sepiaoid, *sf²* safranin-2, *st* scarlet, *v* vermilion, *w* white, *wᵃ* apricot, *wᵉ* eosin.

(From Beadle and Ephrussi.)

make it from. As would be expected, *cn* eyes in *v* hosts remain *cn* in colour.

In the *ca, cm, pᵖ* and *rb* eyes a process must have occurred which not only interrupts the chain of synthesis but which also makes the tissues unable to react to the substances even if supplied, since these types, unlike v and cn eyes, do not acquire normal pigment when implanted into wild-type larvae. There is no obvious relation between the groups of genes classified together in this way and the groups derived by Schultz; perhaps future work will reveal the connection.

The ca substance is of rather a different nature, since no eye can make its own supply but must obtain the substance from some other part of the body. Thus a wild-eye disc implanted into a ca larva does not obtain ca substance and therefore cannot manufacture either v or cn substances and develops a claret colour.

The v and cn substances occur in other insects as well as Drosophila, and extraction experiments have been made on the lymph of *Calliphora* pupae. The active substances are water-soluble and contain nitrogen and are probably related to the purines. These substances seem to affect not only the colour in the eye but also the number of facets formed (p. 184).

Another pigment system which has been fairly fully investigated is that of the coat-colours of rodents. The pigments are melanins formed probably from amino-acids such as tyrosin. Wright[1] has argued that all the colours can be explained by invoking two enzyme reactions: Enzyme I, acting on the fundamental substrate amino-acids (or chromogen compounds), gives a yellow pigment, while enzyme II has no effect alone, but when combined with I forms a sepia pigment, even from concentrations below the threshold for the action of I. Some genes act by suppressing or partially inhibiting the action of I and thus affect all colours (e.g. the albinism series of hypomorphs weakens the action of I while dominant white (neomorph) inhibits it altogether by the presence of an antienzyme), others affect only the black or sepia pigment-formation due to II.

In all the cases so far mentioned, the gene-controlled substances are only observed after a long series of developmental processes has occurred, and it is not by any means clear that these end-products are directly produced by the genes concerned; it is therefore not safe to assume that the nature of the observed substances gives any clue as to the nature of the genes. In a few special cases, however, the developmental path between gene and substance is much shorter. The best cases are in gamophase characters of higher plants.[2] For instance, in maize a plant heterozygous for waxy[3] produces two kinds of pollen, in

[1] Wright 1917, 1918, 1927, for transplantations, cf. Reed 1938.
[2] Cf. Stern 1938.   [3] Brink 1929.

which different starches are laid down very soon after the segregation of the genes. In grains containing waxy, the amylase concerned with the synthesis of the starch is different from the normal enzyme, and the starch formed stains red-brown instead of blue with iodine. The abnormal type of starch is correlated with a slower pollen tube growth, and it is possible that there are many other biochemical differences yet to be detected in the numerous cases known in which pollen tube growth is directly affected by the particular allelomorph contained in the gamete. No attempts have apparently been made to detect enzymes in segregating animal sperm.

3. *Time-effect and Dose-effect Curves*

In the examples mentioned in the last section we have attempted to describe the course of the developmental reactions by which a certain substance is produced. We might summarize the reactions in a given case by plotting the quantity of substance present against the time. The curve which would be obtained might be called the time-effect curve of the gene under investigation. We shall in this section try to generalize this idea so as to make the time-effect curve of a gene summarize all the information which we have about the developmental action of the gene.

In the first place, we must inquire into the relations between the time-effect curve and the dose-effect curves of the same gene. The dose-effect curves which were discussed earlier (p. 164) were obtained by plotting the dose of the gene against the final effect produced in the adult organisms. From a developmental point of view, the final effect of a gene in the adult must either be an asymptote to the time-effect curve, when development slows off gradually as maturity is attained, or in some animals it may be an end-value reached when development is suddenly brought to an end by a metamorphosis. In either case, the end-value of the time-effect curve is the same as the value plotted for that gene on the dose-effect curve. If we have a set of allelomorphs, the dose-effect curve is in fact merely a summary of the end-values of the separate time-effect curves.

The importance of this point is that it shows that certain conclusions about dose-effect curves also apply to time-effect curves. For instance, we have rather little detailed information about the dependence of the time-effect curves on the genotypic milieu, although Ford and Huxley have described the effect of some modifying factors on the time-effect curve of pigment formation in the eyes of Gammarus and a few other cases are known. But we have much more evidence from dosage compensation, cases like that of shaven, etc., which shows that the dosage curves are dependent not solely on the particular gene under investigation, but rather on the balance between that gene and the whole of the rest of the genotype. We can now see that this effect of the genotypic milieu on the dosage curve must be a result of its effect on the time-effect curves, and we can thus give a much stronger basis to the important conclusion that the time-effect curve is a function of the whole genotype.

The simplest type of time-effect curve is that in which we summarize certain developmental processes which are directly observable, such as the deposition of pigment in the eyes of Gammarus or the skins

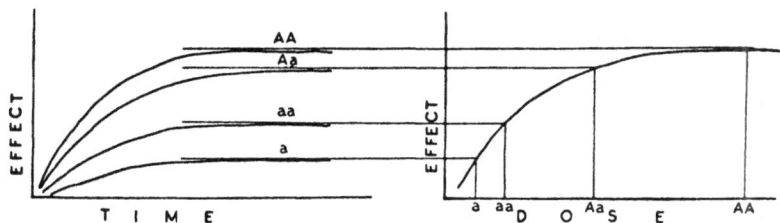

Fig. 85. The Relation between Time-effect and Dose-effect Curves.

of Lymantria caterpillars. But the investigations on eye pigments in Drosophila, for instance, clearly show that the observable process is only the final reaction in a whole series of changes which lead from the gene to the pigment. We can, ideally, expand the idea of the time-effect curves to cover not only the progress of the final observable process but also that of the earlier processes, about which we usually know very much less. For instance, if a pigment is formed from a precursor, we can not only plot a curve showing the speed of formation of pigment, but we can include an earlier curve which gives the rate of formation of the precursor.

If we attempted to formulate this in a strict way, we should find that for every new substance whose concentration we wished to plot we should have to introduce a new dimension of space, and this soon becomes rather alarming to non-mathematicians who are not used to creating universes to their own specifications. But even without any complication of multi-dimensional space, it is very easy to grasp the essential points which emerge when the time-effect curves are generalized in this way.

In the first place, we find cases in which the effects produced differ merely in quantity and vary continuously over a certain range; then all that we can deduce about the time-effect curves is that the rates of the processes concerned can vary continuously, so that different quantities of the end-product are produced when development ceases. Perhaps more important are the cases in which there are several fairly sharply demarcated and alternative developmental processes, which can only be represented by a system of branching lines. For instance, we have seen that in Drosophila there is a period of development when the normal vermilion substance is essential for normal eye pigmentation. If vermilion substance is absent, the pigment-forming substances will change so as to form vermilion pigment; if vermilion substance is present, they change so as to produce normal pigmentation, but come to another branching point where the presence or absence of cinnabar substance decides in which direction they will proceed. In such a case we have a mixture of reacting substances, say two or three enzymes and some substrates, and at a branching point there are two alternative possible ways in which the mixture can change, according as the vermilion substance is present or not; for instance, the vermilion substance might inhibit the most active enzyme and allow a less active one to work. We do not in fact know any of the details about the processes involved; all we know is that we are dealing with a system with alternative possible ways of changing.

If we want to consider the whole set of reactions concerned in a developmental process such as pigment formation, we therefore have to replace the single time-effect curve by a branching system of lines which symbolizes all the possible ways of development controlled by different genes. Moreover, we have to remember that each branch curve is affected not only by the gene whose branch it is but by the whole genotype. We can include this point if we symbolize the developmental reactions not by branching lines on a plane but by branching valleys on a surface. The line followed by the process, i.e. the actual time-effect curve, is now the bottom of a valley, and we can think of the sides of the valley as symbolizing all the other genes which co-operate to fix the course of the time-effect curve; some of these genes will belong to one side of the valley, tending to push the curve in one direction, while others will belong to the other side and will have an antagonistic effect. One might roughly say that all these genes correspond to the geological structure which moulds the form of the valley. Genes like vermilion which have their main effect at certain branching points are like intrusive masses which can divert the course of the developmental processes down a side valley.

This attempt to symbolize the developmental reactions may seem unduly picturesque and too abstract to be of much value. Its abstractness, however, must be blamed on the fact that we know so little about the actual processes concerned. The two important, but abstract, facts

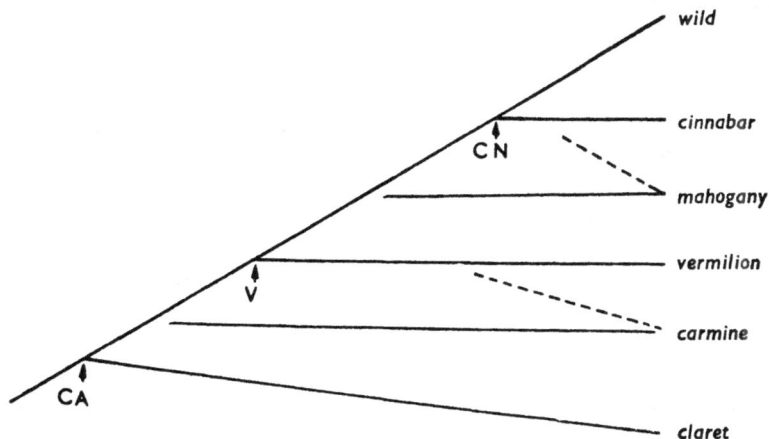

Fig. 86. **Diagram of the Developmental Processes of Pigment Formation in the Eyes of Drosophila.**—The developmental process moves from left to right along the branching tracks. The points marked *ca*, *v*, and *cn* symbolize the alternatives dependent on whether the claret, vermilion, and cinnabar substances are produced. There must be very many more tracks which we cannot yet fit into the picture. Thus it is known that carmine has no, or little, vermilion substance, and mahogany little cinnabar substance; but we do not yet know whether the developmental processes in flies homozygous for these genes branch off from the normal track before the vermilion and cinnabar forks, or are secondary branches, as indicated by the dotted lines.

which are expressed in visual form by the valley model are, firstly, that the course of any developmental process is determined by many genes, and secondly that these genes often define alternative courses along which the reactions may go.

This same scheme may be used to describe the development of characters which are not simple substances. For instance, a similar history of successive reactions has been invoked to explain the developmental effects of the Bar gene in Drosophila which has been worked out in some detail by studying the effect of temperature on the number of facets formed in the eye.[1] Unfortunately, we only see the end-result

[1] Cf. Margolis 1935, Margolis and Robertson 1937.

"frozen" by the occurrence of metamorphosis. It is found that there is only a certain period during development (just before the facets actually appear) during which temperature changes affect the number of facets formed in Bar flies. The interpretation suggested is that during early development a facet-forming "substance" is formed and that the Bar gene sets going reactions which break down this substance; then a third set of processes determine that facet formation shall actually begin, and a number of facets are formed proportional to the amount of substance still present. The temperature effects, which only occur when the Bar gene is present, presumably affect the break-down process for which Bar is responsible. In this example the end-product which is the "effect" in the time-effect curve is not a single substance like a pigment, but is a relatively complicated tissue, the eye facets. The facet-forming substance, however, may be a single chemical compound, since Ephrussi[1] has recently shown that the number of facets formed in Bar eyes may be increased by the injection of suitable extracts from normal pupae. We know nothing about the mechanism of action of the substance; it might either increase some sort of inductive stimulus to facet formation or lower the threshold of reaction to such a stimulus. The substance is probably related in some way to the vermilion substance, and may be identical with it; it is known that Bar inhibits the formation of vermilion substance in the eye itself, though not in the rest of the body.

In considering development from an embryological point of view we can, as with Bar, not yet express the characters in which we are interested in terms of quantities of definite substances, but must talk instead of histological types such as neural tissue, eye facets, etc. But experimental embryology leads to the formulation of exactly the same kind of system of alternative possibilities as we have had to develop to describe the genetical results. For instance, the ectoderm of the amphibian gastrula has two alternative methods of change open to it; it may become epidermis, or, if the evocator is added to it, it may become neural tissue. The case is exactly parallel to that of the pigment system in Drosophila at one of its branch points. Both the methods of approach to the study of development formulate the main problems in the same kind of way, and we may hope that genetics and embryology can collaborate in finding the answers.

We must now consider certain genetical problems in the light of the scheme of thought which has just been developed.

[1] Ephrussi, Khouvine and Chevais 1938.

4. Dominance

Dominance of hypo-hypermorphic genes is a question of the dose-effect curve. Stern, Muller, and others have shown that this curve has an approximately hyperbolic form, rising rapidly for small doses, but flattening out and approaching an asymptote as the doses become larger. Wright[1] has argued that this is a simple consequence of the ordinary chemical dynamics of the gene-reaction system, assuming gene-quantity to correspond to the quantity of enzyme or catalyst.

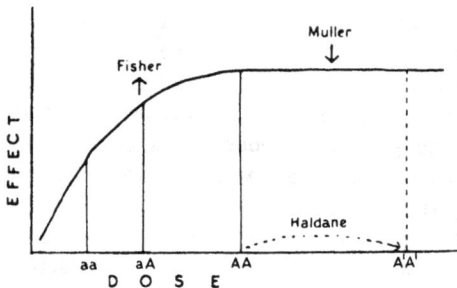

Fig. 87. The Evolution of Dominance.—Fisher suggests the selection of modifiers which push up the early part of the dose-effect curve; Muller suggests a similar pushing down of the later part; Haldane suggests the adoption of a more efficient wild-type allelomorph.

An allelomorph A is dominant to a if AA lies on the horizontal part of the dose-effect curve and if the compound Aa also comes in the horizontal part of the curve and thus shows the same effect as AA. If AA and Aa lie in any other part of the curve they will show some sort of incomplete dominance, i.e. the difference between AA and Aa will be less than that between Aa and aa (because the slope of the curve falls off). Thus some degree of dominance is to be expected as a general rule, particularly for large mutations, but it is probably necessary to provide some explanation of the great frequency of complete dominance; i.e. some mechanism by which AA is pushed so far along the horizontal part of the curve that Aa is also on it. (The problem is exactly the same as that of dosage compensation in the X, p. 166). The alternatives are (1) direct selection of modifiers which make Aa similar to AA, i.e. steepen the early part of the dose-effect curve so that the flat part occurs early (Fisher);[2] (2) selection of normal wild type genes

[1] Wright 1934.　　　　　　　　　　　　[2] Fisher 1928, 1931.

which lie far along the horizontal part of the curve (Haldane);[1] (3) pushing down the horizontal part of the curve, thus preventing large doses from having too great an effect and making the flattening out occur earlier (Muller).[2] (Cf. p. 297.)

The "effects" in the dose-effect curves mentioned above are the limiting, asymptotic values of the processes caused by various doses of genes. Goldschmidt[3] has given a theory of dominance in which the gene reactions are supposed to proceed at a constant speed not tending to any limiting value. But the same principle of explanation is introduced in another form: the dominance is explained by supposing that what we actually observe is not the gene reaction itself but another reaction which, when plotted against the dose, gives a curve which rises steeply and then becomes horizontal. Goldschmidt supposes that this second process goes on more slowly as time passes, and that a given dose of the gene corresponds to a constant interval of time during which the reaction is allowed to proceed: thus unit of dose is the same thing as unit of time, so that if the visible reaction when plotted against time gives an exponential curve, so must it when plotted against dose. In this scheme the gene-reactions, which now merely control the length of time for which the second visible reaction proceeds, have become completely hypothetical.

5. Mimic Genes

In a species, certain types of morphological variation may occur particularly frequently, and from diverse genetical and environmental causes. In terms of the "geological" model of gene reactions, we could say that the landscape defined by the genotype includes certain definite valleys branching downwards, so that any slight variation in the upper part of the main valley may divert the gene reaction into one or other of these already existing side channels. A very good example of this type of behaviour is the Minute group of genes in D. melanogaster.[4] The Minutes are all dominant (and lethal when homozygous), and they all shorten the bristles to various extents and have a characteristic effect on the eyes, wings, etc. Their loci are scattered throughout the chromosomes, but although they are definitely different genes, they all produce the same syndrome of effects, though with different strengths. Schultz showed that in compounds (double heterozygotes) two Minutes do not reinforce one another, and thus do not perform the same primary

---

[1] Haldane 1930, 1932b, cf. Wright 1929.   [2] Muller 1932b.
[3] Goldschmidt 1932.   [4] Schultz 1929.

reaction. Each Minute has a different effect on the early part of the valley along which the bristle-forming process is moving, but any of these effects is sufficient to divert the course of the process out of the main valley into the Minute branch valley. The course of the branch valley is not absolutely fixed by the rest of the genotype, but is also effected to some extent by the particular Minute used, since the final effect can be more or less extreme. Schultz discovered some of the modifiers which determine the course of the Minute branch valley; these are factors (Delta, Jammed, etc.) which will increase or decrease the effects of Minutes, and which modify all Minutes in the same way, and therefore must affect the Minute valley itself and not the different primary effects which originally diverted the process into that valley.

Abnormal Abdomen in *D. funebris*,[1] like the Minutes, is a character brought about by several different genes, and it can also be caused by chance mishaps during development. Here again the side valley is fixed by the genotype, and several things may divert the stream into it.

6. *Plus-Minus Variations in Gene Effect*

The vertical dimension in our geological model represents probability, so that the valley bottom is really a representation of an equili-

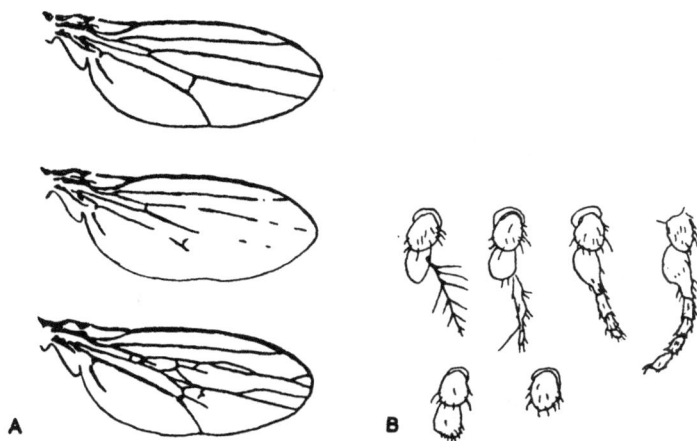

Fig. 88. Plus-Minus Variation in Gene Expression.—*A* A gene in *Drosophila funebris* which causes either an increase (below) or a decrease (middle) in venation of the wings; the normal wing is shown above for comparison. *B* Similarly *aristopedia ssᵃ* in *D. melanogaster* may cause the aristae to become larger and footlike (upper row) or to be reduced and even to disappear (lower row); the normal arista is shown at the left of the upper row.   (After Timofeeff-Ressovsky.)

[1] Timofeeff-Ressovsky 1932*a*.

brium. This is rarely apparent in genetic experiments, but is much more clear in experimental embryology, where an experiment usually consists in carrying the point representing the state of a process into some unnatural place in the landscape defined by the genotype, and from this place the process may run down into the normal valley (i.e. equilibrium is restored, regulation occurs). The equilibrium aspect of the valley only appears in genetics if by chance the valley bottom becomes flat, i.e. there are several equally probable values for the process. Then the gene which causes the flattening is observed to lead to variable results. Timofeeff-Ressovsky[1] has described a good case where the only effect of a gene seems to be to flatten the bottom so that the phenotype varies equally on both sides of the normal; flies of D. funebris with the gene may develop either more or fewer bristles than the wild type. Another similar gene affecting wing-venation is illustrated in Fig. 88.

7. The Variability of Gene Effects

The effects produced by a gene vary in dependence on both the rest of the genotype and the environment. Timofeeff-Ressovsky[2] has proposed discussing these variations in terms of the concepts of expressivity, penetrance, and specificity.

(a) Expressivity.—The expressivity is a measure of the amount of effect shown by the gene. If we are dealing with a quantitative character it can be measured by the quantity of substance produced, either in the appropriate real units or in terms of some arbitrary unit. We have already seen that the expressivity of a gene depends on the rest of the genotype, since modifying genes affect the degree or quantity of the effect produced. It also depends on the environment in many cases. Thus if we have Bar in a given genotypic milieu in Drosophila, the number of facets formed depends on the temperature during the sensitive period (p. 184).

The dependence of expressivity on environment is particularly important in connection with the problem of determining how far the differences between two organisms are hereditary and how far they are caused by different external conditions during development. If we have two Bar flies reared at different temperatures, the ratio of the number of facets in the two types gives a measure of the effect of the environment on the expressivity of Bar; and if we have Bar and not-Bar flies

[1] Timofeeff-Ressovsky and Timofeeff-Ressovsky 1934.
[2] Timofeeff-Ressovsky 1931.

reared at the same temperature, we can measure the effect of the Bar gene at that temperature. But if we simply have two flies, one Bar reared at 25° C. and the other not-Bar reared at 15° C., we cannot directly compare either the effect of the Bar gene as against not-Bar, or the effect of the different temperatures, because the change from 25° C. to 15° C. may have different effects on Bar and not-Bar.[1]

For instance, one can imagine there being two genes A and B which had exactly the same phenotypic expression at 15° C. while at 25° C. the expression of A remained the same while B gave a phenotype different, say, by 10 units. No comparison between two of these forms could

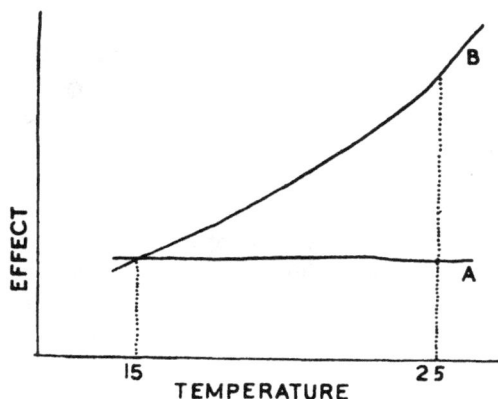

Fig. 89. **Diagram of the Dependence of Expressivity on Environment.**— Two (imaginary) genotypes produce the same phenotype at 15° C., but different ones at higher temperatures.

reveal the true state of affairs; if we compared A at 15° with B at 25° we might think the whole difference was hereditary until we discovered that A and B gave the same effect at 15°, or we might choose to suppose the whole effect was environmental till we found that A and B at 25° differed by exactly the same amount. Clearly there is no basis for either opinion, and neither is correct. We can only find out the true state of affairs by having two genetically identical stocks of A, one reared at 15° and one at 25°, and similarly two identical stocks of B reared at these two temperatures. The special importance of this conclusion is with reference to human genetics, since there we have to deal with populations brought up under very different circumstances; but we very rarely have genetically identical stocks which can be reared under different conditions and thus provide a test of the effect of the environment as opposed to that of the hereditary factors. Almost the only

[1] Cf. Hogben 1933, Haldane 1936b.

chance to obtain this information is by investigating identical twins, which are formed from one fertilized ovum and therefore have identical genetical constitutions (p. 337).

(b) *Penetrance.*—The penetrance is the frequency, measured as a percentage, with which the gene shows any effect at all. Most of the genes usually worked with in genetical experiments are chosen for having a high or complete penetrance; every organism homozygous for a recessive factor shows it. But for many genes, which tend to get rather neglected in experiment, this is not so. Only some of the homozygotes show the effect. The frequency of effectiveness depends both on the environment and on the genotype. A very good case of environmental influence is the gene giant in *D. melanogaster*.[1] In a stock of this mutation, the percentage showing the effect is dependent on the amount of food; under conditions of severe larval competition very few giants emerge. The effect, as its name implies, is an increase in size, caused by a delay in pupation; Gabritschevsky and Bridges suggest the analogy with castes[2] in social insects, where again there are two sharply distinct phenotypes formed as a response to different quantities of food.

In the giant stock, the expressivity hardly varies at all; all giants are of much the same size relative to normal. With other genes expressivity and penetrance may both vary, but they often do not vary together. Thus lines with high penetrance may have low expressivity and vice versa.

Penetrance is a statistical idea, and it presumably expresses a statistical variation in developmental processes. It seems easiest to picture it if we suppose that a gene with low penetrance does not determine the whole course of the developmental reaction producing the organ or character which the gene affects, but only acts for a certain short period. Then we can think of penetrance in the same terms as were used in discussing the Minutes (p. 186). The rest of the genotype may be taken to define a "landscape" along the main valley of which the character-producing reaction moves; but there are side valleys, and a gene acting at the right time may push the reaction out of its normal course into one of these. A gene with low penetrance provides a push which only occasionally carries the reaction out of the main valley; one with high penetrance is usually or always successful in this.

According to this suggestion, the potentiality for the altered phenotype, for instance, the development of giant larvae in the example quoted above, is already given by the normal phenotype. We are

---

[1] Gabritschevsky and Bridges 1928.   [2] Wheeler 1937.

reminded of the "alternative reaction systems" invoked to explain how a single gene can, in some cases, decide between two complicated types of sexual differentiation. But if the genotype really provides alternative methods of reacting, or alternative valleys along which the developmental processes may go, it should be possible to reveal these potentialities without using the "realizer" genes which normally bring them into action. This requirement can be fulfilled. Goldschmidt,[1] in particular, has shown that if normal flies are subjected to severe environmental influences (sub-lethal temperatures, etc.) a small percentage of modifications appear which exactly parallel the changes produced by genetic factors. These so-called phenocopies are evidence that the alternative valleys are given even in the normal genotype. They can be said to provide evidence of positive values for the expressivity of genes which are not actually present! They are not, of course, inherited, though other heritable changes may be produced by the same agencies (p. 279). Their importance is that they reveal the underlying "landscape" of the normal genotype; unfortunately, the methods known for producing them are as yet fairly unspecific and do not reveal much about the nature of the "push" required to carry the developmental reactions into the side-valleys.

(c) *Specificity*.—The last variable to be discussed in relation to gene-effects is the specificity, which is the name proposed for the variations in the actual qualitative nature of the effects. That is to say, it is the variation in the course of the side-valleys, giving a variation not in the amount of the final effect but in its kind. This again is controlled largely by the rest of the genotype, partly by the environment. The variations which have been worked on are mainly variations in pattern, and are discussed on p. 200 in connection with other pattern effects.

[1] Goldschmidt 1935, 1938, Friesen 1936.

CHAPTER 9

The Genetic Control of Pattern

1. *The Nature of Developmental Patterns*[1]

Morphological patterns, such as we find in living organisms, are arrangements of different substances or tissues in definite relative positions in space. These relative positions may not remain constant, but may change as development proceeds, but the changes, if they occur, follow a regular course. If we symbolize the pattern as it is at any single instant by a point, the whole pattern as it changes would have to be expressed as a line plotted against time. During development, there is always some period when the tissue tends, after experimental disturbance, to "regulate"; that is to say, the disturbed tissue gets back to the normal course of the pattern-development and finally produces a normal organ. The normal pattern-development is therefore, at least during this regulatory period, an equilibrium state to which the mass of tissue tends to return.

We have very little information about the nature of the forces which are concerned in such equilibria. They might be intermolecular forces of the kind responsible for liquid crystal formation; probably diffusion forces are important, and several other suggestions are possible. But whatever their nature, they could only give rise to an equilibrium which is a morphological pattern if in the first place they proceed from different points in the mass of tissue. However far we can analyse the development of a pattern, we shall therefore always be left with an initial heterogeneity to account for. The basis of such a heterogeneity might be (1) local differences in chemical forces at different points on molecules formed by pattern-genes; (2) local differences in the cytoplasm of the egg; (3) local differences between different parts of the chromosomes. Differences of these three kinds certainly exist. But it is difficult to see how local chemical differences within a molecule could give rise to sufficiently complicated patterns of the right size. Moreover, there is no evidence that the linear arrangement of genes has anything to do with the developmental patterns; in fact, the evidence of translocations, inversions, etc., is directly against this. We are left with local differences of the cytoplasm as the immediate source of

[1] *General references:* Henke 1933, 1935.

the whole pattern of the animal. During development new substances and tissues are produced by the interaction of different regions of the egg, and in this way the pattern gradually becomes more complex. If the nature of the reacting substances is altered by a gene substitution, the equilibrium which they attain will be altered and a new pattern produced.

As we have seen (p. 143), the case of Limnea gives some support to the hypothesis that the pattern-properties of the egg cytoplasm are themselves dependent on the genes of the mother in which the egg was formed.

2. Genes Affecting the Manifestation of Patterns

Some genes alter the patterns which can be seen in the adult organism, but on analysis are found merely to have changed the expression of a pattern which itself remains fundamentally the same. The simplest case is when a gene alters the type of pigment which is distributed in a pattern. The fact that the actual pattern is not altered by changing the colour of the pigment is so obvious that we should hardly be tempted to call such a gene as a pattern gene at all. But other cases are rather more subtle. For instance, many colour patterns are based on a fundamental pattern of readiness to receive pigment during a certain period of development. If a gene causes an increased amount of pigment to be formed, the size of the pigmented patches may be increased. Goldschmidt[1] has explained some melanotic forms in Lepidoptera in this way. Another possible mechanism, to which Goldschmidt has also drawn attention, is as follows. Suppose that the deposition of pigment is dependent on the diffusion from a centre of a pigment precursor, which can be converted into pigment only during a certain limited period of development; then if a gene causes the precursor to be formed earlier than usual, the diffusion will cover a larger area before the pigment-forming reaction is brought to an end, and the coloured areas will be larger. In the Lepidopteran cases described by Goldschmidt, the precursor substance which diffuses out from the centre of the future spot seems actually to be something which slows up development of the scales. Pigment formation happens at a later period, and can only occur in the relatively undeveloped scales affected by the diffusing substance, the rest of the scales being too advanced to be coloured.

One of the most fully analysed cases of pattern formation in the

[1] Goldschmidt 1927.

Lepidopteran wing has been worked out by Kühn[1] and his students in the mealmoth *Ephestia kuhniella*. The essentials of the pattern on the forewing are a central mid-field which is bounded by two bands, each consisting of a white stripe between two dark stripes, while proximal to this there is a basal field and distal to it an outer field. The mid-field is an element of the pattern which is said to belong to the "symmetry system," and a similar element can be recognized in many Lepidopteran patterns.

The mid-field is produced by two streams of some agency, probably

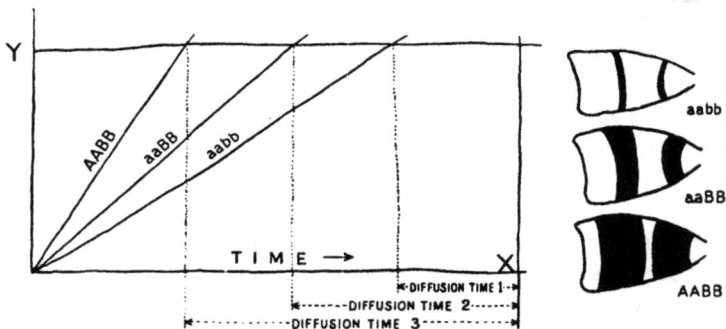

Fig. 90. **Diffusion and Pattern.**—A diagram of a case in which the pattern is determined by the diffusion of pigment from two bands crossing a Lepidopteran wing. Diffusion begins when the gene-controlled processes (indicated by the lines *AABB*, *aaBB*, and *aabb*) produce a threshold concentration Y of pigment or precursor, and continues until some definite stage of development X. The extent of the dark bands depends on the length of time during which diffusion of pigment occurs, and that again depends on the rates of the reactions *AABB*, *aaBB*, and *aabb*. (Modified from Goldschmidt.)

of some substance, which start from the anterior and posterior edges of the wing and then spread out from the middle towards the sides. The stream is proceeding during the period between the 24th and 72nd hour after pupation, at 18° C. If the pupal wing is wounded during this period by burning with a cautery, the stream appears to be frozen in the position which it has reached when the burn is made. By burning at different ages, a series of pictures of the stream can be obtained. If the burn is made earlier, the stream proceeds to completion, but is checked by the dead tissue.

We are still ignorant of the nature of the streaming agent or substance, and even do not know exactly what it does. The first effect which can be detected is an alteration in the growth rate (measured by

[1] Kühn 1932, 1936.

the number of cells in mitosis). The mitotic rate is increased in those areas which will later become darker, that is, the edges of the mid-field.

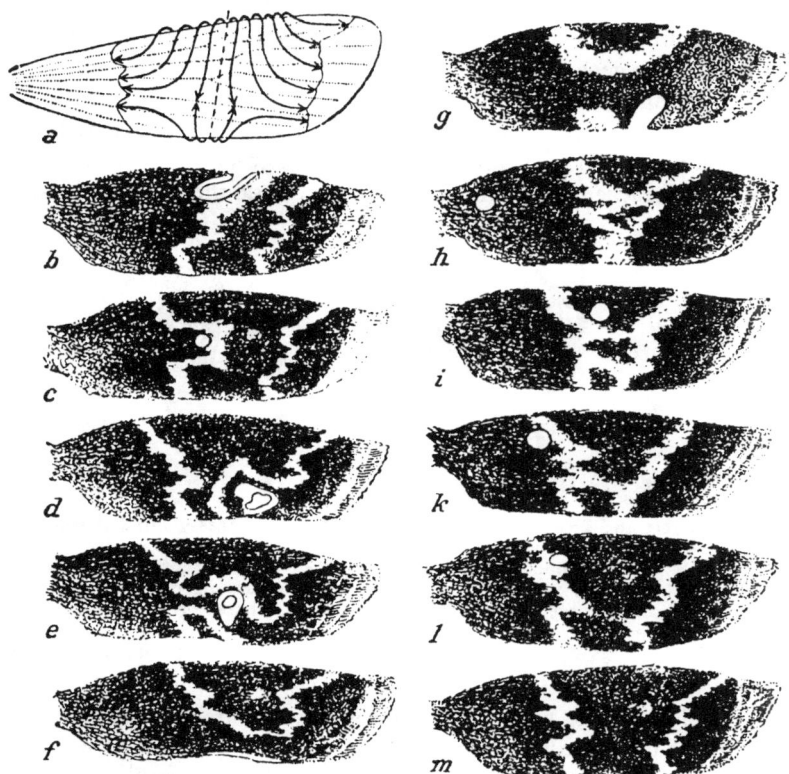

Fig. 91. **The Development of Wing Pattern in Ephestia.**—At the left above (*a*) is a diagram of the diffusion streams which lead to the formation of the midfield. *b* to *f* show the results of wounds made in the first 24 hours of pupal life; the streams have proceeded with their normal intensity, but have been checked by the wounded spots. In *g* to *l* the wound was made later (24–72 hours) and the streams have been halted at the point they had reached when the wound was made, earliest in *g* and latest in *l*. An unoperated wing is shown in *m*.

(From Kühn.)

The final effect involves not only the colour but also the shape of the scales and the detailed distribution of pigment in them.

Genes are known which affect the pattern of the wings in Ephestia by altering the equilibrium pattern which is finally attained by the determination stream. The dominant *Sy* (recessive lethal) causes the stream to halt at a stage which is attained and surpassed in normal development. It may be that the stream flows more slowly in *Sy* moths, and therefore does not spread so far in the time available, or it may be that there are forces opposing the stream and these are intensified by *Sy*. A recessive gene *syb* has a similar effect.

The final spread of the determination stream can also be affected by high temperatures (45° C.) applied for three-quarters of an hour during the pupal period. These temperature shocks do not fix the stream at the stage it has reached at the moment of shock, as do the cauterizations,

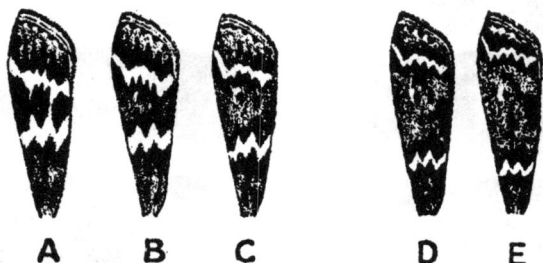

Fig. 92. Genetic and Environmental Control of Wing Patterns in Ephestia. —The drawings show wings from pupae which were heated to a moderate temperature (45° C. for 45 minutes) at various stages of development. *A* and *B* are modifications in moths with the constitution *Sysy*, whose normal appearance is *C*; but *C* can also be obtained as a modification of *sysy*, whose normal appearance is *D* or *E*.

(From Kühn.)

but actually alter the final equilibrium which the stream attains after diffusion. If the stimulus is applied during the first 36 hours of pupal life, the stream is reinforced and the mid-field broadened; a *Sysy* genotype can be caused to develop like a normal *sysy*. Between 48 and 60 hours the effect is reversed, and the mid-field contracted. Later the reactivity of the wing falls and the stimulus produces no great effect. When genetic types are imitated by environmental agencies such as temperature the results are known as phenocopies (p. 191). Unfortunately, it is still quite obscure in the above case how the temperature actually affects the determination process and we cannot yet draw any conclusions about the mode of action of the genes.

By the study of temperature effects it is possible to analyse complicated wing patterns into the physiological elements of which they are composed. In *Vanessa urticae*,[1] for instance, there is a general tempera-

[1] Köhler and Feldotto 1935.

ture-sensitive period during the first two days after pupation; but this general period can be broken up into a set of shorter periods during which the stimulus affects one particular element of the pattern. The individual parts of the pattern which can be identified in this way can be recognized not only in related species but in quite distant ones; possibly it will be found that comparatively few such elements are responsible for all Lepidopteran patterns. By comparative studies it may be possible to say that a certain element which might be expected is missing in a certain species; for instance, the central part of the mid-field in the upper side of the forewing of *Vanessa urticae*; and in this case the missing element, which is as it were squeezed out of the normal wing, can be produced by a heat stimulus applied at the appropriate time. The stimulus has apparently made a new element in the pattern, but clearly the potentiality for this element must have been implicit in the genotype of the moth, although normally unexpressed.

The genes which have so far been studied all seem to affect the degree of realization of the fundamental patterns which, as we have said, are common to large groups of Lepidoptera. There must be other genes which determine the underlying potentialities for these patterns, but we know very little about them.

Goldschmidt[1] has recently described another case of the control of degree of realization of a pattern. The allelomorphs of the vestigial locus in *D. melanogaster* produce wings which look as though little pieces had been cut off from the tips. The study of the development of the various forms shows that this is almost literally true. With the less extreme allelomorphs, at least, the wings develop quite normally until after the imaginal discs have been everted, but then undergo a degeneration of parts of the distal ends. Goldschmidt could not decide between two alternative explanations; either some substance necessary for development diffuses into the wing from the proximal end, and in the mutant phenotype does not reach the tip, or some substance which causes degeneration is produced in the tip and diffuses from there towards the proximal end. In either case, if one arranges the wings in a series of increasing degeneration, one obtains a picture of the course of the diffusion stream of the hypothetical substance (which would of course proceed in opposite directions according to the two alternative hypotheses). The various allelomorphs can easily be arranged in a series of increasing effectiveness, and their relative dosage value

---

[1] Goldschmidt 1937, 1938.

obtained, and Goldschmidt discovered modifiers which affect the dose-effect curves of the mutants and thus alter their dominance relations compared to wild type (he has coined the word "dominigenes" for such dominance modifiers). Moreover, just as in the Ephestia case, the phenotypic effects can be produced as phenocopies by suitable temperature shocks.

Lillie and Juhn[1] have shown that the mechanism of pattern-formation in the feathers of fowls also depends on an underlying pattern of developmental rate like that discovered by Goldschmidt in Lepidoptera. They stimulated pigment formation by injections of thyroxin or female sex hormone into Leghorns in which the females are more darkly coloured than the males, and observed the deposition of pigment in new feathers regenerating from follicles from which the original feathers had been plucked. Different parts of the individual feather differ in growth rate, and so do feathers from different regions of the body. In both cases, the slower the rate of growth the lower the concentration of hormone which is effective in stimulating pigment formation. But these differences in thresholds are not the only factors involved, at least in the individual feather. Although the slow growing regions (those near the shaft of the feather) react to a lower concentration of hormone, they also have a longer latent period between the time when their threshold is reached and their reaction begins. If a large dose is given, and is gradually absorbed, the time taken for the concentration in the feather germ to reach the threshold of the slow-growing parts, and then for the slow-growing parts to begin reacting, may be just as long as the whole time necessary to reach the high threshold of the fast growing parts which begin reacting almost at once. In such a case, all parts of the feather will begin depositing pigment at the same time. By varying the dosage, one can vary the relative importance of threshold and reaction time, and in this way obtain different patterns of pigment. It is then fairly easy to see how some pattern-controlling genes might work, though none have yet been fully worked out.

The pattern of feather growth rates in different parts of the body is genetically controlled, and becomes determined quite early in development, so that pieces of skin from newly hatched chicks, when transplanted to other sites in the body, develop the same kind of feathers as they would have done in their place of origin.[2] Wright[3] has suggested

---

[1] Lillie and Juhn 1932, Juhn and Fraps 1936.
[2] Danforth 1929. [3] Wright 1917.

that this is really a phenomenon of gene-autarchy, and that the pattern of a bird with, say, black wings and a light body, is really a pattern of something which causes a colour gene to mutate to a black-determining allelomorph in the wing areas. It is known that mutable genes may mutate with particular frequency in some tissues, but Wright's hypothesis is not yet proved and would in fact be difficult to test. It is clear that some irregular spottings are produced in this way (p. 372).

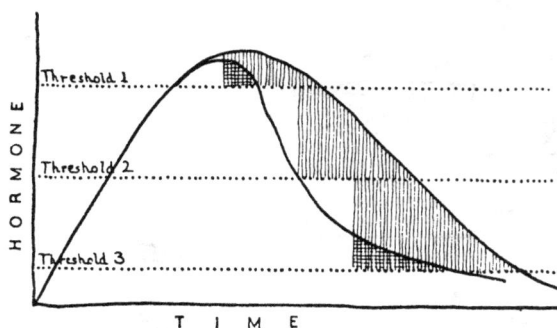

Fig. 93. **The Formation of Bars in Feathers of Fowls.**—The curves give the concentration of hormone (e.g. thyroxin) during its absorbtion and excretion after a single large dose. Different parts of the feather have different thresholds below which they show no reaction in pigmentation; the threshold is lowest in the central, axial, region (Threshold 3). Regions of low threshold have, however, a longer latent period before they begin reacting. If the dose is excreted as in the upper curve, pigment is formed in all parts of the feather (as shown by the shaded areas) and gives a transverse bar; but excretion as in the lower curve gives no deposition of pigment in the intermediate regions, and causes the formation an axial and a marginal spot.

(Modified from Lillie.)

Transplantation work[1] has recently shown that the determination of feather colour in different breeds of fowls takes place even earlier than was previously realized; pieces of skin behave autarchically as regards feather-colour when transplanted between chick embryos of three days incubation. At this time the type of feather (neck, body, hackle, etc.) has not yet been determined, and the feathers appearing in the region of the graft are determined by the region of the host's body and not by the region from which the graft was taken. Probably, however, the actual feathers are not formed by the graft, which seems to sink underneath the ectoderm and give rise only to the pigment.

Another example of a colour pattern which is dependent on an

[1] Willier, Rawles, and Hadorn 1937.

G*

underlying pattern of thresholds is the Himalayan rabbit,[1] which is white except for black patches on the paws, nose and tail. The gene concerned is an allelomorph of the albino locus. Its action is to determine a pattern of temperature thresholds above which the formation of tyrosinase is impossible. Over most of the body the threshold is such that the body temperature is too high to allow the enzyme to be produced, so that the skin remains white. If the animals are kept at a low temperature, pigment may be formed. In the exposed regions at the

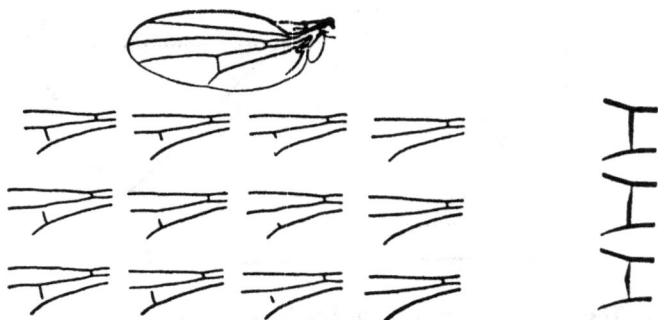

Fig. 94. **Variations in the Manifestation of the Gene** v_{ti} in *Drosophila Funebris*.—Above to the left is a normal wing, and beneath it three rows of figures showing the manifestation of v_{ti} in three different inbred stocks. In each row the degree of expression (expressivity) increase from left to right. At the right are three diagrams of the transverse vein, which is drawn thinner the more readily it disappears in the stock concerned.

(From Timofeeff-Ressovsky.)

extremities the temperature of the skin is lower and the enzyme is formed. The threshold in these regions is also lower, since if body skin is transplanted to the extremities it does not produce black hair.

In some cases the underlying pattern of thresholds can be modified by selection. Timofeeff-Ressovsky[2] has described different inbred stocks of *Drosophila funebris* which all carry the gene v_{ti} (*venae transverae incompletae*) but in each of which it has a particular type of manifestation (or specificity). In one line, the feebler grades of expression of the gene have a break in the lower end of the transverse vein, in another line the break occurs at the upper end, while in the third the vein may be broken at both ends. In all three lines the higher grades of expression lack the vein altogether. Timofeeff-Ressovsky considers this in terms of a "vein-breaking" stimulus and the resistance of the vein; one might

[1] Cf. Engelsmeier 1935, Danneel 1938.   [2] Timofeeff-Ressovsky 1931.

compare such a formulation to the ideas of evocators and competence.

Another method by which a fundamental pattern can be modified in its expression is by effects on relative growth rates. Sinnott and Dunn[1] have described some good cases in squash plants, in which there are different genetically determined fruit shapes. In some cases the different shapes are determined very early, in the fruit primordium, which then grows uniformly in all directions. But in other cases, the fruit primordia may have the same shape, and the differences between the fully grown fruit may be due to different growth rates in different directions. Similar effects on relative growth rate are exerted by whole groups of genes acting together in the trisomics of Datura, each of which produces fruit of a characteristic shape. The genes responsible for facial conformation in man presumably work in the same way. Goldschmidt[2] pointed out that even such a striking pattern-abnormality as brachydactyly in man can be explained as a result of abnormal relative growth rates. In brachydactyly, the second row of phalanges develops very late and grows very slowly so that it is extremely small in the adult and may in extreme cases be reabsorbed.

3. Genes which Affect the Fundamental Pattern

In all the cases we have described above, the action of the genes can be described as modifying an underlying pattern which is essentially unaltered. Particular elements in the pattern may be expanded or contracted, but the changes are quantitative rather than qualitative. One can, however, find examples of genes whose effects seem to be more radical. Again, we are not dealing with a sharp division of genes into two categories, but with a gradual increase in the importance of the gene effects. A clear-cut distinction between the two sorts of genes could only be made if we had a definite idea of what constitutes a qualitative change in a pattern; this is a task for topology, but so far the mathematicians have not provided a scheme suitable for application to biological material.

(a) *Disruption of the Pattern.*—As a first group of genes with radical effects on patterns we may consider the numerous genes which prevent any pattern being formed. They break down the old pattern but do not substitute any other definite arrangement for it. For instance, the short-tailed gene in mice[3] causes, when homozygous, a profound disturbance of all the anatomical relations in the posterior end of the

[1] Sinnott and Dunn 1935.   [2] Goldschmidt 1927.   [3] Chesley 1935.

animal. The gene affects the mesoderm as well as the other two layers, and perhaps one might suggest that its primary effect is on the individuation pattern within the organization centre. Another similar case is that of the rumpless gene in fowls.[1] Similar malformations of pattern may be caused by genes which affect the pattern-forming properties of the competent tissues. Little and Bagg[2] have described a strain of mice in which a gene or chromosome rearrangement causes the production of an excess of cerebro-spinal fluid. This escapes from the neural tube and forms blisters under the skin of the embryo. Very often a blister comes to lie in the region where the limb buds are forming, and the pattern of the limb becomes distorted, probably by a simple mechanical effect.

If pattern formation depends on the attainment of an equilibrium between several different processes, we should expect it to be very sensitive to rather unspecific conditions. This seems to be true. Lehmann[3] has recently shown that several poisons can affect pattern formation in particular regions of the amphibian embryo, and has drawn attention to the parallel between such phenomena and the effect of genes like the short-tailed gene in mice. Similarly the effect of the rumpless gene can be imitated by incubation of eggs at a low temperature.[4] Many years ago Stockard[5] pointed out that there are crucial periods in development when important events are taking place and the embryo is particularly sensitive to non-specific harmful conditions. The time of gastrulation, when the organization centre is active, is one such period and there are many others. A general inhibition, such as a slowing up of growth rate may, if it occurs at the crucial period for a certain pattern, entirely prevent the pattern being formed.

Perhaps the best example of this type of phenomenon is that of the creeper fowl.[6] The mutant gene (or deficiency?) is a dominant. The heterozygote shows a condition like human chrondrostrophy, that is, the long bones are malformed, shortened and bent, which gives the animal a characteristic gait from which the name is derived. The homozygote usually dies at an early stage; if it survives it shows still more marked abnormalities of the limbs. There may be a complete lack of calcification or periosteal ossification, resembling human phokomelia. The pattern of the limbs is completely disorganized. These changes seem to be brought about by a general retardation of growth

---

[1] Dunn 1925, Landauer 1928.   [2] Little and Bagg 1924, Bonnevie 1934.
[3] Lehmann 1936.   [4] Danforth 1932.
[5] Stockard 1931.   [6] Landauer 1932, 1933.

rate which begins quite early, at about the 36th hour of incubation, and increases to a maximum at the time of limb-formation, when death usually occurs. All the processes of differentiation which are proceeding most rapidly at this period are affected, and since many of these processes are essential for pattern formation in different organs, the characteristic disorganization is produced. The conditions have been most fully studied in the limbs, though other organs (e.g. the eyelids) also develop abnormally. Fell and Landauer[1] showed that if normal limb buds are cultivated in vitro in optimal conditions for

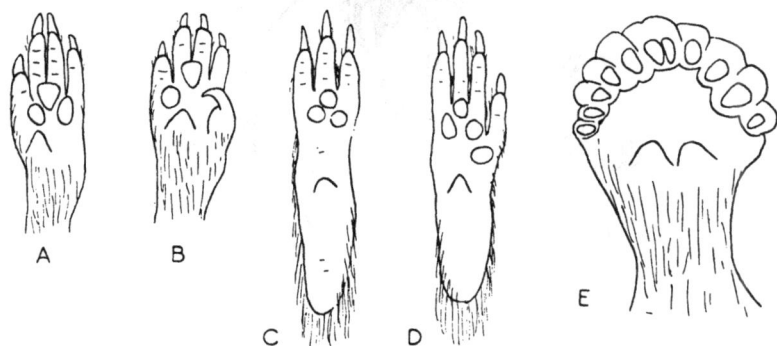

Fig. 95. **Polydactyly in Guinea-Pigs.**—A forefoot of normal, B of heteroyzgote, C hindfoot of normal, D of heterozygote, E forefoot of homozygote embryo.
(After Wright.)

growth, quite good differentiation is obtained, but that if the cultivation is made in specially growth inhibiting conditions the resulting limbs show the same type of abnormality as is found in creepers. The general inhibition of growth first prevents the formation of hypertrophic cartilage, and this secondarily prevents periosteal ossification, which normally seems to be induced by the hypertrophy of the cartilage.

(b) *Reorganization of the Pattern.*—Sometimes a gene, which in an ordinary genotype produces merely a disorganization, can in a suitably selected geneotypic milieu produce constant and orderly effects which are worthy of being called a new pattern. Wright[2] has described a polydactyly gene in guinea-pigs, which was homozygous when lethal, the rare survivors having eight to eleven toes on each foot. The normal guinea-pig has four digits on its forelimbs and three on the hind. By selection Wright built up a race in which the homozygote still had many

---

[1] Fell and Landauer 1935   [2] Wright 1935a, Scott 1937.

too many digits, but the heterozygotes usually had five toes on the forefoot and four on the hind. The case is particularly interesting for several reasons. The substitution of a five-toed for a four-toed condition is as clear-cut an example as one could wish for of a pattern effect as opposed

Fig. 96. **Comb Types in Poultry.**—A single, B pea, C rose, D walnut. Pea and rose are each due to a separate gene, both of which are dominant to single, which is the double recessive. Walnut is the double dominant.

(After Morgan after Bateson.)

to a substance effect. If a pattern is the result of an interaction between different forces, it is reasonable to find that the formation of an orderly pattern requires the co-operation of several genes if not of the whole genotypic milieu. Finally, the production of a pentadactyl limb can be regarded as a reversion or atavism, and the fact that it is only possible by a change in the whole geneotypic milieu may explain why evolutionary changes are in general not reversible.

Finally, there are genes which normally produce an orderly new pattern. Well-known examples are the genes controlling the shape of cocks' combs.[1] The recessive form is the single comb; one dominant factor *R* converts this to the rose comb, while another *P* gives the pea comb. The combination *RP* produces the walnut comb. Each comb shape is a definite and distinct pattern. We have not yet any full embryological data on this case. Many other genes with clear-cut effects

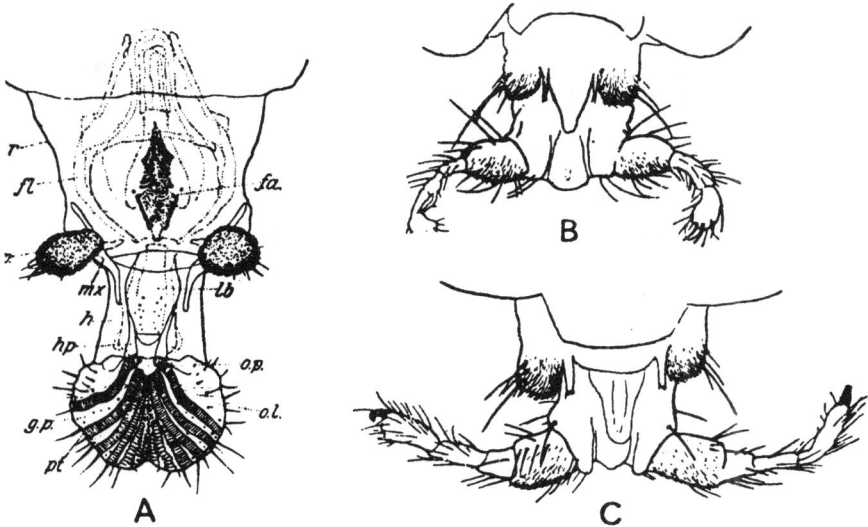

Fig. 97. **Hereditary Homoösis.**—*A* Proboscis of normal Drosophila, frontal view, extended position; *B* and *C* the same for proboscipedia, showing the tarsus-like characteristics. *f.a.* filter apparatus, *f* fulcrum, *g.p.* gustatory papillae, *h* haustellum, *hp* hypopharynx, *o.l.* oral lobes (labellae), *lb* labrum, *m.p.* maxillary palp, *mx* maxilla, *o.p.* oral pit, *pt* pseudotracheae, *r* rostrum.

(From Bridges and Dobzhansky.)

on the underlying patterns will probably be found among genes determining colour varieties, such as the different types of banding in snails.

The fact that a gene changes an underlying pattern and does not merely modify its expression is perhaps most obvious for genes which alter the number of elements in an organ. All genes determining meristic segmentation are of this kind. Good examples are the factors four-jointed and dachs which cause the formation of four instead of five tarsal joints in the legs of Drosophila.

Rather similar are the factors which cause hereditary homoösis, that

[1] Punnett 1923.

is, the formation in a particular region of an appendage which does not belong there but would normally be found in some other part of the body. Aristopedia[1] and proboscipedia[2] in Drosophila cause the aristae and mouth parts respectively to develop into leg-like appendages. Again, no experimental study of these genes is yet available; it would be very interesting if one could determine whether they act by causing the wrong evocator to be formed in a certain place, or whether they alter the type of reaction to the normal evocator.

The foregoing account has dealt almost exclusively with pattern genes in animals. Many are known in plants, affecting both colour patterns and the shapes of organs such as leaves and flowers. But so little is known of the causal factors involved in plant morphogenesis that there is at present little hope of analysing how such genes act.[3]

[1] Balkaschina 1929.   [2] Bridges and Dobhansky 1933.   [3] Harder 1934.

CHAPTER 10

Sex Determination[1]

It was realized in the very early days of Mendelism that there is some connection between sex determination and Mendelian heredity. Even before the rediscovery of Mendel's[2] work, Bateson in 1894 pointed out that sex is an example of discontinuous variation, and included it in the body of facts which, presented as *Materials for the Study of Variation*, nearly brought him independently to the idea of discontinuous hereditary units. Correns[3] was the first to give an actual demonstration that the connection was a real one. He showed that in Bryonia the male is heterozygous for sex factors, and produces two classes of pollen, while the female corresponds to the homozygous recessive. This was soon followed by the discovery of unequal sex chromosomes by McClung.

Since that time, the subject of sex determination has developed into a detailed and fairly comprehensive part of genetics. A wide view of sexuality in the biological realm raises considerable difficulties of a kind which previously could only be considered from a philosophical point of view. What in fact do we mean by sexuality? In its common use the word sex is applied to diploid animals and plants which elaborate haploid gametes which unite in fertilization. If we consider that the essential characteristic of sex is its subservience to a particular kind of reproduction, it would seem logical to define it to apply to the actual gametes whose union is the fundamental reproductive act. This, however, runs counter to the commonsense usage. The difficulty of choosing a suitable meaning is increased when we attempt to deal with sexuality in lower plants which have an equal alternation of generations, since here we frequently find that the diploid phase is asexual, while the haploid phase is sexual in the sense of bearing sexual organs and elaborating gametes, which latter are sexual in the sense of actually uniting in fertilization.

In this account, a discussion of these possibilities will be postponed

[1] *General references:* Bridges 1925, 1932, Crew 1927, Goldschmidt 1923, 1931*b*, Hämmerling 1937, Mainx 1933, Wettstein 1936, Witshic 1929, *Plants*, Allen 1932, Correns 1928*a*, L*r Plants*, Hartmann 1929*b*, Kniep 1928.
[2] Mendel himself considered the question. See his letters to Nägeli, published in Correns 1924.　　　　　　　　　　　　　　[3] Correns 1907.

to the end, when the facts will be available on which a discussion should be based. Meanwhile the word sex will be used in three senses: (1) for the differentiation of the gametes into different kinds; (2) for the differentiation of the body of the haploid organism into kinds bearing different sexual organs; (3) for the differentiation of the diploid organism in the same way. Organisms, whether diploid or haploid, which bear both types of sexual organs and produce both types of gametes, will be called bisexual, or monoecious or hermaphrodite, and here the concept of sex is applied to the individual organs rather than to the body as a whole.

The second and third types of sexuality have been defined in terms of haploid and diploid organisms. This requires modification. The alternation of generations in a plant is essentially an alternation between two morphological phases and is not absolutely coupled with a definite chromosome cycle. Thus gametophytes can be produced, in mosses, for instance, which are diploid instead of haploid (p. 213). Similarly, the sex of a male bee, which is haploid, is of the same order as the sex of the diploid female, since both obviously belong to the same morphological phase in spite of their difference in chromosome number. Mainx has therefore proposed the terms Gamophase and Zygophase to replace haploid phase and diploid phase above, and this suggestion will be adopted here. The first kind of sexuality mentioned above will be called Gamete sexuality.

A. GAMETE SEXUALITY

In organisms with zygophase sexuality, the question of whether a cell in a sexual organ develops into a male or female gamete or gametophyte seems never to be decided by its own genetic constitution, but always by influences coming from outside the cell. It is a particular case of differentiation during development and is decided like other differentiations. Thus in vertebrates it is the result of a process of induction: any primordial germ cell can develop either into an egg or a sperm, and the way it does develop is determined by whether it is acted on by the cortex of the gonad, which is highly developed in females and induces eggs, or by the medulla, which is developed in males and induces sperm.[1] Humphreys has been able to demonstrate this directly by grafting male gonad-forming mesoderm over the presumptive germ-cells of a female in the Axolotl; in spite of their female

[1] *Rev.* Witschi 1934.

genotype, the germ cells developed in the grafted male gonad into sperm. Similarly, in Drosophila,[1] a case has been found in which after X-raying a female gave X and no-X eggs; clearly one was lost from some of the presumptive germ-cells, leaving them XO, genotypically male, but in spite of this they developed as eggs since they were in an ovary.

Gamete sexuality in animals is therefore not a genetic question but an embryological one. So is the further question (which cannot yet be answered) of whether all cells of an embryo are potentially capable of

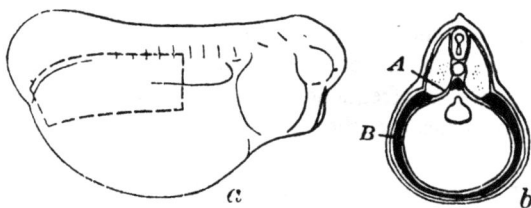

Fig. 98. **The Induction of Gamete Sexuality.**—*a* is a figure of an embryo of *Rana sylvatica* in the tailbud stage in lateral view; *b* is a transverse section. The region within the dotted lines of *a*, and between A and B in *b*, was exchanged between different embryos. The germ cells at this stage lie in the dorsal ridge of endoderm (black in *b*) and were not moved. In some cases the sex of the gonad on the operated side was different to that on the normal side. In these the grafts must have been made between embryos of different sexes. The germ cells always developed according to the sex of the gonad in which they found themselves, and therefore sometimes at variance with their genetic constitution.

(From Witschi, after Humphrey.)

being differentiated into gametes, or whether the competence for this differentiation is from the beginning confined to a certain series of cells constituting a definite germ-track.

In organisms in which the sexuality is confined to the gamophase, it is probable that the differentiation of the actual gametes is also often an embryological rather than a genetical question (e.g. in the diploid hermaphrodite moss gametophytes, p. 214). Only in those organisms with an asexual zygophase and a sexual gamophase which consists simply of the gametes themselves, does the gamete sexuality seem to be determined directly by the factors contained in the gametes (e.g. Fungi).

Hartmann[2] has shown that in some of the lower plants (e.g. in the

[1] Muller and Dippel 1926.   Hartmann 1931, 1932, 1934.

Alga Ectocarpus) the strength of the "male" and "female" fertilization producing tendencies of the gametes may vary, so that one can distinguish strong and weak male and strong and weak female gametes; the strength can be measured by statistical studies of the frequency of conjugation when gametes of two different sorts are brought together. The most important result is that gametes which react as weak males towards female gametes (i.e. conjugate with them in a low percentage) may act as weak females towards a strong male gamete. Hartmann speaks of this as relative sexuality, and interprets it by expressing the strengths of the male and female tendencies in arbitrary units. Thus, denoting maleness by negative numbers and femaleness by positive ones, we might have the gamete-sorts $a = -20, b = -10, c = +10, d = +20$, when the male and female combinations would react with the strengths or frequencies $ad = 40, ac = bd = 30$ $bc = 20$, while the two males or the two females would react with the strength 10.

In Ectocarpus, where these facts were first discovered, the gamete sexuality is not determined by genetic factors, but is dependent on the sexual constitution of the plant producing the gametes, which is itself determined phenotypically by unknown environmental factors. Thus a plant acquires a sexuality of a certain strength which then characterizes all its gametes. Hartmann has also applied a similar scheme of interpretation to other cases, such as the multipolar sexuality of fungi (p. 214), where the gamete sexuality may be directly determined by genetic factors in the haplophase.

Some authors[1] have questioned whether it is yet shown that the relative sexuality of the gametes in such cases as Ectocarpus is not really the result of different degrees of "ripeness," which hypothesis, if correct, would largely destroy the importance of the facts for a general theory of sex. But this criticism can hardly apply to some of the other cases recently described by Hartmann and his pupils. Thus Moewus[2] has investigated two species of Chlamydomonas, *C. paupera*, which has three races, strong, middle, and weak, and *C. eugametos*, also with different races. In the hybrid between these two species, non-disjunction of a sex chromosome took place, as could be shown both by the inheritance of a factor lying in the chromosomes and also cytologically. Of the haplophase individuals derived from the F1, those from which a chromosome was lacking because of the non-disjunction were inviable, but the others, with both the sex chromosomes, survived and had a

---

[1] E.g. Mainx 1933.   [2] Moewus 1935, 1936.

sex-potency which was the sum of those of the gametes which had originally been crossed. Thus crossing a *paupera* $+$ 3 with a *eugametos* $-$ 1, gave an F1 which formed non-disjunctional gametes with a value of $+$ 2; and the process could be repeated so that crossing this F1 with, say, a *paupera* $-$ 1 gave an F2 with a value of $+$ 1. The parallelism

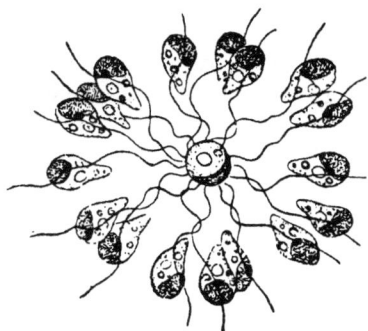

	♂ 33	♂ 35	♂ 38	♂ 40	♀ 31	♀ 32
♂ 33	–	1	–	–	3	3
♂ 35	1	–	1 ˣ	–	3	2
♂ 38	–	1 ˣ	–	–	3	3
♂ 40	–	–	–	–	3	2
♀ 31	3	2	3	3	–	–
♀ 32	3	2	3	2	–	–

	♀ 34	♀ 37	♀ 39	♀ 41	♀ 43
♂38	3	2	3	2	3
♂35	1	2	1		2

Fig. 99. **Relative Sexuality in Ectocarpus.**—The drawing on the left shows the conjugation, with many male gametes swarming round a female gamete. The tables on the right indicate by numbers from 1 (weak) to 3 (strong) the "strength" of the conjugation between gametes from different plants. The lower table gives a comparison between the males Nos. 35 and 38; it is apparent that 38 is strong, while 35 is weak. In the upper table it will be seen that the weak 35 can act as a female towards the strong 38 (and also towards 33).

(After Hartman.)

between the chromosome behaviour and the potencies of the sex factors here makes it quite clear that we are dealing with a genetic determination of sex, and further that the sex factors act quantitatively and can be added and subtracted.

Moewus has reported other results with Chlamydomonas and Polytoma which he interprets by a hypothesis which involves crossing-over in the 2-strand stage. This is supposed to bring the $+$ and $-$ factors, which are not allelomorphs, into the same chromosome. This interpretation cannot yet be regarded as convincingly proved.

Hartmann has generalized the concept of relative sexuality derived from observations of this sort into a general theory of sex determination. We shall find that in the higher organisms the "sex factors" can vary in potency in the same way as the sexual tendencies discovered by Hartmann and the factors which he postulates to control them. But in the higher forms the variations have nothing to do with the fertilization-producing tendencies of the gametes; they concern the degree of realization of morphological sex differentiation in the zygophase. Thus the sexuality which is relative in the higher forms is not at all the same sort of thing as the sexuality whose relativeness provides the basis of the theory, and it appears unjustifiable to homologize them. One can, perhaps, draw attention to the fact that in both cases the genetic factors on which the sexuality is based vary in a quantitative manner, but this amounts to no more than pointing out that in both cases the factors are of a kind which can have hypo-hyper-morph allelomorphs, which is true of many factors besides sex factors, and is therefore not very enlightening.

B. GAMOPHASE SEXUALITY

We can classify the possible types of gamophase and zygophase sexuality as follows:

1. Zygophase asexual, gamophase bisexual or with separate sexes environmentally determined, e.g. monoecious mosses with bisexual gametophyte.
2. Zygophase asexual, gamophase separately sexed by genetic determination, e.g. dioecious mosses.
3. Zygophase bisexual or with environmental determination of sexes, gamophase sex determined embryologically during development in the zygophase, e.g. hermaphrodite animals, higher plants hermaphrodite during the zygophase.
4. Zygophase separately sexed by a genetic determination, gamophase sex determined embryologically as before, e.g. separately sexed animals and higher plants.

Roughly one may say that what we should ordinarily call sex is associated with the gamophase in 1 and 2, and with the zygophase in 3 and 4. It is usually considered that the evolution of sex determination has been along the series 1, 3, 4, with 2 as an offshoot from 1. The intermediate stage between 1 and 3 is probably seen in heterosporic ferns.

1. Bisexual Gamophase

The first type of sex determination is essentially non-genetic and therefore does not concern us here. One must imagine that the possibilities of producing male or female gametes are present in all the gamophase organisms of this type; and the potency of the gametes may be subject to variation as in Ectocarpus. The determination of which type of gamete is actually developed is performed by environmental or developmental conditions.

2. Separately Sexed Gamophase

The second type of sex determination is found in many lower plants. Marchal,[1] working with dioecious mosses, was able to show very clearly that the sex factors segregate at meiosis. The haploid spores grow into gametophyte plants which are haploid and are either male or female, and remain so under all experimental conditions. The factors determining this maleness or femaleness are genetic but have no effect in the zygophase or sporophyte, which produces spores but shows no differentiation into sexes or even any characters which could lead one to call it bisexual. It is in fact asexual. But nevertheless it contains the male and female factors, which can be wakened into activity if the sporophyte is converted experimentally into a gametophyte. This conversion of phase takes place on regeneration; a small part of the diploid sporophyte regenerates as a gametophyte, which is still diploid and therefore still contains both the sex factors. It is found to be bisexual and bears both male and female organs; since the male organs are formed first during development, it can be spoken of as a protandric hermaphrodite. It is sometimes also called an intersex, though this is perhaps not strictly accurate since the plant is not intermediate, but is both fully male and fully female at the same time.

More complex hermaphrodites could be made by first obtaining diploid gametophytes with narcotics by suppressing a cell division during spore formation. Representing the male factor by M and the female by F, these gametophytes obtained by producing double spores were MM and FF, and were apparently normally male or female. By crossing them tetraploid $MMFF$ sporophytes were obtained and from

[1] Marchal and Marchal 1911.

these tetraploid gametophytes were made by regeneration. These *MMFF* plants were also hermaphrodite, but even more protandric than the *MF* diploid ones obtained in the first experiment, and the female organs did not appear till the end of development. On the other hand, *MFF* gametophytes, got by regenerating a sporophyte derived from a cross between diploid *FF* and haploid *M*, were protogynic, the female organs appearing first. Similar diploid gametophytes, heterozygous for the sex factors, have been obtained in other organisms, and show other dominance relations. In Sphaerocarpus[1] the female factor is completely dominant, in Oedogonium[2] partly dominant. The relations between these hermaphrodites and zygophase intersexes is discussed later (p. 232).

The most interesting fact about gamophase sexuality is that it is not restricted to a bipolar scheme as is zygophase sex. Particularly in fungi, sporophytes are found which give rise to four types of gametophytes which fall into two pairs within which copulation is possible.[3] The sporophyte can be represented as heterozygous for two factors, so that it is *AaBb*. It gives the four haploid "sexes" *AB, Ab, aB, ab*, which show no morphological differences but among which *AB* can only conjugate with *ab*, and *Ab* with *aB*. This so-called multipolar sexuality is further complicated by the occurrence of local races containing different allelomorphs of the *A* and *B* loci, gametophytes containing different allelomorphs being able to conjugate.

It is perfectly clear that the multipolar sexuality in the gamophase is something very unlike the bipolar zygophase sexuality which we usually mean by the word sex. In fact, it has been suggested that it is better to consider the *A* and *B* factors not as sex factors but either as incompatibility factors, or, inversely, as factors positively leading to conjugation. This is tacitly done even in those fungi which have only two sexes, which are commonly referred to as + and − rather than male and female. However, from the point of view urged later on in this discussion, that sex is not the same thing genetically in different groups of organisms, it becomes unnecessary to deny that name to the phenomena in the fungi while allowing it to all the other phenomena which concern the differentiation of the gametes. It might, however, be better to consider the multipolar sexuality of Fungi as a gamete, rather than a gamophase, sexuality.

Some data have already been obtained on the physiological mechan-

[1] Allen 1932, Knapp 1936.
[2] Mainx 1933.   [3] *Rev.* Kniep 1928, 1929.

ism of the A and B factors.[1] In Ustilago the A factor is strictly a sterility factor or lethal factor, the homozygote being inviable; mycelia (gameto-phytes) with the same A factor but different Bs can perform a limited copulation giving inviable sporophytes. The B factor directly controls conjugation, which does not take place, even in rudimentary form, unless the B factors in the two mycelia are different.

C. ZYGOPHASE SEXUALITY

We know considerably more about the genetic control of zygophase sexuality than that of gamophase. The existence of sex chromosomes leads one to postulate a simple pair of factors homozygous in the homogametic sex and heterozygous in the other; thus in most animals and plants, where the heterogametic sex is the male, the factor in the X chromosome would be a femaleness factor. But this simple scheme is unable to deal with sex reversals and intersexes and hermaphrodites, which show that both males and females contain the potentialities for both types of sexual differentiation. It is necessary then to postulate some underlying bisexual potencies which are set in action or restrained by the controlling genetic mechanism.

There are two main variants of this hypothesis.[2] According to Correns[3] all cells of both sexes contain a set of factors called AA and GG which respectively give the potentialities for male and female differentiation. These underlying complexes are controlled by the "realizers" aa and $\gamma\gamma$ which are contained in the sex chromosomes and act specifically on AA and GG. By making various assumptions as to the strengths of the factors a, γ, A and G, this hypothesis can be used to explain the inter-mediate types of sexuality found in intersexes. The other variant of the hypothesis is mainly due to Goldschmidt,[4] who contrives to formulate the difference between the sex factors in the heterogametic sex as a quantitative difference instead of a difference between the qualitatively different factors a and γ. This can be done by invoking two pairs of sex realizer factors FF for femaleness and MM for maleness. In any given species, one of these pairs is always homozygous in all members of the species, while the other is heterozygous in the heterogametic sex, the two allelomorphs being related as hypo-hyper-morphs, so that the difference in effect can be expressed quantitatively. Thus in male

---

[1] Bauch 1930, 1931.   [2] *Rev.* Mainx 1933.
[3] Cf. Correns 1928*a*.   [4] Goldschmidt 1923, 1931*b*.

heterogamety, the female is *MMFF* and the male *MMFf* (often the *f* = *O* as in Drosophila, where the *Y* is empty). These two pairs of factors act by stimulating the underlying potencies for the two types of differentiation, which are considered by Goldschmidt as two reaction-types of the system as a whole rather than two definite complexes of genes. The stimulations due to the *M* and *F* factors are antagonistic to one another, and the result is dependent on whether the *M* or the *F* factors are stronger; thus in the case given above the *FF* factors are stronger than the *MM* in the female, but the *Ff* in the male is weaker. Goldschmidt's hypothesis is perhaps more flexible than Correns's and its quantitative character, even though perhaps rather hypothetical, since we do not know exactly what the quantities actually consist of (but see p. 226), nevertheless makes possible more accurate prediction than can easily be attained by Correns's symbolism; it is commonly adopted by zoologists at the present day, but botanists often still retain Correns's formulation.

1. *Lymantria*

Goldschmidt developed his theory from his work of the Gypsy moth, *Lymantria dispar*.[1] The moth occurs throughout northern Europe and Asia, and different local races, when crossed, sometimes produce intersexual offspring, as Standfuss first discovered. The moths have female digamety, so their formula in Goldschmidt's notation is ♀ *MmFF*, ♂ *MMFF*. The value of the *m* factor is *O* in all the races, and the *FF* factors, which lie in the autosomes, have the same strength in all races; they therefore have no noticeable effect except in the triploid intersexes, analogous to those in Drosophila (p. 219), with which we are not concerned here. The intersexes arising in crosses between local races depend on variations in the strengths of the *M* factors and of another female tendency which is not the same as the autosomal *F* factors mentioned above, but is inherited through the cytoplasm. It is symbolized as $\boxed{\text{F}}$.

We must therefore rewrite the formulae of the males and females in terms of the effective factors as $\boxed{\text{F}}$ *MM* and $\boxed{\text{F}}$ *M*. In the pure races, *MM* overcomes $\boxed{\text{F}}$, which itself overcomes *M*. The intersexes arise when the balance between $\boxed{\text{F}}$ and *M* is such that neither completely overcomes the other.

Goldschmidt was able to assign different strengths to the *M*, and

---

[1] Goldschmidt 1931*a*, 1934, Hämmerling 1937.

$\boxed{F}$ factors in the different races and thus obtain a consistent explanation of the whole series of hybridizations. The geographical distribution of the races is fairly regular and suggests that the strength of the sex factors may actually be an adaptive character related to the length of the seasons, since a particular strength is probably an expression of a particular speed of development (p. 272).

In discussing the nature of the intersexual animals and the mechan-

Fig. 100. **Intersexes in Lymantria.**—The intersexes arise in crosses between local races which differ in the strengths of their cytoplasmic $\boxed{F}$ (female) and chromosomal M (male) factors. Some of the combinations are shown below; the male is homogametic.

Cross	F1 Males	F1 Females
weak ♀ × strong ♂	$\boxed{\dot{F}}_w\ M_w\ M_{str.}$ males	$\boxed{F}_w\ M_{str.}$ males
half-weak ♀ × very strong ♂	$\boxed{\dot{F}}_{hw}\ M_{hw}\ M_{v.\ str}$ males	$\boxed{F}_{hw}\ M_{v.\ str}$ near-male intersexes
half-weak ♀ × strong ♂	$\boxed{F}_{hw}\ M_{hw}\ M_{str}$ males	$\boxed{F}_{hw}\ M_{str.}$ medium intersexes
half-weak ♀ × medium ♂	$\boxed{F}_{hw}\ M_m\ M_m$ males	$\boxed{F}_{hw}\ M_m$ low-grade intersexes
neutral ♀ × any male ♂	$\boxed{\dot{F}}_{neut}\ M\ M$ males	$\boxed{F}_{neut}\ M$ females

Note that the first cross gives only males in the F1, the females being completely transformed. From the last cross, with neutral ♀ × weak ♂, intersexual males appear in F2. These are $\boxed{F}_{neut}\ M_w M_w$, when the $\boxed{F}_{neut}$ is too strong to be balanced by the weak M factors.

ism of their development, it is first necessary to make a distinction between intersexes and gynandromorphs. The latter (cf. p. 85) are animals in which different parts have different genetic constitutions as regards sex. They may be formed, for instance, by the loss of an X chromosome from a female, perhaps at the first division of the egg, giving a bilateral gynandromorph, or in later stages, and even several times in the same animal, giving a more complex mosaic. An intersex differs from a gynandromorph in that all its cells have the same genetic constitution; it is not a mixture of genetically male and female parts, but is genetically wholly of an intermediate character.

Goldschmidt's first discovery about the nature of the intersexes was that although they are not genetic mosaics, they are phenotypic mosaics.

The development of each organ is partly male and partly female. Closer investigation showed that the intersexes were of two types, males transformed towards females, and females transformed towards males. In the former type, the male intersexes, the first formed organs or parts of organs are purely male in type, while the later formed ones are female. Thus the phenotypic sex mosaic is really a time mosaic, a result of early male and later female development. The same thing holds, *mutatis mutandis*, for female intersexes. The time in develop-

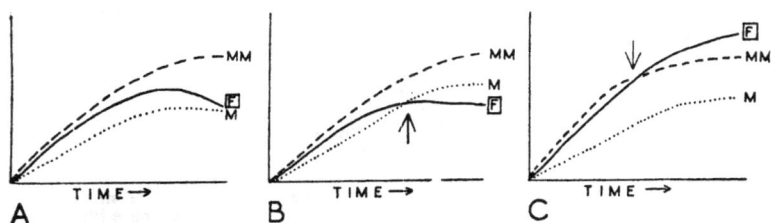

Fig. 101. **Intersexuality and the "Switch-over."**—The curves show the production of female substance (full line) by all individuals under the control of the cytoplasmic [F] factor, and of the male substance (dotted and dashed lines) under the control of *MM* in males and *M* in females. A conditions in the normal sexes; B with female intersexuality (M comes above [F]); C with male intersexuality ([F] comes above MM). The "switch-over" points are indicated by arrows.

(After Goldschmidt.)

ment when the change from male to female development occurs is the "Drehpunkt" or switch-over.[1]

The sex factors M and F can be supposed to produce substances which control the underlying alternative development-potencies of the tissues in such a way that the tissues develop in either a male or a female direction according as the male or the female substance is in excess. We can make a diagrammatic representation of the production of these substances, plotted against developmental age. A transformed female intersex will show the curve for the female substance at first above that for male substance, but gradually overtaken and surpassed by it, the point of intersection of the two curves being the switch-over.

The grade of intersexuality can be measured by the time at which the switch-over occurs; thus, with a certain female factor, a strong

[1] For criticism of the theory see Baltzer 1937, and reply of Goldschmidt 1938a.

male factor will produce male substance more quickly and overtake the female stuff sooner than a weak one, and will therefore produce a more male type intersex.

Goldschmidt suggested that, since the effects of the different allelomorphs of the sex factors, determining the different strengths, differ only in a quantitative way so that they can be added and subtracted, the genes themselves differ only quantitatively, the different allelomorphs representing different quantities of some substance; and further, since each allelomorph corresponds to a definite time-curve of production of the male or female stuff, the gene substance may be an enzyme. This suggestion is to be regarded as a working hypothesis subsidiary to the main theory. It is by no means clear that hypo-hypermorphs always differ only in quantity since it is easy, for instance, to imagine two enzymes which differ in chemical nature but which catalyze the same reaction with different efficiencies, so that their effects would differ quantitatively though they differed qualitatively themselves.

2. *Drosophila*

The hypothesis that sex determination in the zygophase is the result of an interaction between male and female factors is not based solely on Lymantria but also follows directly from a study of intersexes in Drosophila.[1] These intersexes are modified triploids; similar animals occur in Lepidoptera and were described and interpreted by Standfuss,[2] but the data available for Drosophila are much more extensive. The intersexes have three of each sort of chromosome except of the X chromosome of which they have only two; representing the haploid set of non-X chromosomes (autosomes) as A, the intersexes can be given the formula $3A\ 2X$. They differ from normal females ($2A\ 2X$) only in the presence of an extra set of autosomes, which therefore must carry the male factors which cause the intersexuality. Since we know that the addition of an X chromosome to a male ($2A\ 1X$) will convert it into a female, the X chromosomes must carry female factors. *A priori* there are two possibilities: either the intersexuality depends on the difference between the $3A$ and $2X$ or on the ratio between them. Actually it is probable that the second alternative is the true one, since tetraploids, triploids, and haploids with 4, 3, or 1 X respectively are all females

[1] Bridges 1932.
[2] Standfuss 1908, Goldschmidt and Pariser 1923. For triploid intersexes in a plant (Rumex), see Ono and Shimotomai 1928, Ono 1935.

and the same as the normal $2A$, $2X$, whereas, of course, the differences $4A - 4X$, $3A - 3X$ and $A - X$ would be different. (The haploid material is only known as patches of tissue in mosaic flies.) Further,

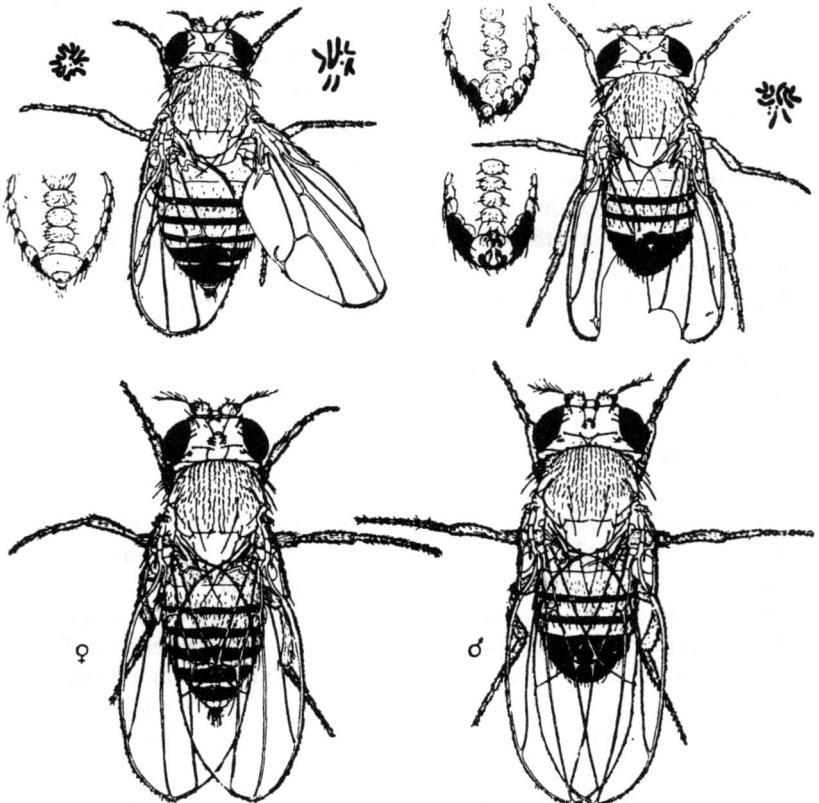

Fig. 102. **Drosophila melanogaster.**—Below are the wild-type female and the wild-type male. Above on the left is a female type intersex, and on the right a male type. Both are triploids with only two X chromosomes; there may be only two IVths, and a Y is present in the chromosome group on the left.

(After Bridges.)

the presence of Y chromosomes makes no difference to the sexuality either of ordinary males and females or of intersexes; although parts of the Y are necessary for fertility in males they have no effect on sexual differentiation.

The sex of an individual is therefore determined by the ratio between the autosomes, carrying a tendency towards maleness, and the X chromosomes carrying a tendency towards femaleness. We have the ratios $X/A = 1$ for females, $X/A = 2/3$ for intersexes, and $X/A = 1/2$ for males. Other ratios can be obtained in (1) diploids with an extra X chromosome and (2) triploids with only 1 X. The first gives a ratio of $X/A = 3/2$; the flies only survive in a few cases and are sterile and malformed females known as superfemales. The second type gives a ratio of $X/A = 1/3$ and again give sterile and more or less inviable flies, this time of male type known as supermales. If the balance theory of sex, which we have just been discussing, is really the whole of the story, these animals should be, as their names suggest, more extreme sexual types than normal males and females. But it is not at all clear that this is the case. The intersexes of Drosophila, like those of Lymantria, are complex mixtures of male and female parts, which seems to show that during development there are only two alternative ways in which sexual differentiation can occur. There is no evidence for an intermediate type of development, and thus it is perhaps unlikely that there are other more extreme types. It is possible that the supermales and superfemales have merely followed the normal male and female types of differentiation, modified by their extra or deficient X chromosomes exactly as they might have been modified by any other duplication or deficiency which was not connected with sex.

The intersexes themselves are very variable in grade. They range from nearly pure males to nearly pure females. It can be shown[1] that their anatomical peculiarities can be explained in the way suggested by Goldschmidt for Lymantria;[2] they start development as males and, after a switch-over, continue it as females, and their grade of intersexuality depends on the relative lengths of the two periods. Kerkis,[3] in fact, has given evidence that the same phenomenon occurs in the normal sexes. The gonads start growing at the same rate in both sexes and in the male the growth rate changes slowly and continuously throughout larval life, while in the female there is a sudden change of rate at an early period which probably corresponds to the switch-over.

Many genes are known which affect the anatomical development of the sex organs in Drosophila. Thus there is a gene which prevents the formation of the penis and claspers in the male, another which removes the sex-combs, etc. Similarly the anatomical development of the inter-

[1] Dobzhansky 1930.
[2] But cf. Bridges 1938a, where he adopts a point of view more like that of Baltzer 1937.   [3] Kerkis 1934.

sexes and even of normal males and females is affected by duplications and deficiencies of parts of the X. A controversy has developed between Goldschmidt[1] and Bridges[2] as to whether these genes should (Bridges) or should not (Goldschmidt) be considered as true sex genes; Goldschmidt wishes to argue that there is only one primary F gene in the X chromosome. It is probably true, as Bridges points out, that there is at present no way of classifying the genes which affect the anatomical development of the sex organs into true sex genes and others. But

Fig. 103. Time of Development of Sexual Characters in D. melanogaster **and their stability in intersexes.**—The table gives the age in hours of pupae in which various stages in development are reached. Since the intersexes all start development as males and later switch over to female differentiation, male characters are less likely to be normal if they develop late (after the switch-over) while female characters are more likely to be normal if they develop late.

Character	Pupal Age of Pupa	Stability.
Male—		
Anal plates lateral	50–52	Unstable
Penis normal in shape	44–46	
Testes spiral	30–32	
Testes connected with vas deferens ..	26–28	
Penis present	24–26	
Male ducts present	22–24	
Genital arch present	18–20	Stable
Female—		
Anal plates dorso-ventral .. ∴	44–46	Stable
Chitinous spermathecae present ..	42–44	
Vaginal plates present	32–34	
Vagina present	24–26	
Gonads ovarian in structure	2–4	Unstable

Goldschmidt is also right in so far that there is a possible way in which such a classification might be attempted. If we accept Goldschmidt's hypothesis that sex differentiation is controlled by male and female substances, these substances presumably act by stimulating the developing organs, and the actual results of development will depend on the nature of the stimulated rudiments as well as on the nature of the stimulant. For instance, it may be that the sex substances act as evocators, when the character of the organs developed would also depend on the competence of the reacting tissues. If any reaction of this kind occurs, it should theoretically be possible to classify genes affecting sexual differentiation into those sex genes which affect the production of the evocating sex-stuffs, and those genes which affect the competence of the organ rudiments. The distinction is essentially the same

---

[1] Goldschmidt 1935a.   [2] Bridges 1932.

as that between the "realizers" and the alternative reaction systems, A and G in Correns's notation, which we shall discuss again later. The theoretical possibility of such a distinction does not, of course, by any means prove Goldschmidt's point that there is only one F gene in the X chromosome of Drosophila. In fact it seems much easier to interpret the results of Dobzhansky and Schultz,[1] who studied the effect of duplicated fragments of the X on intersexes, by admitting their suggestion that the fragments really changed the time of the switch-over and therefore contained true sex genes affecting the production of the male and female substances. The effects of the fragments were of a fairly low order, however, as they had little or no effect on the sexual differentiation of normal males and females, which have probably got rather more margin of safety than the intersexes; it is possible therefore that there are one or more intense sex genes in some other part of the X. The location of male factors in the autosomes has been very little studied in *D. melanogaster*, but in *D. simulans*, Sturtevant[2] has found a factor in the IIIrd chromosome which has a considerable effect in the male direction, even converting normal females into intersexes.

3. *Other Organisms*

In certain other organisms, the differential sex factor does seem to be a single gene. Thus in the fish Lebistes,[3] there is normal crossing-over between the X and the Y, so that there cannot be a large number of sex factors in the X as they would be separated by crossing-over and the mechanism disrupted. But even here there are minor factors which influence sex determination, since Winge[4] has been able to select a race which was homozygous for the normal sex-differential gene and thus all XX and by rights female, but in which the sex determination was taken over by another gene pair whose effects are normally obscured by the rest of the genotype. This was heterozygous in the female, although the original race had male heterogamety.

A case of sex determination depending on a single differential gene pair has been produced artificially in maize, which is normally monoecious or hermaphrodite. The two races described by Emerson[5] depend on the factors barren-stalk, which suppresses the formation of female flowers, and tassel-seed, which converts male flowers into female ones. One race contains the allelomorphs barren-stalk-1 and tassel-seed-2,

---

[1] Dobzhansky and Schultz 1934, cf. Patterson, Stone, and Bedichek 1935.
[2] Sturtevant 1921.   [3] Winge 1923.
[4] Winge 1932.   [5] Emerson 1932, Jones 1934.

both of which are recessives; the female is homogametic with a constitution $ba\ ba\ ts_2\ ts_2$, while the male is heterogametic $ba\ ba + {}^{ts_2}\ ts_2$. In the other race a dominant allelomorph Ts_3 is used, and in this case the male is homogametic $ba\ ba + {}^{Ts_3} + {}^{Ts_3}$ and the female heterogametic $ba\ ba\ Ts_3 + {}^{Ts_3}$. In both cases tassel-seed is the differentiating gene, and we see there is not much difference between male and female heterogamety. We need not be surprised, then, to find that in the family of fish to which Lebistes belongs, in which sex determination also appears to be dependent on a single differential gene, some species like Lebistes itself show male heterogamety, while others, such as Platypoecilius,[1] show female heterogamety.

4. *Phenotypic Sex Determination and Hermaphrodites*

Sex determination in most of the organisms we have spoken of is a fairly efficient mechanism; the genetical differences between males and females are sufficient to ensure that intersexes and other aberrant forms only occur rarely. In fish, however, we have seen that the determination is not very stable, so that it is possible to alter its basis by selection. In these circumstances it is not surprising that environmental conditions may become important, and in Xiphophorus, nearly related to Lebistes and Platypoecilius, the genetic sex determination is very weak, or even absent, and is commonly overruled by external agents. The classical example of this type of behaviour, type 3 of our classification (p. 212), is the Gephyrean worm *Bonellia viridis*, in which the larvae develop into females if kept isolated, but into males if they become attached to the body of an adult female.[2] The mechanism of the change is still being investigated. There is some evidence[3] that male development is provoked by a specific substance secreted by the female, but other authors[4] claim that the same effect can be produced by several external agents (e.g. acidity of the sea water, various inorganic salts, lowered oxygen consumption). If we postulate sex realizers at all in this case, we must suppose that they are homozygous in both sexes, or at least that the two allelomorphs in the heterogametic sex are very nearly alike. The variation in degree of intersexuality which can be obtained in larvae treated in the same way (e.g. by a stay of a certain length of time in contact with a female) is sufficient to suggest that there may be some genetic variation in sexual tendencies between different individuals, but there seems to be a considerable range of

[1] Cf. Kosswig 1935.
[3] Nowinski 1934.

[2] *Rev.* Baltzer 1937.
[4] Herbst 1935.

variation, such as would be explained by many randomly assorting genes, rather than a clear division into two types.

In Bonellia, there must be two alternative reaction systems which are activated by the external conditions. In other organisms, true hermaphrodites, these two systems are present, but there is no differential effect of external conditions, and no antagonism between the two types of sexual differentiation, so that both are realized in the same animal, sometimes even in the same organ (e.g. Mollusca). In some cases there are separate male and female organs existing contemporaneously, in others there is a definite time sequence, the animal functioning as a male first in protandric types, as a female first in protogynic types, or finally there may be a regular alternation between male and female phases.

We do not know much about the genetics of sex determination of these complete hermaphrodites, but we understand fairly fully the conditions in some animals intermediate between hermaphrodites and animals with efficient sex determination. The best known case is in frogs.[1] The female is homogametic $MMFF$ and the male heterogametic $MMFf$, but the difference between the X and the Y chromosomes cannot be seen cytologically. In some races, the so-called undifferentiated races, the F and f allelomorphs are nearly of the same strength and the frogs show evidence of hermaphroditism, there being a period in early life when even genetic males develop as females. It seems that all the animals start development as females, and in males the development is later switched over into the male direction at a time determined by the relative strengths of the MM and Ff factors; this depends chiefly on the strength of the f, as the M and F factors are of nearly the same strength in all races. In Drosophila we have seen that development always starts in the male direction, in Lymantria in either the female direction, giving female intersexes when switched over, or in the male direction giving male intersexes when switched. The strengths of the f factors in frogs have been calculated by Witschi, on the hypothesis that the sex determination is due to the difference between the male and female factors rather than on their ratio; the data are hardly sufficient to allow a final decision between the two hypotheses in this case. As in Lymantria, it is found that the strength of the factor is vaguely correlated with climatic conditions, weak f genes, with efficient sex determination, being absent in northern regions and at high altitudes.

In the Amphibia, which lend themselves to embryological experi-

[1] *Rev.* Witschi 1934, cf. Witschi 1937, Burns 1938.

ments, a considerable amount is known about the developmental mechanisms by which the sex factors produce their effects. As was pointed out earlier, the actual germ-cells develop into eggs or sperm according to the type of gonad in which they find themselves, and the determination of their development can be regarded as an embryonic induction. The somatic constituents of the gonad consist mainly of the outer layer or cortex and the central medulla, both of which are present in the larval organ. In the female, the medulla largely disappears and it is the cortex which is the inductor of egg-development, while in the male things are reversed, the cortex vanishes and the sperms are induced

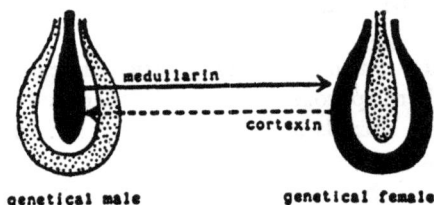

Fig. 104. Differentiation of the Gonads in Salamanders.—In the male the central medulla becomes fully developed and produces "medullarin" (the male substance), while in the female it is the cortex which becomes highly developed. The arrows indicate the suppressing action of the sex substances in heterosexual parabiotic pairs; the action of medullarin in suppressing cortical development is the stronger of the two.

(From Witschi.)

by the medulla. As well as the balance of the M and F factors, certain environmental conditions influence the relative development of these two parts; for instance, in frogs kept at high temperatures the cortex is inhibited and the medulla may become more highly developed and females converted into males. Cold has the opposite effect. In the toad no medulla is ever formed in the anterior part of the gonad and in the male this part cannot develop into a proper testis but remains a purely cortical organ, more or less like an ovary, known as Bidder's organ.

The cortex and medulla perform their inductions by substances which are capable of some diffusion. If a male and female are grafted together, there is antagonism between the cortical and medullary substances. In toads there is little effect, probably because the substances are not easily diffusible. In frogs grafted side by side the male medulla will suppress the cortex of the ovary nearest to it, but the diffusion is not sufficient to have much effect on the further ovary. In urodeles the effect spreads more widely and both gonads are affected; again it is the male which suppresses the development of the female, unless a very small male is grafted with a very large female, when it may be the female which takes the lead.

This beautiful work brings the male and female substances, postulated by Goldschmidt, out of the realm of the hypothetical into a sphere where they should be capable of biochemical analysis. The two main gaps remaining in the embryological part of the story are: (1) can the

Fig. 105. Sexual Development of "Parabiotic Twins" in Amphibia.—The animals are grafted together in early embryonic stages and allowed to develop until the sexual differentiation is well advanced. In pairs of toads, partners of different sexes have no effect on one another. In frogs, a male suppresses the differentiation of the ovary nearest to it, and may convert part of it into a testis-like organ. In salamanders and newts, the male dominates if it is larger or the partners equal in size; usually it entirely suppresses the ovarian development of its "co-twin"; it may itself suffer some retardation in the development of its testes. If a male of a small species is grafted on to a female of a large species, the female dominates and converts the male organs into ovaries. In the diagram testes are represented black, ovaries white. Note Bidder's organ (ovarian) in the male toad.
(After Witschi.)

somatic part of the gonad induce neutral cells to become germ-cells or can it only decide what type of development shall be followed by cells which are already determined as germ-cells? (2) the mesodermal part of the gonad is itself presumably induced by the primary organization centre of the gastrula; if one placed non-gonad mesoderm of a male into the gonad forming region of a female it would almost certainly form a gonad, but we do not know whether this would be a male or a female gonad.

In mammals and birds the mechanism of sex determination is probably essentially similar, depending on the stimulation of the cortex or medulla by the F or M substances. The differentiation of the sex organs other than the gonads (ducts, external genitalia, etc.) and of the secondary sex characters is, however, not directly under the control of their genetic constitution but depends on influences proceeding from the gonads. These influences are the well-known sex hormones; a detailed study of their action belongs rather to physiology than to genetics. Here we may mention that there are two theories under consideration. According to one, which may be called the classical theory, both hormones are active stimulating agents of secondary sexual differentiation during the later stages of embryonic life. According to the other theory, recently advanced by Wiesner,[1] only one of the hormones is an active agent in development, namely that of the heterogametic sex, male in mammals, female in birds. Differentiation towards the organs of the homogametic sex is supposed to go on independently of hormonal stimulation, a reaction to the homogametic hormone being not shown until the organs are fairly fully developed.

The relation between the hormones and the medullary and cortical (M and F) substances is still not clear. Certainly M and F substances are found in insects, and in insects sex hormonal influences must be weak, since quite small patches of tissue can develop their own sex in mosaics; even in Drosophila, however, there is some inhibitory influence of a gonad on the development of ducts of the opposite sex.[2] In the Amphibia, Witschi[3] regards the medulla and cortical substances as different from the male and female hormones, since he has not been able to obtain sex reversal by injection of oestrin. In the chick, on the other hand, it has recently been reported that injection of large quantities of hormone may influence the earliest stages of gonad development.[4] In mammals, hormones certainly affect the development of the gonads in fairly late stages (at birth and after). In earlier stages the position is not so clear; in unlike-sexed twins with joined placental circulations in cattle, the female gonad is transformed into a testis, the resulting intersex being known as a freemartin, but there is no proof that the M substance responsible for this is the same as the adult testicular hormone. On the other hand, even male embryos contain large quantities of oestrin derived from the mother, which has no effect on their sexual development, but it is probable that this oestrin is inactivated by combination with protein.

[1] Wiesner 1935.

[2] Cf. Bridges 1932. For an interpretation of Lymantria gonads in terms of cortex and medulla, see Sato 1932.

[3] Witschi 1937. In more recent work (unpubl.) positive results were obtained.

[4] Willier, Gallacher and Koch 1937.

5. The Hymenopteran Type of Sex Determination

In this type of sex determination, which is found in Rotifera, Acarina, Thysanoptera, and some Hemiptera as well as in most Hymenoptera,[1] the females are diploid and those of their eggs which are fertilized develop into diploid females, while the other eggs develop parthenogenetically into haploid males, which form haploid sperm by a failure of the reduction divisions. It has always been very difficult to bring this method of sex determination into line with the theory of balance

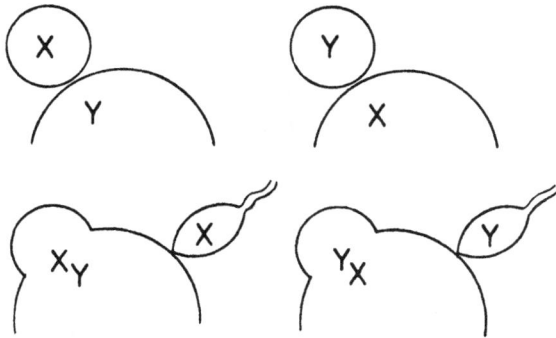

Fig. 106. **Sex Determination in the Wasp Habrobracon.**—The female complex is in two parts, carried on the X and Y chromosomes. The upper two figures show the separation of X and Y in the formation of the second polar body, leaving haploid eggs with either X or Y which develop by parthenogenesis as haploid males. Fertilization, if it occurs, happens before the formation of the second polar body. If the fertilizing sperm carries the X chromosome, the egg passes its own X out into the polar body; similarly, if the sperm carries Y, it is the Y which is lost in the polar body (lower two figures).

between male and female factors such as we have discussed up to the present. Darlington[2] has suggested that the system may have evolved from a protandric hermaphrodite form in which the development of the haploid eggs was affected by the slowing up which is often seen in haploid and in which the whole life therefore came to be passed in the first (male) stage of the hermaphrodite succession. The need for this and other similar ingenious theories may perhaps be removed if certain facts discovered by Whiting[3] in the parasitic wasp Habrobracon turn out to have general importance. In this form, the female factors are located in two chromosomes, which we may call X and Y. The haploid eggs are either X or Y and develop into males. Fertilization takes place

[1] *Rev.* Schrader and Schrader 1931—For sex determination in other parthenogenetic organisms, see Morgan 1915, 1926.

[2] Darlington 1932a.   [3] Whiting 1935.

before the extrusion of the polar bodies, and a differential maturation occurs; if the fertilizing sperm contains the X chromosome, the egg pushes out its X into the polar body, so that the fertilized egg becomes XY, and similarly if the sperm contains Y, it is the Y which gets eliminated from the egg. Thus fertilized eggs normally contain the whole F complex, shared between the X and the Y. Occasionally the differential maturation mechanism breaks down and diploid XX or YY individuals result; they are exceptional diploid males.

6. *Progamic Sex Determination*[1]

In a few animals, it has been suggested that sex determination is dependent not on genetic factors but on the size or physiological condition of the eggs. This is often spoken of as progamic sex determination. Thus in Dinophilus, a worm, there are two classes of eggs, large and small, and the large ones develop into females and the small into males. The interpretation of these facts in terms of a determination dependent on a balance between genetic factors, is still obscure. In frogs, where a genetic basis is certainly present, it has been shown by Hertwig and Witschi that the rates of development of the male and female substances can be affected by the physiological state of the egg when it is fertilized;[2] overripe eggs develop as males even if they have a female genotype. Again, little is known of the chromosome cycles of organisms with progamic sex determination. In Dinophilus, there are reported to be ten chromosomes in both eggs and sperm, but it is possible that there is an XY pair, whose members are not distinguishable cytologically, and that the dimegaly of the eggs is correlated with a differential maturation, so that in the large eggs the Y is passed into the polar body while in the small eggs it is the X which is extruded.

D. THE THEORY OF SEXUALITY[3]

1. *The Two Sexes*

Differentiation into two sexes is an extremely common property of living organisms. Several theories have been put forward to account for this. One of these theories can now be dismissed. Bütschli suggested that sexual union is necessary to rejuvenate living substance, basing the idea on the alleged fact that cultures of infusorians cannot be kept

---

[1] *Rev.* Whiting 1935a.   [2] Cf. Witschi 1934.
[3] *General references:* Darlington 1932a, 1938, Hartmann 1931, Mainx 1933, Muller 1932a.

going indefinitely by vegetative division, but require occasional periods of sexual conjugation. This fact, however, has failed to stand experimental testing;[1] for instance, Woodruff[2] has kept the *Blepharisma undulans* for fifteen years without conjugation, while even the cells of higher organisms can grow and multiply indefinitely under certain conditions, as is shown by Carrel's famous strain of chick fibroblasts, now over twenty years old. Mortimer[3] has shown that parthenogenetic Cladocera such as Daphnia can be kept indefinitely without sexual union if the environmental conditions are correct.

Weissmann suggested that the importance of sexuality is that it leads to the crossing of individuals and the production of new varieties. The development of genetics has demonstrated the correctness of this view and allows us to put it in more precise form. Crossing does not lead to the production of new hereditary units, but it does enable the already-formed units to be combined in new ways. There must certainly be an enormous evolutionary advantage in this;[4] for one thing a vegetatively reproducing species can only acquire two new characters when the two mutations happen one after the other in the same clone, while in a sexually reproducing population favourable mutations are spread as it were through one another without hindrance.

It is clear, then, that sexuality, in the sense of conjugation of two haploid nuclei, is of great evolutionary advantage and will be preserved whenever it occurs. Moreover, as Darlington has pointed out, the mechanism which has actually been evolved is just sufficiently modified from mitosis to give recombination by crossing-over as well as by random assortment of chromosomes; if the prophase of meiosis were slightly more precocious than it is, the splitting of the chromosomes which now occurs in pachytene would be postponed till the division was finished, there would be no formation of diplotene loops, no chiasmata, and no crossing-over. Thus in its details as well as in its broad outline, the sexual process is admirably adapted to give recombination of characters.

Recombination is only important in hybrids. If self-fertilization were the rule, and most organisms were therefore homozygous, the advantages of the sexual cycle would be lost. Clearly then the function of the differentiation between the sexes is to discourage self-fertilization and encourage crossing. The evolution of hermaphroditism, which sometimes goes so far as to allow self-fertilization, has been discussed by

---

[1] *Rev.* Robertson 1929. [2] Woodruff 1935.
[3] Mortimer 1935. [4] Cf. Wright 1931. Muller 1932a

Altenburg[1] in terms of the most economical way of performing the "work" necessary for reproduction.

The considerations given above suggest why evolution should have produced sexuality, but they are not adequate to explain exactly how sexual differentiation is caused. For that we require a physiological theory. Bütschli and later Schaudinn put forward what is known as the Sexuality Theory, which has been thus expressed by its main modern exponent, Hartmann:[2]

> "Every Protist or sex-cell, in fact every cell at all, is hermaphrodite or bisexual and has the full germs or potencies of the male and female sex. By the preponderant development of one or the other potency, a cell becomes male or female in comparison with other cells in which the opposite potency has been realized. In this way the cells acquire a male or female tendency."

Hartmann has developed this point of view into a comprehensive theory of sexuality. Using Correns's symbolism, he postulates for all cells an underlying alternative reaction system AG, where A represents potentialities for male development, G those for female development. These he represents twice in a diploid cell $AAGG$ and once in a haploid cell AG, but he considers that they are of the same nature throughout, and that sexuality, to some extent at least, is the same thing in the gamo- and zygo-phases. Acting specifically on these underlying complexes, Hartmann postulates "realizers" a and γ, which exist in allelomorphs of different strengths.

2. The Alternative Reaction Systems

This theory can be taken as the starting point of our discussion. The first point which needs elucidation is the nature of the alternative reaction systems A and G. We cannot suppose that the first living matter was endowed with such dual potencies. Moreover, there are forms of life at the present day in which a bipolar sexual differentiation is apparently absent; for instance, the so-called isogamous Algae in which the gametes all seem to be equal. It is difficult to distinguish two sexes in organisms like Spirogyra or Ciliates, and impossible, without elaborate subsidiary hypotheses, in Fungi with multipolar sexuality. Even in those organisms which show two alternative sexes, it is difficult to say how the so-called male in one group resembles the male in another. Hartmann avoids the difficulty by saying that the fundamental

[1] Altenburg 1934a.   [2] Hartmann 1931.

nature of the sexes is unknowable, but it is difficult to be content with a conception of such a mystical kind.

On the other hand, one cannot avoid some hypothesis of underlying complexes, on which the alternative sex-differentiation in any given species depends, since, for instance in fish, one finds the choice between the two types of differentiation, with all their complicated effects, made by a single gene, or even by external conditions. We must suppose that the complexes, like all other attributes of living organisms, have been evolved, and they must therefore have been, originally at least, of the nature of genes or complexes of genes.

It may seem remarkable that two sets of genes should have been evolved, which control rather complicated developmental reactions and give rise to complicated organs, but such that quite trivial factors, such as slight variations in the environment, can cause one set to be active and the other inactive. Yet this is exactly what we must postulate for phenotypic sex determination. The strangeness of the idea disappears, however, if we remember that this is exactly what always happens with all characters, though usually the differential conditions occur within one animal body instead of between different animals. For instance, the cells of the ectoderm of a gastrula contain two complexes of genes, or two ways of reacting, one of which causes the development of the central nervous system, the other that of skin; and the choice between them is made by a single simple condition, the presence or absence of the evocator. Probably in all such systems, it is only one master reaction which is directly affected by the deciding stimulus; thus in an organism whose sex can be determined phenotypically by the *pH* of the medium (e.g. *Bonellia viridis*), it is unlikely that each of the gene reactions in the sex-development is itself affected by the acidity, but more probable that the direct effect is confined to some master reaction which then affects certain other reactions and so on through a whole hierarchy of secondary and tertiary effects. We know examples both of genes which do not have any effect except in combination with certain other genes (e.g. modifiers), and of genes whose activity is dependent on the external conditions. If we regard *A* and *G* as symbolizing two alternative types of reaction of the genotype, based on genes of this kind, they become less difficult to accept than they would be if we had to regard them as inherent tendencies of living matter.

It also becomes clearer why we rarely find more than two sexual types, since it would obviously be more difficult to build up three or four alternative methods of reaction than two. In multipolar sexuality, which is more or less confined to the Fungi, the sexual differentiation is very slight, and the four alternative complexes are very little developed.

In different types of sex determination, the alternative between the A and G types of reaction is decided at different stages in the life cycle. The fundamental reason for the decision, or rather its evolutionary advantage, is that it ensures that fertilization shall be between gametes from different animals. Probably, then, the original mechanism was an alternative mode of reaction in the gamete itself. This survives in some Algae, in which sex determination is performed by environmental agencies working on the gamete. Usually, however, the time at which the alternative is decided is pushed back in the life-cycle, probably on a safety-first principle.[1] Eventually, in the higher animals and plants, the sex determination of a gamete has been pushed back to the fertilization of the zygophase before.

How far can one say that sex is the same thing in an organism with determination of sex in the gamete and one where it is decided much earlier? This is really the question of how far the A and G complexes are the same in the two cases, since the A and G merely symbolize the alternative modes of reaction which constitute the sex difference. Hartmann seems to take maleness and femaleness as unanalysable concepts, which we can naturally recognize as analogous in different groups of organisms. This is a very abstract point of view. We can perhaps get a more definite answer by considering an example. Suppose that in a certain organism the sex of the gametes is determined by whether they contain a substance a, which can stimulate the A complex to develop and the cell to become a sperm; otherwise it becomes an egg. Suppose further that we are dealing with a case of phenotypic sex determination and that a comes from the external environment. Now in the next step in evolution, perhaps some of the organisms in the species succeed in producing a in their gamophase cells before the actual gametes are formed; these organisms will then be males and all their gametes will contain a and will develop as sperm. But now the actual sex determination depends on whether the organism forms a in its gamophase cells or not, and this alternative clearly will depend on quite a different genetic basis to that which determined the reaction of the gametes to the presence of a. Thus the sex determination is now founded on a new alternative reaction system $A'G'$. The old A and G must of course still be there to provide a basis for the effect of a,

[1] Cf. Haldane 1932a.

so that the species has now actually got an alternative reaction system $A + A'$ and $G + G'$.

This example has been chosen so that the evolutionary change takes place entirely within the gamophase. The change from gamophase to zygophase sexuality will also involve a difference in or addition to the alternative complexes, but there is no obvious reason why it should be a difference of another kind to the differences which cause the evolutionary pushing back of sex determination into the gamophase itself.

In some organisms, when the sexual differentiation is slight, we may be tempted to neglect the alternative system altogether and think of the sex determination as entirely controlled by the realizers. For instance, in tetrapolar sexuality Mainz has suggested that it is unnecessary to invoke more than the A and B factors. But, since we believe that any differentiation always involves many factors, if not the whole genotype, we must even here admit that the A and B factors can only be active in co-operation with other genes, and in so far as the A and B factors have different effects, the other genes must be regarded as some sort of an alternative reaction system, only one in which the alternatives are not very highly developed.

3. The Differential Sex Genes or Realizers

The other half of the sex determining mechanism are the differential sex genes or "realizers."

The complexity and efficiency of the differential sex genes varies very much in different groups. In the fish, as we have seen, there is a highly complex alternative reaction system controlled probably by a single gene or even by external conditions alone. In most cases the differential is more complicated. In the first place, it may involve both male and female factors, which build up the male and female substances which are the immediate stimulants of the alternative reactions. Secondly, either or both of the male and female differentials may contain many factors. Since one sex must always be heterozygous for one of these sets of factors, a differential system containing several genes can only be stable if crossing-over is suppressed between them. It is a very general rule, therefore, that crossing-over is reduced in the heterogametic sex, and this suppression is often, as Haldane has pointed out, not confined to the section of the chromosome where the differential factors lie but affects all the chromosomes. Crossing-over within the actual differential set may, however, be more efficiently suppressed by collecting the heterozygous genes together into a certain part of the X chromosome and allowing no crossing-over between this part and its partner in the Y. This complete suppression can only be caused by the genetic control of the position of chiasma formation. The non-crossing-over part of the Y presumably originally carried hypomorphic allelomorphs of the factors in the X. We find this condition in frogs, where the F factors in the X and the f in the Y are nearly equal in strength in those races with a considerable hermaphrodite period. In more highly specialized types, a clearer decision between male and female is obtained by diminishing the importance of the Y allelomorphs. This comes about as a natural consequence of the fact that the Y is never homozygous, so that hypomorphic mutations (and most new mutations are hypomorphic) will always be masked and cannot be eliminated from it by adverse selection, as they would be in any other region of the chromosomes. Similarly the Y will lose any factors it may have originally have had tending to produce the heterogametic sex and the autosomes will take over the function of balancing against the Xs. Thus the Y eventually becomes empty of genes and may disappear.

4. *The Meaning of "Sex"*

The discussion has now reached a point where we can try to define what seem to be the most important concepts. In the first place, the important concept from the functional point of view is that of sexual reproduction in the sense of reproduction by single cells in such a way as to encourage recombinations of factors. In its fully developed form, this involves the mechanism of fertilization following on reduction of chromosome number by meiosis, with consequent segregation. There are, however, less efficient methods, but those known to us can probably be regarded as degenerate types derived from fully sexual reproduction. One form of degeneration is found in diploid apomixis by the production of restitution nuclei; a slight amount of recombination is still possible by crossing-over in the four strand stage (p. 61). Another degeneration is complementary to this; the suppression of crossing-over with the retention of chromosome segregation, as happens for the Y chromosome or the autosomes of male Diptera.

From a morphological point of view, we can define sexual differentiation in its broadest sense as anything which tends to increase the chance of crossing of different individuals. The type of differentiation involved is very different in different organisms. It is the result of the activation, by a differential factor or group of factors, of one or the other of two alternative reaction systems of the genotype. These systems were probably first evolved in the gamophase, and during evolution their action gets gradually pushed back in ontogeny. This must entail the addition of factors which act earlier, but not necessarily the loss of the factors which act at the later stages. The morphological sex-differentiation evolved in this way may lose its primary function, as it does in hermaphrodites with well-differentiated sex organs.

In fact, if we regard sex from a morphological point of view, it becomes an extremely ill-defined and imprecise concept. The important difference between reproduction with or without crossing-over becomes irrelevant, while there seems, from this point of view, to be a fundamental distinction between fully developed sexuality such as we see it in a strip-tease artist and mere self-sterility and incompatibility mechanisms. But this is a distinction which is very difficult to define; on which side of the line shall we place the multipolar sexuality of Fungi or the relative sexuality of *Ectocarpus*?

PART THREE

Genetics and Evolution

That biological organisms have evolved has long ceased to be a theory, and has become a generally accepted fact, which itself requires an explanation. In the formation and testing of an explanatory hypothesis, genetics must collaborate with systematics and ecology. There is no space here to summarize the relevant data from the two latter sciences, but in the first chapter a short account is given of the most definite evidence concerning the nature of evolutionary change. The next chapter deals with the analysis, in genetical terms, of taxonomic differences; this is perhaps the most valuable contribution which genetics has yet made to the study of evolution. These two chapters pose the problem which a theory of evolution has to solve; the attempts to solve it, mainly by means of Darwin's theory of natural selection translated into modern terms, are discussed in the last chapter.

CHAPTER II

Processes of Evolution[1]

Evidence of the fact of evolution can be found in most fields of biological study. But facts which give some indication of the genetic mechanisms involved are much fewer. Most of them fall into one of two classes, which demonstrate two important theses. Firstly, those fossils whose evolution can be followed in a continuous series through some considerable period of time give conclusive evidence that evolutionary change can be by gradual transitions, which, moreover, progress in a single direction. Secondly, there is evidence, perhaps not quite so strong, but nevertheless very cogent, that some evolutionary changes can be by sudden and discontinuous jumps. This has emerged most clearly from studies on the distribution of plants.

1. The Palaeontological Evidence[2]

Palaeontology provides evidence about the evolution of the hard parts (skeleton, shell, etc.) of organisms which were common enough to be frequently preserved as fossils. For the theory of evolution there are two fundamentally important classes of evidence, which concern quite different time-scales and which must therefore be sharply distinguished. On the one hand, we can describe the general outline of the course of evolutionary change in a large group of organisms, in some cases even in a whole phylum. We are then dealing with phenomena which last throughout many geological periods, probably through times of the order of some hundreds of millions of years. On the other hand, we have data as to the gradual alteration of one species or small group of species, a process which usually takes place within one of the more recent geological periods, over a period of a few million years. In both cases students of variation among living animals will be surprised by the high degree of orderliness which palaeontogists ascribe to the processes concerned. On the large scale, whole orders and phyla seem to pass through an orderly series of changes which have been described as "programme evolution"; while on the small scale it has been shown that a group of "species" may evolve progressively along a certain line

[1] *General references:* Haldane 1932b.
[2] *General references:* Bather 1927, Davies 1937, Lang 1923, Swinnerton 1923, 1932.

of change which has been called a "trend." The evidence for the small-scale trends is, however, considerably more compelling than that for the large-scale programme evolution. We will discuss the two phenomena separately, taking the small-scale trends first.

(a) *Trends.*—One of the classical examples of evolution in trends is provided by the Gryphaeas of the Jurassic.[1] During that period several different stocks of oyster-like Lamellibranchs belonging to the genus Ostrea independently evolved along a certain line of change which converted them into forms classified in the genus Gryphaea. The essential change was the development of a very thick shell with highly-curved umbones. At any given time, represented by a certain geological horizon, the population of Gryphaeas belonging to one of these stocks shows a certain amount of variation in these characters. Part of the variation is due to age differences; old specimens have thicker shells than young ones, and are also larger in area. Variation due to age differences can therefore be eliminated by considering only shells of the same size. Within a group of shells of the same size, the variation is distributed in the normal frequency curve. In a higher, i.e. later, horizon, it will be found that the whole of this curve of variation is shifted towards greater thickness and curvature. By such insensible changes of frequency distribution the population gradually comes to consist of forms which would be classified in a different genus. Clearly the usual concept of species cannot be applied to such a case, and palaeontologists usually speak of such assemblages as a gens.

In the Gryphaeas, the evolutionary changes seem at first sight to carry forward in a general way the developmental changes which occur in each life history. We can regard an early, flat form as consisting of a short piece of a spirally coiled shell, while the later form with a curved umbo simply consists of a larger part of the same spiral and is derived from the flat form by mere continuation in its spiral growth. Phenomena in which a later evolutionary form is produced by continuing a developmental process which stopped short in its ancestors is known as palingenesis, and we shall see that such a process is described as occurring in the large-scale programme evolutions which will be discussed later. It is therefore of importance to note that the apparent palingenesis of the umbonal curvature of the Gryphaeas will not bear critical examination. The evolutionary sequence does not involve merely the development of a larger amount of the same spiral, but the actual tightness of the coil increases in the later forms.

[1] Trueman 1922, 1930.

The spiral coiling of the Gryphaeas, like most molluscan spirals, is a particular example of "heterogonic growth."[1] In heterogonic growth, two organs or parts of an animal grow at different rates (measured by increase relative to the mass which is already present) but their growth

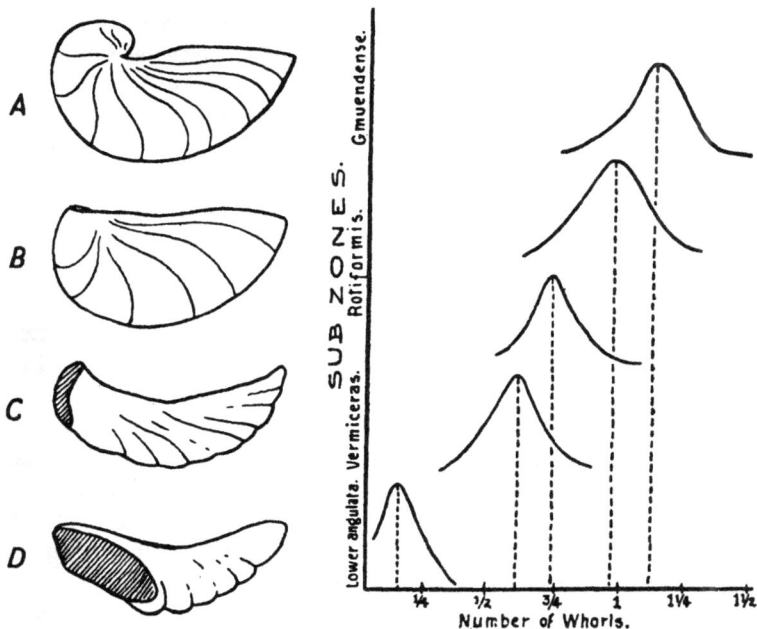

Fig. 107. **The Evolution of Gryphaeas.**—On the left are drawings of the left valves of Gryphaeas at successive levels in the Lower Lias formation, to show the increase in coiling from the older types (below) to the more recent (above). At the right are shown the frequency curves of variation in coiling in successive populations during a short part of this period. The different levels in the geological sequence (sub-zones) are given distinguishing names, and the zero lines of the frequency curves have been lifted up to show the level of the population in the sequence of beds.

(From Swinnerton, after Trueman.)

rates remain in a constant ratio to one another. In the case of spiral coiling, the two parts whose growth rates are in a constant proportion are the inner and outer edges of a coil. It is remarkable that this same phenomenon seems to be involved in several cases of evolution in trends. Thus the horns of Titanotheres are heterogonic organs; their growth rate is in constant proportion to that of the body as a whole.

[1] Cf. Huxley 1932.

And we find that several different lines of Titanotheres underwent trend evolution, in which the horns became larger and larger.[1] It is not known whether this connection of heterogonic growth and trend evolution is mere coincidence or not.

Not all trends concern heterogonic growth. A very beautiful example may be taken from the Chalk Micrasters,[2] which were Echinoderms rather like the modern heart urchin. We find here several slow continuous processes of evolutionary change affecting many different parts of the shell; the general body outline, the position of the mouth and apical disc, the depth of the anterior groove and the cross-sectional shape, granulation and suturing of the ambulacral grooves all change slowly in definite directions. Many other examples could be given. One of the most famous is that of the reduction of toes in the horse, pictures of which will be found in most textbooks. But we have nowhere nearly so complete a series of overlapping forms for any land animal, even so wide-ranging and common as the horse, as we have for marine invertebrates, and it is the latter, therefore, which provide the most critical evidence.

Two further facts about trends require noticing. In the first place, a trend may continue so far that it appears to lead to the extinction of the gens. The latest representatives of any of the Gryphaea gentes show such a curvature of the umbones that the opening of the shell appears mechanically difficult, and it is plausible to suggest that this was a reason for the dying out of the line. Similarly, the latest representative of a Titanothere gens may have such unwieldy horns as to appear definitely ill-adapted. This appearance can never be confirmed by actual observation and must remain a hypothesis. But if it is true it presents a particular difficulty to any theory which explains evolution as a result of natural selection; and it is of course quite unreconcilable with any theory of evolution in terms of the inheritance of adaptive acquired characters.

The second point is that we find examples of two or more related stocks which start off on several different trends, and in such cases gens *A* may progress rapidly along trend *P* and more slowly along trend *Q*, while gens *B* moves slowly along *P* and rapidly along *Q*. The trends have then a certain independence. We shall see that this independence has been elevated into a general principle for dealing with the programme evolution on a larger time scale.

(b) *Programme Evolution.*—Programme evolution is most strikingly

[1] Cf. Robb 1935.   [2] Rowe 1899.

seen in groups whose entire evolutionary history lies in the past and is now available for study. The group for which this theory was first advanced, and which is still the most famous example of it, is the Ammonitoidea, which were cephalopods flourishing from the Silurian to the Cretaceous. They are now only represented by the allied stock of Nautiloids.

Hyatt (1867–1903) claimed that the earliest representatives of the ammonites had straight conical shells, and that the early stages of their evolution consisted in the gradual acquirement of an increased curvature by which the shell became coiled into a tight spiral, with the outer whorls eventually overlapping and partly enclosing the inner whorls. At the same time there was an increasing elaboration of the "suture," which is the line along which the outer surface of the shell is met by the septa which divide the coiled up cone into chambers. After reaching a maximum of coiling and sutural-elaboration, the ammonites at the end of their evolutionary course retraced their steps and became straight with more simplified sutures. Having thus attained a racial second childhood, the so-called "senescent" stage, they finally died out.

The basis of observation on which this somewhat anthropomorphic account is erected has recently been severely criticized, particularly by Spath.[1] He points out that the actual evidence for the temporal sequence of forms showing progressively increasing coiling is very inadequate. Actually there appear to have been straight, slightly coiled and tightly coiled forms coexisting from the earliest times. The suture line however was simpler in the early members of the group and became more complicated later. During the greater part of the time when the group was in existence, far the commonest form had a more or less tightly coiled shell and elaborate suture, but loosely coiled forms were always present. Shortly before the extinction of the ammonites in the Cretaceous, an enormous evolutionary radiation occurred, and very many new types were produced. Among these were straight shells with reduced sutures, but there were also species which cannot be fitted into the simple scheme, for instance, spirally coiled types. There is thus no adequate evidence of an orderly evolution from simple through complex back to simple forms. In fact, Spath states that if we examine in more detail the particular sub-groups whose evolution can be followed with certainty through fairly short periods, the evidence suggests an evolution from tightly coiled forms to straight ones rather than the reverse.

[1] Spath 1933.

Hyatt placed great emphasis on palingenesis in evolution; that is to say, on the theory that the early development of an organism summarized the complete evolutionary history of its ancestors (recapitulation), and that the evolutionary advance appeared as something

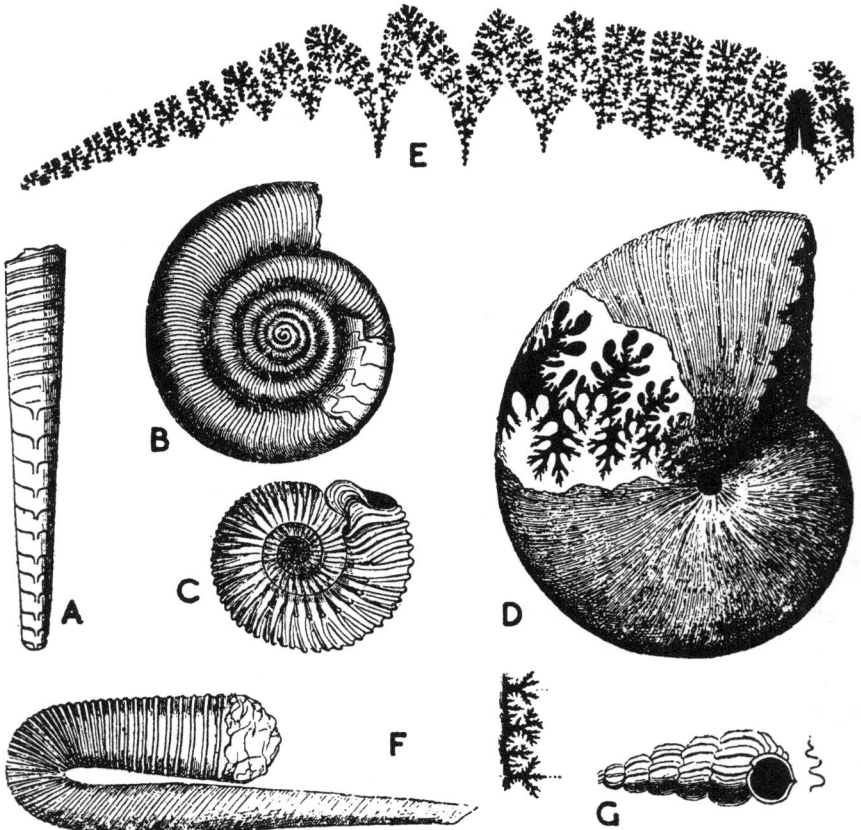

Fig. 108. **Some Ammonites to Illustrate "Programme Evolution" in the Group.**—*A.* An early straight form with simple suture line (Bactrites, Upper Devonian). *B.* Early loosely coiled, simple sutured form (Clymenia, Upper Devonian). *C.* A typical middle form (Stephanoceras, Jurassic). *D.* Tightly coiled form with complex suture (Phylloceras, Jurassic). *E.* The suture line at its most complicated; the space between two successive lines has been painted black. The drawing shows only one side of the shell, the mid-line being to the right (Pinacoceras, Trias). *F.* An uncoiled late form with simplified suture (Hamulina, Chalk). *G.* A form with helical coiling such as is found in the latest species of the group; the form figured, however (Cochloceras), is from the Upper Trias, and shows that these types occurred in the heyday of the group and not only in its "senescent" phase.

(After Zittel.)

added on to the end of this. Spath points out, however, that Hyatt used this principle in classifying his ammonites into series of ancestors and their descendants, and has therefore no right to use the phylogenetic lines thus obtained to prove the correctness of the principle.

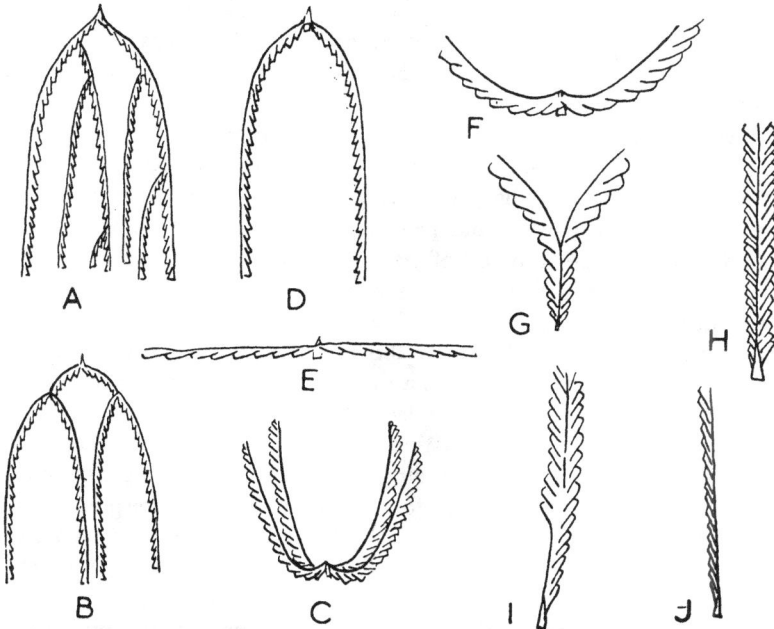

Fig. 109. Semi-diagrammatic drawings of **graptolites** to illustrate some of the **"orthogenetic trends"** in the evolution of the group. The time sequence runs from *A* to *J*. *A* Pendent, many-branched form (Bryograptus); *B* pendent four-branched form (Tetragraptus); *C* scandent four-branched form (Tetragraptus); *D* and *E* pendent and scandent two-branched forms (Didymograptus); *F*, *G*, and *H*, transition from scandent two-branched to single-branched forms; *I* and *J*, transition to one-sided, one-branched forms (Dimorphograptus to Monograptus). Note the two main trends, from pendent to scandent growth, and reduction in number of branches; and note further that they are to some extent independent (compare *B* and *C*; *D*, *E*, and *F*).

Another famous example of programme evolution is that of the Graptolites.[1] These were colonial organisms, probably allied to the Coelenterates, but perhaps more safely placed in a phylum of their own. The colonies consisted of branches hanging from an attachment organ which probably adhered to seaweed; along the branches the individual polyps were arranged, each in a small cup or theca. There is

[1] Elles 1922, Bulman 1933.

no question in this case of a strictly palingenetic type of evolution, in fact, new characters are said to appear in the first formed thecae of a colony. Nor is there any evidence of a circular evolutionary course. But the whole of graptolite evolution can, it is claimed, be described in terms of a very few progressive series of changes, of which the most important are (1) reduction of number of branches; (2) change from the pendent form to a form with branches sticking upwards; (3) elaboration of the shape of the thecae.

Evolution normally proceeded in a definite direction along each of these lines, although some examples of reversal of direction, i.e. regression, are known. The progress along each of the three different lines was to a large extent independent; one phylogenetic stock might rapidly reduce the number of its branches while retaining simple thecae, another might develop elaborate thecae while still having a large number of branches. Each of the series of changes is often compared with the trends discussed above, and is often spoken of as a trend; but the graptolite trends on the whole took much longer to run through than those of the Gryphaeas for example, and the evidence for them comes from the seriation of comparatively rare specimens, and not from collections so numerous that we can follow a continuous process of change. There is therefore room for errors of interpretation; the orderly programme of evolution may perhaps represent the neatness of the systematists' mind rather than anything to do with the graptolites themselves. While such a doubt is still possible, it is better to distinguish between the large-scale and more doubtful programme evolution and the small-scale undoubted trends. Further palaeontological research is necessary before we can judge how far such doubts are justified. Meanwhile the geneticist should note the alleged phenomenon of programme evolution as one which may require his attention; at present genetics has very little to say about evolution over such long periods of time.

2. *The Sudden Origin of Species: Age and Area*

Examples have been given (p. 256) of the sudden origin of new species by the formation of tetraploids in plant hybrids, both in nature and in the laboratory. There is other evidence that many plant species, which may not always be polyploids, have a sudden origin. Willis,[1] working on the flora of Ceylon, pointed out that a rare species tends to be confined to a very small area in a certain locality. He showed that

---

[1] Willis 1922.

there is no ground for supposing that this limitation in habitat is, in general, caused by an adaptation to a very narrow range of conditions which are realized in that spot and nowhere else, since if that were the case we should expect to find that if the territories of two rare species overlapped at all they would be exactly the same, which is not true. Further, it is difficult to believe that these rare species all represent the last remnants of species which were formerly much commoner and more widespread, since then there should be a fair number whose range consists of several isolated patches, which again is not true.

These considerations led Willis to suggest that the rare species occurring in a single isolated area had in fact been newly formed. He

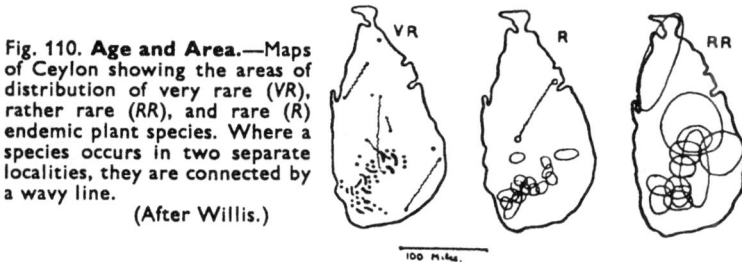

Fig. 110. **Age and Area.**—Maps of Ceylon showing the areas of distribution of very rare (VR), rather rare (RR), and rare (R) endemic plant species. Where a species occurs in two separate localities, they are connected by a wavy line.
(After Willis.)

put forward the following hypothesis, known as the theory of "age and area":

"The area occupied at any given time, in any given country, by a group of allied species at least ten in number, depends chiefly, so long as conditions remain reasonably constant, upon the ages of the species of that group in that country, but may be enormously modified by the presence of barriers such as seas, rivers, mountains, changes of climate from one region to the next, or other ecological boundaries, and the like, also by the action of man, and by other causes."

Roughly, then, in a statistical way, the older the species, the greater the area it occupies.

The importance of this extremely simple, but within its limits plausible, hypothesis for our present discussion is this. The narrowly restricted range of some of the rare species which the age and area hypothesis identifies as new species indicates that these species have had a sudden and local origin in the recent past. The evidence strongly suggests that the species did not originate by the gradual transformation of the gene-complex of a population, but that the new species appeared quickly in a few individuals.

3. The Local Origin of Important Groups of Plants

A species which has originated suddenly and in a localized area in the way postulated by Willis will then spread and undergo further evolution and may give rise to a large and varied group of descendants. Vavilov[1] pointed out that the greatest number of variants should be found in the region in which the group first originated, where it has had longest to produce new types. The places of origin of such groups can therefore be identified as the regions where the greatest number of variants can be found. Vavilov and his associates have conducted very extensive collecting expeditions to many parts of the world, chiefly in connection with attempts to find new races of crop plants which may be used to improve the yield of agriculture in the U.S.S.R. Although an enormous number of specimens have been collected, including for instance, fourteen new species of potato, only one species of which was previously known in Europe, there are still very many gaps in our knowledge; we have so far had to rely, in this field, almost solely on the efforts of the one country which has serious economic reasons for trying to improve its crop yields.

On the basis of what is known at present, it seems likely that most of the important crop plants originated in a few regions. The conditions which caused each of these regions to be the birthplace of several large groups of important crops are not understood. Vavilov distinguishes seven such centres:

(1a) South-West Asia, including North-West India, for soft wheats, rye, several Leguminosae and many fruit trees.
(1b) India for rice, sugar cane, and many tropical crops.
(2) East China for fruit trees and soya bean.
(3) Abyssinia for hard wheats, barley, and leguminous crops.
(4) Mediterranean regions for the olive and certain forage plants.
(5) South Mexico and Central America for maize and upland cotton.
(6) Peru and Bolivia for the potato.

More detailed study shows that some of these primary centres can be

[1] Vavilov 1928, 1932.

further broken up. Thus for maize, which is probably the most wide-spread crop in the world, ranging from 58° N. to 40° S., with an enor-

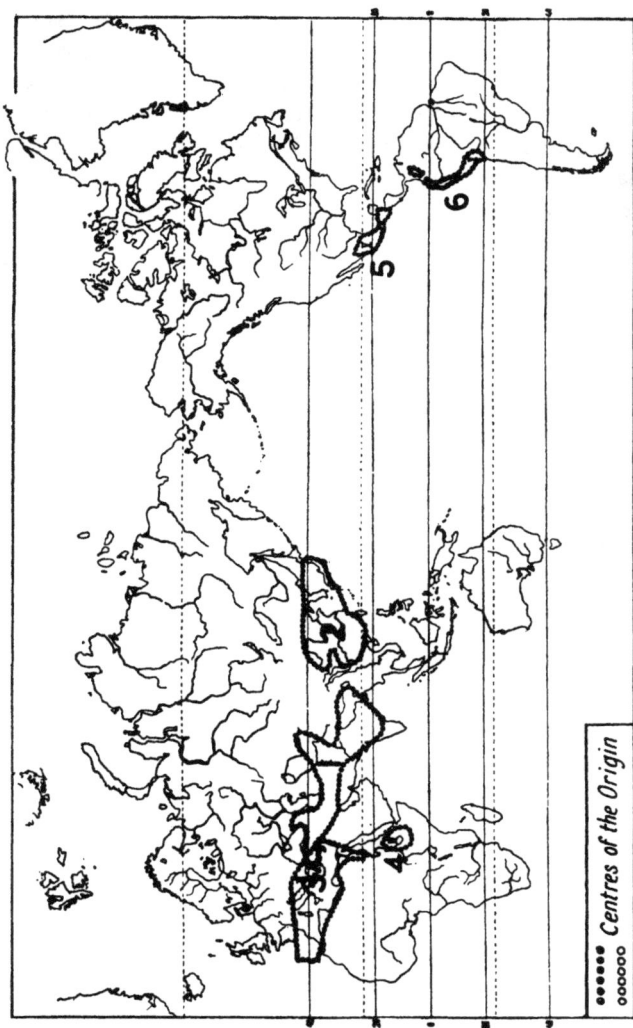

Centres of the Origin

Fig. 111. Centres of Origin of Crop Plants (from Vavilov).—1. S.W. Asia and India. 2. S.E. Asia. 3. Mediterranean. 4. Abyssinia. 5. Central America and Mexico. 6. Peru and Bolivia.

mous number of variants, some of which ripen in seventy days while others require eleven months, and a range in size between 60 cm. and 6 cm., subsidiary centres of diversity can be distinguished for the subsidiary groups, e.g. soft maize is most variable in Peru and Bolivia, hard maize in South Mexico and Central America.[1]

Vavilov's theory is, of course, only generally true and may be expected to break down in some cases. For example, Turesson[2] has pointed out that the centres of diversity of some Eurasian plants is in Finland and Sweden, which were probably beneath the polar ice cap (during the Ice Age) when the species originated. He suggests that their present great diversity of forms may be due to a climatic similarity with the original centres of origin.

[1] Kuleshov 1933.   [2] Turesson 1931.

CHAPTER 12

The Genetic Nature of Taxonomic Differences[1]

The genetic analysis of taxonomic differences is still in its infancy; by far the greater majority of animals and plant species are genetically unknown, though cytologists have already advanced rather farther in the comparative study of their chromosomes, particularly in plants. However, the comparatively small number of well-analysed cases are already enough to show us that species, and other taxonomic groups, may differ in any of the ways genetically possible. We find a rough, but only a rough, correspondence between the magnitude of the genetical differences and the taxonomic interval between different groups. The imperfections of the correlation are perhaps only to be expected when we reflect that the word species covers groups which are very different in genetical status. Darlington[2] lists six types of species, which will be affected in different ways by the evolutionary mechanisms of variation and inheritance. They are given in order of increasing hybridity (i.e. genetical inhomogeneity) of individuals.

(1) Simple diploid monoecious or hermaphrodite species.
(2) Polyploid species.
(3) Mixed species which include several different races, which may differ by polyploidy, or by chromosome translocations, inverversions, etc.
(4) Diploids with a sex chromosome mechanism in which one sex is permanently heterozygous. These are clearly near to (1) and (5).
(5) Complex heterozygous species such as those of Oenothera (p. 110).
(6) Clonal species, which reproduce vegetatively or by apomixis; they are frequently aneuploid, i.e. contain chromosome sets which are unsuited to performing regular reduction, as in triploids, trisomics, etc.

It will be convenient to discuss polyploid species first. At the end of the chapter we shall consider whether there may be species differences which are not dependent on the usual chromosome mechanisms.

[1] *General references:* Dobzhansky 1937a, b, Haldane 1929, 1932b.
[2] Darlington 1932a.

A. CHROMOSOME DIFFERENCES

1. *Polyploid Species*

The fact of polyploidy may be suspected if an organism is found to have a chromosome number which is an integral multiple of that of a related species. Judged on this criterion alone, polyploids seem to be extremely common among angiosperms; probably a quarter of the known species have multiple chromosome numbers of this kind. It has been suggested that the evolution of angiosperms from lower plants was due to polyploid formation.[1] Among the lower plants, the evidence is not extensive, but polyploidy seems to be less important, and indeed in some large groups all the species have the same chromosome number (e.g. nearly all gymnosperms have a haploid number of 12). In animals polyploidy is very rare; haploids are known in Ascaris, and simple tetraploids in Artemia[2] and a few other crustacea[3], while triploid and tetraploid Drosophila have been obtained in the laboratory. The chromosome numbers make it possible to suggest that polyploidy has played a part in species formation in hermaphrodite Mollusca and Annelids, but nothing definite is known about this. This absence of polyploid animals is peculiar;[4] it may be correlated with a disarrangement of the sex determining mechanism, since the few known polyploids are parthenogenetic or hermaphrodite; but on the other hand, it may be that only parthenogenetic or hermaphrodite polyploids can survive.

Among angiosperms, certain genera, e.g. Antirrhinum, Orchis, have a characteristic chromosome number which is found in all species; in others, e.g. Carex, the species have different chromosome numbers but these do not form a series of multiples; while in those genera in which polyploidy occurs, the different multiples are usually correlated with differences between species or larger groups, though in a few cases the polyploid forms are classified only as races or sub-species.

2. *The Experimental Production of Polyploid Species*

The general principles of the formation of polyploid species have been most clearly revealed in investigations which have led to the artificial production of polyploids worthy of being ranked as new species.

Polyploids originate by a partial failure of cell division.[5] If the failure

---

[1] Anderson 1934. [2] Gross 1932. [3] Vandel 1927*b*. [4] Muller 1925.
[5] This was originally suggested on theoretical grounds by Winge 1917.

occurs in somatic tissue, the two daughter nuclei may unite and constitute a single nucleus with the tetraploid chromosome number. If a shoot originates from such a cell, it will also be tetraploid, and bear flowers which give diploid gametes (if the reduction is regular, cf. p. 69). Failure of cell-division may occur in germinal tissue shortly before the reduction division, in which case diploid gametes will be formed immediately in the descendants of the affected cell; this process is known as syndiploidy. The meiotic divisions themselves may also fail, either by an imperfect separation of the daughter nuclei, which gives a so-called "restitution nucleus," or by a failure of zygotene pairing accompanied by a suppression of the second division; the latter abnormality occurs particularly in hybrids. Both these types of failure of meiosis give rise to diploid gametes, which will give tetraploids on selfing or triploids when crossed with normals. Similar processes occur in the formation of unreduced gametes in organisms with diploid parthenogenesis (p. 63).

These processes of doubling were until recently not under experimental control. In mosses[1] it is possible to suppress one of the meiotic divisions by the use of narcotics such as chloral hydrate, and thus to obtain diploid gametophytes; recently Blakeslee[2] has found that in the higher plants anaphase separation in mitosis can be rather regularly inhibited, with the formation of tetraploid restitution nuclei, by treatment with colchicine. Failure of somatic mitosis can sometimes be stimulated by various sorts of ill-treatment (e.g. extreme temperatures) and processes of this kind may turn out to be applicable to animals. A failure of mitosis is rather common in the callus tissue growing over wounds in Solanaceae,[3] and many tetraploids have been produced from shoots originating in such tissue. A special case of polyploid formation occurs in some diplohaplonts, in which portions of the diploid sporophytes may grow into plants which are gametophytes, although they retain the diploid number.[4]

Polyploids originating within a single species usually differ rather little in appearance from the original diploids, although they are often somewhat larger and more vigorous. Their fertility is usually lower because of the irregularity with which the reduction takes place (p. 70). More important for the study of natural polyploids are those which

[1] Marchal and Marchal 1911. *Rev.* effects of temperature etc. Sax 1937.
[2] Cf. Blakeslee and Avery 1937, Nebel and Ruttle 1938. For an attempt to produce tetraploid animals in this way, see Pincus and Waddington 1939.
[3] Winkler 1916, Jorgensen 1928.   [4] Wettstein 1927.

I

have arisen from hybrids. A hybrid between species A and B may be almost completely sterile if the A chromosomes cannot pair with those from B. Such a hybrid may form a few diploid gametes, either by a chance migration of all the chromosomes to one pole, by syndiploidy or suppression of the second division or finally by a somatic doubling. If diploid gametes are formed in any of these ways, triploids or tetraploids may be formed on selfing; and the latter, being $AA\ BB$, will be able to breed true, but give infertile triploids when crossed to the original parents.

Certain naturally occurring species have been synthesized in this way, by the chance occurrence of chromosome doubling in hybrids; and some new forms, deserving the name of new species, have been created. As examples we may take the following:

(1) *Nicotiana "digluta."*[1]—This "synthetic species" was probably the first whose origin was fully understood. It arose in the FI of a cross between *N. glutinosa* ($2n = 24$) and *N. Tabacum* ($2n = 48$). On selfing, all the FI was sterile except for one plant, which gave hexaploid ($2n = 72$) offspring, and must itself have had this somatic number. Presumably it arose from a seed in which the chromosome number was doubled soon after fertilization.

(2) *Primula kewensis.*[2]—This new species arose by somatic doubling in a hybrid between *P. floribunda* and *P. verticillata*, both of which have $2n = 18$. The original hybrid was nearly sterile, but the tetraploid branch bore flowers which were fairly fertile on selfing, though giving infertile triploid hybrids with the parent species.

(3) *Raphano-Brassica.*[3]—The intergeneric cross between the radish *Raphanus sativa* ($2n = 18$) and the cabbage *Brassica oleracea* ($2n = 18$) is nearly completely sterile, since there is a complete lack of pairing between the chromosomes in pachytene. Occasionally unreduced gametes are formed by syndiploidy, and on selfing the hybrids can therefore give a few rare tetraploids. These are quite fertile on selfing, though giving unfertile hybrids with the parent species. By suitable crosses with the parents, many different infertile triploid, pentaploid, etc., forms have been built up. These show very clear examples of balance between the radish and cabbage characters, depending on the dosage of radish and cabbage genes. (Fig. 81, p. 171.)

(4) *Galeopsis tetrahit.*[4]—Müntzing crossed *G. pubescens* ($2n = 16$) with *G. speciosa* ($2n = 16$). The chromosomes of the hybrid paired

---

[1] Clausen and Goodspeed 1925.   [2] Digby 1912, Newton and Pellew 1929.
[3] Karpechenko 1924, 1928.   [4] Müntzing 1932.

heterogenetically (cf. p. 72), and a small F2 was obtained on selfing. Among this F2 was a triploid apparently formed by the fertilization of an unreduced diploid ovule by a haploid pollen grain; the latter presumably contained a mixture of speciosa and pubescens chromosomes, so that the triploid can be symbolized as $pb + sp + \dfrac{pb + sp}{2}$. The triploid was then crossed again with pubescens, and gave a tetraploid

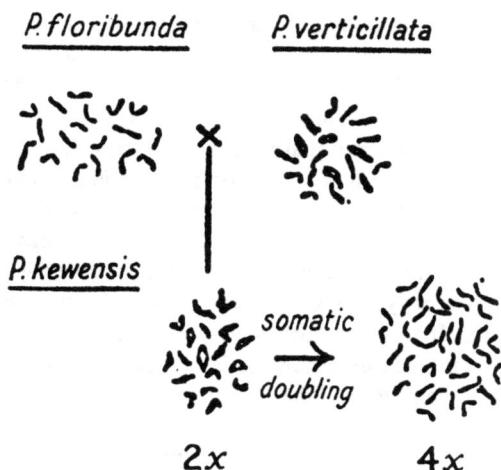

Fig. 112. **The Origin of Primula Kewensis.**—The drawings show typical mitotic figures, those of P. verticillata and the diploid hybrid being early anaphases, the others metaphases.

(From Darlington, after Newton and Pellew.)

by fertilization of an unreduced triploid gamete. This tetraploid would then contain two sets of pubescens chromosomes, one set of speciosa, and one set of mixed speciosa and pubescens. It exactly resembled the natural species G. tetrahit, and was perfectly fertile when crossed with it or when selfed, but was infertile with either G. pubescens or speciosa. It is clear that the crosses had synthesized the species G. tetrahit.

3. The Analysis of Natural Polyploid Species

It is usually not possible to reverse the process by which artificial polyploid species are made, and to break down a species into two components. This has, however, been done by Wettstein[1] in the moss

[1] Wettstein 1932.

Physcomitrium ($n = 36$) which can be shown to consist of two separately viable sets of chromosomes ($n = 18$). In most cases, the analysis of a polyploid species has to be based on the pairing of its chromosomes in hybrids with the constituent diploids or with other related polyploids. Thus if we cross a tetraploid with a diploid and obtain a triploid in which two-thirds of the chromosomes pair[1] at the reduction division, we may argue that the tetraploid contained a diploid set of chromosomes homologous with those of the diploid used in the cross. One of the first observations of this kind was made by Rosenberg,[2] who found pairing in ten bivalents and ten univalents in a hybrid between *Drosera longifolia* ($n = 10$) and *D. rotundifolia* ($n = 20$). He argued that in the hybrid the 10 longifolia chromosomes paired with 10 of the rotundifolia ones, and that the ten unpaired chromosomes were the rest of the rotundifolia set; and concluded that rotundifolia is an allotetraploid in which longifolia is one of the constituents diploids. But clearly this argument in itself is not conclusive; we know that in some cases autosyndesis may occur and it is therefore possible that the twenty rotundifolia chromosomes may pair among themselves, leaving the longifolia set as univalents. If this were the case, one would have to assume that the two diploids making up rotundifolia are more related to each other than either is to longifolia. There is no immediate way of deciding between these two hypotheses.

Arguments from chromosome pairing must therefore be accepted with great caution unless the series of crosses is so complete as to exclude one of the possibilities. For instance, in the complicated relations which Goodspeed and Clausen[3] have demonstrated between the *Nicotiana* species *sylvestris, tomentosa, Rusbyi, Tabacum, glutinosa* and the artificial *digluta*, the fact that no bivalents are formed in the *Tabacum-glutinosa* hybrid indicates that autosyndesis cannot take place in the *Tabacum* set, and this entitles one to argue that the bivalents in the other *Tabacum* crosses are not formed in this way. (Fig. 113.)

The different species of wheat[4] (*Triticum*) form a polyploid series with the basic number 7. The diploids ($2n = 14$) include *T. monococcum, T. aegilopoides*, etc., and are known as the Einkorn group; the tetraploids or Emmer group include *T. durum, T. dicoccum, T. turgidum*, etc., and are often grown for macaroni; the normal bread wheats are hexaploids, known as the Vulgare group after the best known species

---

[1] Usually the occurrence of prophase pairing is only an inference from observations of metaphase association.   [2] Rosenberg 1909.
[3] Cf. Clausen 1928, Sansome and Philp 1932.   [4] *Rev.* Watkins 1930.

T. vulgare. This grouping of the Triticum species was made some time before the chromosome numbers were known, but these completely confirmed the old morphological classification. Crosses between the diploids and either tetraploids or hexaploids are highly sterile, but the hybrids between the Emmer and Vulgare groups have been exten-

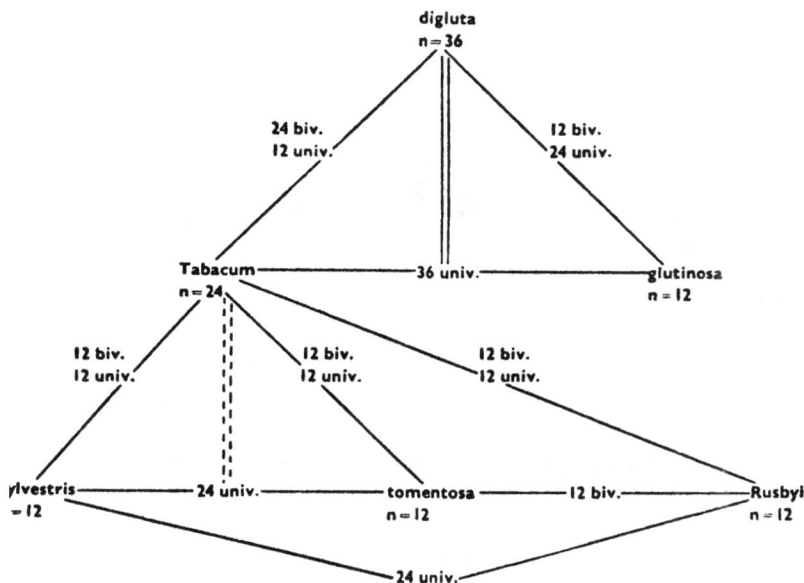

Fig. 113. **Chromosome Associations in Hybrids between Species of** *Nicotiana.*—The numbers of bivalents and univalents formed at meiotic metaphase in the hybrids are given. *N. digluta* is a synthetic species derived as a tetraploid hybrid between *Tabacum* and *glutinosa*. *Tabacum* itself may be a tetraploid hybrid between *sylvestris* and *tomentosa* or *Rusbyi*.

(After Sansome and Philp.)

sively studied. The F_1 has 35 chromosomes, 14 from Emmer and 21 from Vulgare. The meiosis shows 14 bivalents and 7 univalents.

If the F_1 and subsequent generations are selfed, the plants which are obtained fall into two groups: a "diminishing" group in which the chromosome number sinks to $2n = 28$, and an "increasing" group in which the number gradually rises to $2n = 42$. It seems probable that the 14 chromosomes in the haploid set of Emmer are homologous with 14 of the Vulgare chromosomes, the other 7 Vulgare chromosomes being without homologues. The hybrids therefore all form 14 bivalents, which consist of an association between a Vulgare and an Emmer chromosome. Some of the extra Vulgare 7 chromosomes, which we may call *abcdefg*, may be present in addition. The phenomenon of the increasing and decreasing groups seems to be due to the fact that too

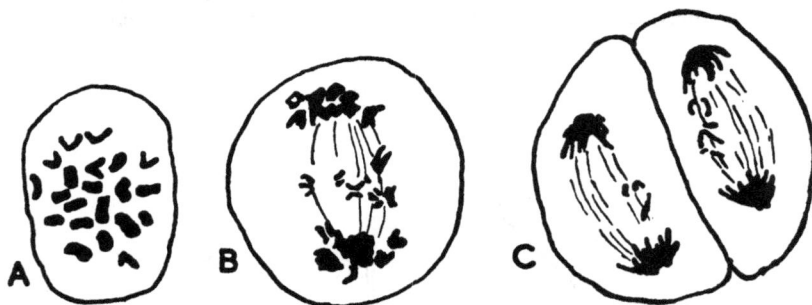

Fig. 114. **Chromosome Behaviour in the F1 Hybrid between Tetraploid and Hexaploid Wheats.**—A is a polar view of the metaphase of the first division, showing 14 bivalents and 7 univalents. In the anaphase B the univalents lag behind on the spindle, and eventually split and the chromatids separate to the two poles. In the second division anaphase C the single chromatids are separated at random to the two poles.

(After Sax.)

great an unbalance produces sterility. Tetraploids are fertile if they have any number of single extra chromosomes (e.g. 14 Emmer + 14 vulgare + *abcef*) but not if any chromosome is represented twice (e.g. 14 *E.* + 14 *v.* + *abccd*); this leads to the loss of extra chromo-

Fig. 115. **Spartina Townsendii.**—Mitotic metaphase plates of *S. Townsendii* (A) and its probable ancestors *S. alterniflora* (B) and *S. stricta* (C).

(After Huskins.)

somes and the resumption of the tetraploid number by the diminishing group. On the other hand, pentaploids with extra single chromosomes are also fertile (e.g. 14 *E* + 14 *v* + *abcdefg* + *abc*) and these will gradually revert to the hexaploid number.

There are several other grasses which are related to the wheat polyploids, the closest relative probably being Aegilops. Some fertile crosses have been made between Triticum and these genera,[1] and some new polyploid hybrids have been synthesized which may have economic importance (e.g the "octoploid" wheat-rye hybrid, derived from doubling in a cross between hexaploid wheat and diploid rye).[2]

In other genera the analysis of the constitution of polyploids has been made, with more or less certainty, purely on grounds of analogy. An example is *Spartina Townsendii*, a species which appeared suddenly in the latter half of the nineteenth century and is intermediate between *S. alterniflora* ($2n = 10x = 70$) and *S. stricta* ($2n = 8x = 56$). Huskins[3] showed that *S. Townsendii* has $2n = 126$, and claimed that it is an allo-octoendekaploid ($18x$) got by doubling in a hybrid between these two species.

There are many other cases in which the taxonomic relations of species have been confirmed by the evidence of chromosome numbers and plausible suggestions of polyploidy can be put forward.

4. *Secondary Polyploids*

Species which are not related as polyploids may differ in more complex chromosomal changes. Perhaps the simplest type of change is the reduplication of one or more chromosomes. Usually trisomic and still more tetrasomic types are relatively inviable, but in higher-multiple polyploids the reduplication of a few chromosomes does not cause such unbalance, and a process of this kind seems in fact to have played some part in plant evolution. Thus apples and pears belong to the family Rosaceae, the greater part of which forms a polyploid assemblage with a basic number of 7. In the Pomoideae section, however, the basic number is 17, and it appears that this number may represent a trebly trisomic diploid, i.e. a diploid with 3 reduplicated chromosomes ($2x + 3$), this whole set being balanced and forming the basic number of a new polyploid series.[4] This phenomenon, which is known as secondary polyploidy, is shown not only by the exceptional ratios obtained with segregating factors, but also by the evidence of the secondary pairing which happens to be very strong in Pomoideae. Secondary pairing, which is also strongly developed in Diptera (p. 74), causes the homologous split chromosomes to lie near each other at metaphase in mitosis, or the two similar chromosome pairs to lie near together in meiosis in a tetraploid. In an apple with 34 chromosomes (i.e diploid in respect of its basic number 17) one finds three groups of three

[1] Cf. Kihara and Lilienfeld 1932.   [2] Meister 1030.
[3] Huskins 1930.   [4] Darlington and Moffett 1930, Moffett 1931.

bivalents and four groups of two, indicating that the haploid set is made up as AAA,BBB,CCC,DD,EE,FF,GG. (Fig. 30 p. 74).

Species in which a chromosome has been lost also occur, again in

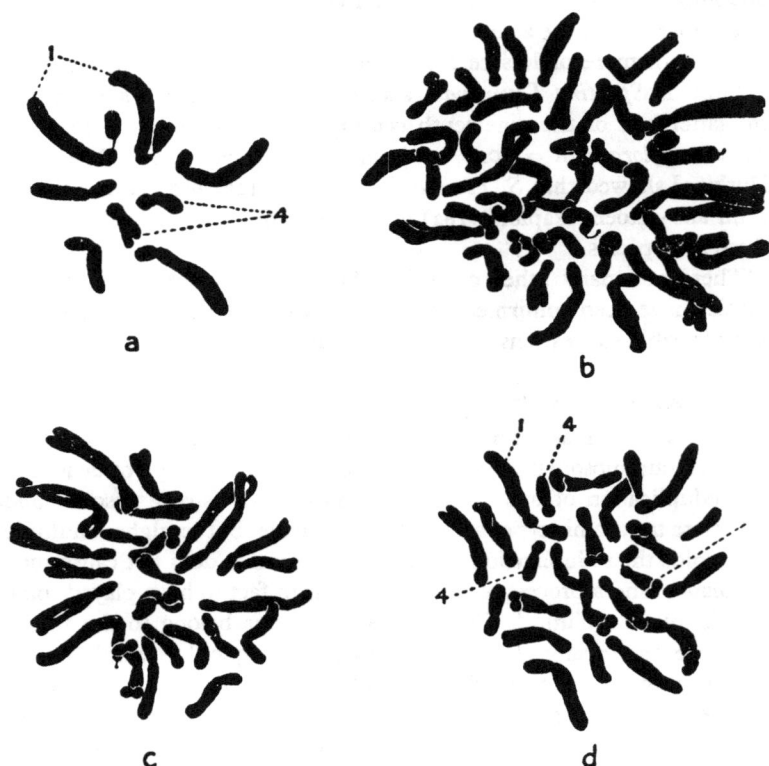

Fig. 116. **Crepis artificialis.**—A. Mitotic metaphase of *C. setosa* $(2n = 8)$. B. *C. biennis* $(2n = 40)$. C. F1 hybrid $(2n = 24)$. D. The true breeding hybrid *C. artificialis* $(2n = 24)$ with 10 pairs of *biennis* chromosomes and 2 pairs of *setosa* ones, which are marked 1 and 4 as in A.

(From Hurst, after Collins, Hollingshead, and Avery.)

conjunction with polyploidy which minimizes the resulting unbalance. The artificial *Crepis artificialis*[1] is an example. It was derived from a cross between *C. biennis* ($n = 20$, probably a tetraploid) and *C. setosa* ($n = 4$, a diploid). In the F1 the 20 biennis chromosomes formed 10 pairs by autosyndesis, but the four setosa chromosomes were unpaired and segregated at random. After continued selfing, a plant segregated

[1] Collins, Hollingshead and Avery 1929.

out in the F4 which still had the 10 pairs derived from biennis, and with four setosa chromosomes which formed two pairs. This enabled the plant to breed true, and it was ranked as a new species and given the name *C. artificialis*; but it is clearly unbalanced in its setosa fraction, two of the setosa pairs being entirely lacking.

In some genera, many polysomic forms occur, which one would suspect of being unbalanced. For instance, in the Melanium section of Viola[1] the chromosome numbers do not fall into a definite polyploid series, but many of the species can be regarded as modified polyploids. Thus we have

Diploid	..	AA	*V. Kitaibeliana* $2n = 14$
Tetraploid	..	$AABB$	*V. tricolor, V. alpestris* $2n = 26$
Hexaploid	..	$AABBCC$	*V. arvensis, V. rothomagensis* $2n = 34$
			V. Kitaibeliana $2n = 36$
Octoploid	..	$AABBCCCC$	*V. nana, V. lutea* $2n = 48$

It is remarkable that *V. Kitaibeliana* has two forms, classified in the same species, but having chromosome numbers which are not even multiples of one another. *Viola canina* is another phenotypically constant species with very variable chromosome numbers, which range from 16 to 25. Probably in such cases many of the chromosomes are more or less completely inert. This is also suggested by the fact that hybrids between such species may be quite fertile although they are obviously aneuploid and give gametes with variable numbers of chromosomes.

5. Non-polyploid Species

Species whose haploid numbers differ by one or a small integer are probably more often related by breakage or fusion of chromosomes, which leaves the balance unchanged, than by complete loss or addition. Straightforward fusion or fragmentation has often been inferred in such cases,[2] and a particularly striking example may be found in the sex chromosome of the Orthoptera. In some groups this is separate, but in others, e.g. in Mermiria, it is fused to an autosome but can still be recognized from the fact that it shows precocious condensation, staining quite deeply in early prophase when the other chromosomes, including the autosome to which it is attached, are still diffuse. The occurrence of perfectly simple fusion, and still more of simple fragmentation, conflict, however, with the large body of evidence dealing

---

[1] Clausen 1927, 1931a, b.   [2] Navaschin 1932.

with the behaviour of the centromeres.[1] Observations of division show that a chromosome fragment which does not possess a centromere does not move to the poles in anaphase, while one with two centromeres, such as might arise from fusion, often gets pulled apart by the movement of the centromeres to opposite poles. It seems necessary to assume that all viable chromosomes possess one centromere. The new "fusion" and "fragmentation" chromosomes are therefore probably always formed in the first place by translocation, and the appearance of forms with an extra or a lacking chromosome is a secondary effect of segregation or recombination (p. 95).

In some genera of plants, chromosome numbers vary continuously over a wide range. Thus in Carex[2] there are species with $n = 28, 30, 33, 34, 35, 37, 39, 40, 42$. Little is known about the exact nature of the differences involved in such cases, but probably polyploidy, secondary polyploidy, "fusion," "fragmentation" and even the occurrence of inert chromosomes are all involved.

6. Translocation, Inversion, etc., in Species Formation

Many nearly related species or local races differ in the shape and size of their chromosomes in such a way as to suggest that deficiencies and translocations have occurred during their evolution. Hybrids between such forms will show the pairing of unequal chromosomes and the formation of unequal bivalents at meiotic metaphase. The Brachystola (Orthopteran) chromosomes described by Carothers (p. 46) are an example, and more complex examples are common in Crepis. Similar changes may often be inferred from a mere comparison of chromosome shape between species which will not cross. The species of Drosophila fall, in their chromosome morphology,[3] into a series of types which can be quite easily derived from each other by a few simple changes involving translocations and "fragmentations" and "fusions" which we have seen to be results of translocation. Careful analysis does not always support the simplest hypotheses which suggest themselves in such cases. Thus in D. Willistoni it might be assumed that one of the rod-shaped chromosomes corresponds to the rod-shaped X of melanogaster, but actually non-disjunction has shown that the sex chromosome is really one of the V-shaped bodies, so that an unexpected translocation must be involved, by which the arm of some other chromosome became joined on to the originally rod-like X.[4]

[1] Cf. Darlington 1937.   [2] Heilborn 1924.
[3] Metz and Moses 1923.   [4] Cf. Morgan, Bridges, and Sturtevant 1925.

Better evidence of inversion, translocation, etc., is found in those species where definite genes or sections of the cross-over map can be identified. Several examples of this are now available in Drosophila, and a beginning has been made with the comparison of maize and related species. The Drosophila comparisons depend on the identification of homologous genes in different species, and a certain caution is necessary in this respect, since we know (p. 186) that in each species there are probably several *loci* capable of producing similar phenotypic

Fig. 117. Diagram of Mitotic Metaphase Figures in Drosophila Species.— The figures are of females, the X chromosome being at the bottom of each figure. (From Metz and Moses.)

effects. When, however, the $F1$ hybrids can be obtained, the allelomorphism of two genes can be tested by examining the double heterozygote, which will be wild-type in appearance if the genes are *not* allelomorphs. Homologous loci have been identified in this way in *Drosophila melanogaster* and *D. simulans,* and the cross-over maps can therefore be compared with some certainty.[1] In general the order of the genes (but not their distances) is the same in both species, but there is one large inversion in the third chromosomes. This can also be clearly seen in the salivary glands chromosomes of the hybrid, which show a characteristic inversion loop. Other minor differences are also present in the salivary chromosomes, and pairing may fail even when no definite structural differences are visible.[2] Relative inversions between species can also be discovered, though not so precisely located,

---

[1] Sturtevant 1929a. Cf. Sturtevant and Tan 1937.　　　　[2] Pätau 1935.

by observations of loop pairing at zygotene in the hybrids, or of the formation of chromosome bridges at anaphase following crossing over.[1] In *Drosophila pseudoobscura*[2] different local races within the species have

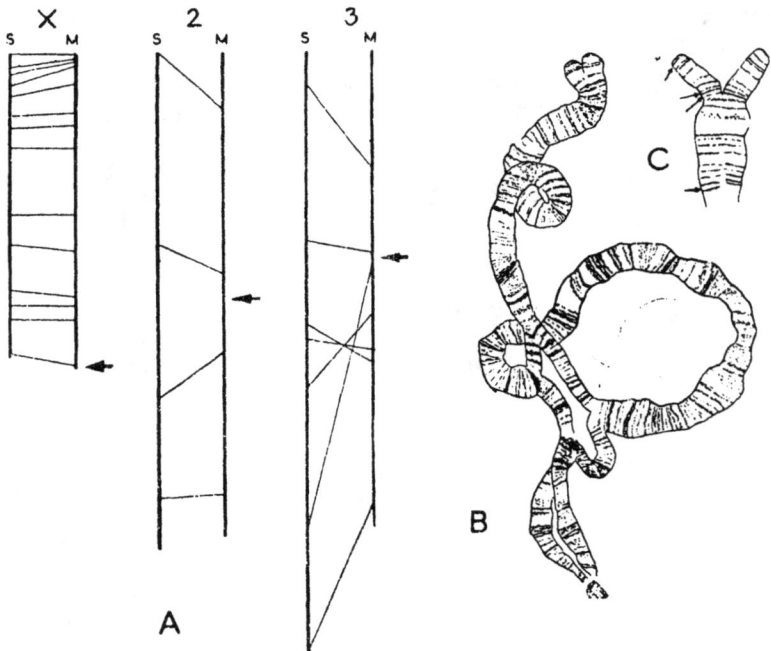

Fig. 118. **Comparison of Chromosomes in Drosophila melanogaster and simulans.**—A shows the comparative crossing over of maps of X, 2nd and 3rd chromosomes (S = *simulans*, M = *melanogaster*); notice the large inversion in the right arm of the third chromosome. The spindle attachments are marked with arrows. B shows the pairing of the inverted region in salivary glands of a hybrid. C shows the left end of the X chromosome in a hybrid; structural differences can be seen at the points indicated by the arrows.

(Ater Pätau.)

different gene arrangements in their chromosomes. These can best be investigated by studies on the pairing in the salivary glands of hybrids. The differences found are mostly inversions, and are sufficiently numerous and complicated to enable one to draw up a phylogenetic tree based on the way in which one arrangement might be transferred into another. Dobzhansky illustrates the method of argument by the following example: suppose we have three gene arrangements *abcdefghi*,

[1] Darlington 1936, Upcott 1937.
[2] Sturtevant and Dobzhansky 1936, Koller 1936, *Rev*. Dobzhansky 1937*b*.

aedcbfghi, and *aehgfbcdi*, it is clear that the second might be derived from the first by inverting the section *bcde*, and the third might then be derived from the second by inverting *dcbfgh*; but on the other hand, there is no simple way in which the third arrangement can be derived directly from the first. It is very interesting to find that the arrangement which one is led to suggest as ancestral in the pseudoobscura stock presents some similarities with that of the distinct species *D. miranda*.

In the salivary glands of hybrids between Drosophila species, large sections of the chromosomes usually fail to pair at all; presumably they have become too radically changed during evolution. An indication of the kinds of changes occurring is given by the comparison of *D. pseudoobscura* and *D. miranda*,[1] where pairing occurs between short homologous sections which have been extensively reshuffled by inversion and translocation; it is worth calling attention to the gross similarity between the chromosomes in these two species, which gives no indication of the very considerable differences which really exist. Arguments of species relationship must be accepted with great caution if they are based only on similarity of metaphase groups. (Fig. 119.)

Detailed investigation of the salivary gland chromosomes within a single Drosophila species often reveals the presence of very small duplicated sections.[2] The duplicated section often lies quite near, or even in contact with, the normal section and is sometimes inverted with respect to it, so that we get sequences of bands like *cdefghfeijk* or *cdeffeghijk*, etc. In the paired chromosomes these small sections may be in contact with their homologues so that the chromosomes are bent round into a sort of knot. Other minute changes would not be so easy to discover, since, for instance, if an inversion is very small there will not be room for the characteristic loop pairing to occur. But the presence of the small duplications makes it very likely that small inversions are also formed, and they may explain why in many regions the chromosomes fail to pair for no apparent reason in hybrids between species such as *D. melanogaster* and *simulans*.

Another type of chromosome change, which, since it leaves the total balance of the chromosome complement unchanged, may be expected to be involved in species formation, is mutual translocation, by which chromosomes *abcd,efgh* become converted into *abgh,efcd*. In hybrids between species differing in this way, multiple associations of chromosomes will form, giving chains or rings. For instance, in the above case we can get an association as a chain *abcd,dcfe,efgh,hgba* and the

---

[1] Dobzhansky and Tan 1936.　　　　　　　　　　[2] Bridges 1935.

last chromosome may be curled round so that the two a's are also paired, giving a ring. Examples of this are common in the species of Oenothera (p. 110), and a similar phenomenon has been found in a few other cases such as in Campanula,[1] Pisum,[2] and Datura.[3]

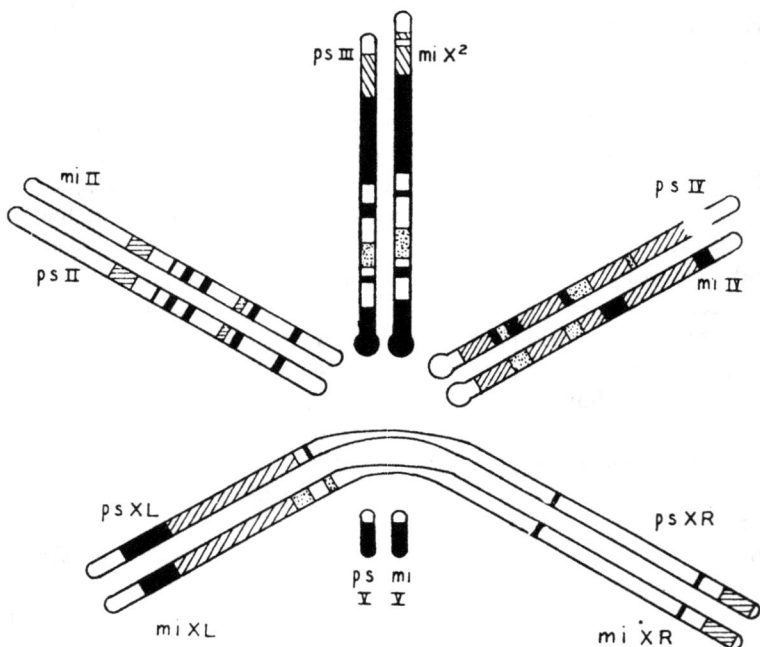

Fig. 119. Comparative Chromosome Maps of *Drosophila pseudo-obscura* and *D. miranda.*—Regions with the same gene arrangements are white; inverted sections, cross-hatched; translocations, stippled; and sections of which homologues are not detectable in the other species, black.   (From Dobzhansky.)

In some cases unbalanced forms, arising by translocation, may be viable and take part in evolution. An artificial case has been described in *D. melanogaster*,[4] in which one of the IV's was translocated to the Y, and then this part of the Y to the X; the male has a supernumerary piece of the Y, but this has very little effect, and similarly the female has two extra pieces of Y. The chromosome number has been reduced from $n = 4$ to $n = 3$.

[1] Gairdner and Darlington 1931.
[2] Pellew and Sansome 1931, Sansome 1932.
[3] Blakeslee 1929, Blakeslee, Bergner, and Avery 1937.
[4] Dubinin 1934, 1936.

More markedly unbalanced forms have been synthesized in Datura.[1] Here the secondary trisomic (cf. p. 172) sugarloaf has an extra chromosome which consists of two of chromosome ends known as 2 united together. By various segmental interchanges, two other types have been produced which also differ from normal in having two extra doses of end 2 which, however, are not attached to each other but to other chromosomes. They are true breeding, and all show much the same effect of their unbalance; the differences between them are presumably due either to the interchanges not being always at exactly the same

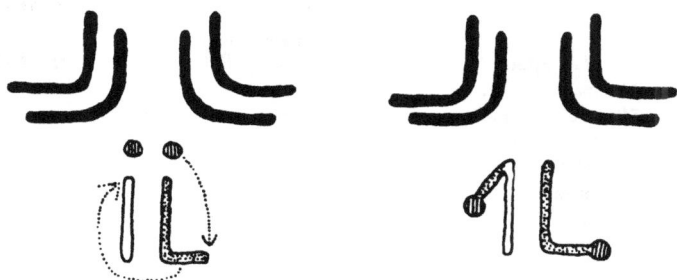

Fig. 120. **The Artificial Reduction of Chromosome Number in Drosophila.**— To the left is a diagram of the normal chromosome complement of a male *D. melanogaster*, showing the steps by which the new race is built up; first a translocation of the IVth on to the *Y*, and then a translocation of part of the *Y* with the IVth to the *X*. At the right is the chromosome set of a male of the new race; the original *X* is white, the *Y* is dotted, and the IVth lined.

(After Dubinin.)

place or to different position effects according to the regions to which the extra ends are attached.

The examples which have been given above show that related groups of organisms may differ by any of the possible forms of structural change, though changes leading to unbalance, such as deficiencies and duplications, are rare. It is not always possible to correlate the magnitude of the cytological changes with the taxonomic interval between the groups. Thus the different races of *D. pseudo-obscura*, which are all classified in the same species, have greater chromosomal differences than do the separate species *D. melanogaster* and *D. simulans*. Sometimes, however, the cytological and phenotypic differences run more or less parallel; this is true, for instance, in the great group of Crepis (Compositae) species, in which the chromosomes show a fairly compli-

[1] Cf. Blakeslee 1932.

cated morphology of trabants, constrictions, etc., which allow of detailed study.[1]

7. Gene Differences Between Species and Sub-species

In all groups of related species which have been carefully investigated, the different species have been found to exhibit the same gene mutations. This would be expected if one of the main factors in species

Fig. 121. Some Homologous Genes in Rodents.—Only allelomorphs of three loci are shown. + indicates that the gene occurs in the normal type, W that it is found in wild races, D that it is found in domesticated animals.

(From Haldane.)

Gene	Effect			Mouse	Norway Rat	Black Rat	Deer-mouse	Cavy	Rabbit
C	Normal	..	..	+	+	+	+	+	+
c^k	Slight dilution ..	..		—	—	—	—	D	+
c^d	Marked dilution	..		—	—	—	—	D	+
c^r	No yellow	..	..	D	D	—	—	D	D
c^b	Himalayan	..	..	D	—	—	—	D	D
c^a	Albino ..	..	..	D	D	—	D	—	D
A^y	Yellow ..	..	..	D	—	—	—	—	—
A^w	Light bellied grey	..	W	+	+	+	—	+
A^g	Grey bellied grey	..	+	—	W	—	+	—
A^r	Ticked bellied grey	..	—	—	—	—	W	—
a^t	Black and tan ..	..	—	—	—	—	—	D
a	Black	..	..	D	D	D	—	D	D
E^d	Black	..	..	—	—	D	—	—	D
E^s	Black	..	..	—	—	—	—	—	D
E	Normal	..	..	+	+	+	+̣	+	+
e^p	Bicoloured	..	..	—	—	—	—	D	D
e	Yellow ..	..	..	? D	—	D	D	D	D

formation has been chromosome rearrangement in the way discussed in the last section. The most complete description of such parallelism is that of Vavilov[2] for the different crop plants; even quite distantly related species of cereals show very similar mutant types. Among animals, one of the most extensive series of data, other than that in Drosophila on which the comparative cross-over maps are based, relates to the coat colours of rodents.[3]

As might be expected, the same locus may be normally occupied by different allelomorphs in different species, so that ordinary Mendelian

[1] Babcock and Navashin 1930, Hollingshead and Babcock 1930.
[2] Vavilov 1922.   [3] Haldane 1927.

segregation takes place in the hybrids. A well-known example is in the guinea-pig;[1] the wild *Cavia rufescens* differs from the domesticated *C. porcellus* in an allelomorph of the agouti locus, which causes the agouti colouration to occur on the belly as well as on the back. Many such cases are also known in plants, e.g. in Antirrhinum[2] and Crepis.

It is perfectly clear, therefore, that specific differences may involve gene differences. There is no reason to suppose, as some authors suggest, that the gene-controlled characters are all trivial. We know genes which, within a species, can cause such important changes as an alteration in the number of segments in the tarsus of a fly, or the number of

Fig. 122. **A Gene with Taxonomically Important Effects.**—A shows the normal two-carpel seed capsule of *Datura stramonium*. B is the three-carpelled capsule produced by the gene Tricarpel. The three-carpel form suggests a relation with the family Polemoniaceae, which belongs to a different sub-order to the Solanaceae in which Datura is classified. The capsules are represented as cut across to show the internal structure.

(After Blakeslee, Morrison, and Avery.)

carpels in the seed-capsule of Datura.[3] These are certainly more important differences than many on which species have been separated. A beginning has been made with the genetic analysis of the important serological differences between related species.[4]

In the past, groups which differ only in a single gene have been described as different species, particularly in plants, but as soon as this was discovered, the species have been amalgamated. In the difference between true species, very many genes are involved. The fullest knowledge we possess on this relates not to fully-fledged species but to local races within a species. One might expect such races to be more alike than two species would be, but even so the differences, in terms of genes, are fairly complicated. We will take two of the best analysed examples.

(a) *Lymantria*.—In a very long series of crosses, Goldschmidt[5] has

---

[1] Detlefsen 1914. For a review of animal hybrids, see Hertwig 1936.
[2] Baur 1932, Lotsy 1911.   [3] Blakeslee, Morrison and Avery 1927.
[4] Irwin and Cole 1936.   [5] Goldschmidt 1934*a*, *c*, 1935*b*.

fully investigated the local races of *Lymatria dispar* (the Gipsy moth) from various parts of Europe and Asia. The data are most complete about the races from the islands of Japan. One characteristic which distinguishes the different races is the strength of the sex factors (p. 216). The male factor which is carried on the *X* chromosome gives rise to a

Fig. 123. **Distribution of Sex Races of Lymantria.**—Notice the fairly orderly distribution, the strength increasing from west to east and south to north. The only exception is the weak race on the northern island of Hokkaido; but the Hokkaido race differs from the others in several ways; there is a considerable change in the whole fauna at the strait between Hokkaido and the main island; and this strait was formed before the main island was separated from the mainland.

series of races which may be called strong, weak, or intermediate, for which Goldschmidt postulates at least eight allelomorphic genes; the strengths cannot be due to combinations of factors at different loci on the *X* since no crossing-over and recombination occurs. The female factors, whose strength in each race is necessarily balanced against that of the male factor, is carried in the cytoplasm.

The rather orderly geographical distribution of the sex races indicates that the strength of the sex factor may be in some way adaptive, but its exact significance for the life of the animal is not understood. Goldschmidt has, however, also investigated some clearly adaptative characters. One of the most important is the latent period before hatching. The eggs are laid in the early summer and immediately develop into young caterpillars, which, however, do not leave the egg till the following spring. The physiological mechanism determining the length of time spent in the egg is complicated, but an essential part is played by a reaction which depends on the summation of all the time spent above a certain temperature. The difference between the local races is a difference in the length of this time which is required before emergence takes place, and the various races are exactly adjusted to the meteorological conditions in which they live. Goldschmidt has shown that this eminently adaptive character is at least partly inherited through genes, though it seems that in each race there are several genes which influence the character, so that the F_2 segregation is not into sharply defined classes but gives the wide variation characteristic of inheritance by multiple factors. Some cytoplasmic differences may also be involved since there is a slight tendency for offspring to resemble their mothers rather than their fathers, and the results of a cross thus differ according to the way in which it is made.

A third character studied in detail by Goldschmidt is the pigmentation of the caterpillars. The types can be roughly classified as light, dark, and those which are at first light and darken later. Closer examination shows that these differences depend on the rate of pigment formation (p. 173). The genetic basis seems to be a set of allelomorphic factors, but again there may be a cytoplasmic factor involved which causes a tendency for matroclinous inheritance.

Each race differs from the other races in all these characters and in many more which have not been fully analysed. Thus the races have characteristic complexes of genes, and the same conclusion must certainly apply to different species.

(b) *Peromyscus*.—Sumner[1] has made a long series of crosses between geographical races of several species of the American deermouse, *Peromyscus*. The differences between the races are mostly quantitative, affecting such characters as relative sizes of parts, depth of colour and size of coloured area, etc. The crosses nearly always give intermediate F_1's and do not show any clear-cut segregation in the F_2's and subsequent generations. There is, however, an increase in variability in F_2 and later, and it is clear that the races differ in many polymeric factors each with only a small effect. The segregation found is usually rather slight, presumably because of the large numbers of

[1] Sumner 1930, 1932.

factors involved. Sumner was at first inclined to doubt whether the race differences really did depend on Mendelian factors at all, and drew attention to the apparently adaptive character of some of the colour varieties. However, experiments in which animals were transferred from one region to another failed to show any direct effect of the environment on colouration, and Sumner eventually was able to satisfy himself that the Mendelian explanation was adequate.

8. *Local Races in Plants*

The investigation of local races in plants reveals a situation essentially similar to that in animals, but since plants are sessile whereas the most fully analysed animals are mobile, there is an even greater multiplication of local genetical types. Particularly extended studies have been made by Turesson,[1] who has given the name of genecology to the combination of genetical, ecological, and taxonomic investigations. He collected samples of plants from many different localities and cultivated them in the uniform environment of an experimental garden. The luxurious conditions of cultivation in some cases allowed stunted forms from unfavourable situations to develop potentialities which had previously been hidden, and apparently uniform populations from such places might reveal considerable hereditary variation. The main result, however, was to show that forms from different localities tended to retain their characteristics, many of which had a clear adaptive significance. The adaptive modifications of wide-ranging species seem therefore to be hereditarily fixed, and the progeny of crosses of different races showed segregation and high variability, often paralleling the mixed populations which can be found in nature in intermediate situations. The differences between the races rarely depended on single gene differences, but usually polymeric multiple genes were involved. Within a given ecological situation, however, the population was usually fairly uniform, so that the genes are not distributed at random throughout the whole species, but are collected into complexes, each complex being characteristic of, and adapted to, a certain set of conditions. It is only in situations intermediate between two localities with well-marked characteristics that the complexes break up and a free mixture of genes is found.

Again, these differences between local races can probably be taken as typical of the differences between fully developed species. Harland[2] has analysed several species of cotton and has shown that, besides a few

---

[1] Turesson 1925, 1930.   [2] Harland 1936.

major gene-differences, the whole genotypic backgrounds of the species are not the same, so that the dominance relations of the same gene, Crinkled Dwarf, are different when it is introduced into the different species (p. 298).

B. CYTOPLASMIC DIFFERENCES

1. *Inheritance through the Cytoplasm*[1]

It has often been suggested that organisms which are far removed from one another taxonomically may differ not only in the genetic characters borne in their chromosomes but also in the nature of their cytoplasm. In the nature of the case it is difficult to test this hypothesis directly, since widely different organisms cannot be crossed. It is, however, certain, if only on embryological grounds, that in many cases the divergence of species during evolution carries with it a divergence in the nature of the egg-cytoplasm. This does not justify, however, the common deduction that evolution is carried out by means of alterations of the cytoplasm and that the differences studied by geneticists are therefore of little importance for it. The evidence, as far as we know it, is that the differences between nearly related species are usually entirely chromosomal, so that one must suppose that the first steps in species-divergence are caused by alterations in the genes, and that the cytoplasmic differences only become developed at later stages in the process.

It is probable, in fact, that in nearly all species the characteristics of the egg-cytoplasm are entirely dependent on the activities of the chromosomes. It is only in comparatively few cases, which will be reviewed below, that we can discover cytoplasmic factors which can be perpetuated in the absence of the appropriate chromosomes.

Cytoplasmic differences are revealed in the comparatively rare cases in which the result of a cross depends on the way it is made; that is, when the offspring resemble the mother, whichever species the mother may be. One of the best examples in higher organisms is in the willow herb Epilobium,[2] in which, even after fourteen generations of crossing *E. luteum* ♀ × *E. hirsutum* ♂ the influence of the luteum plasma could be clearly seen. Another example is the inheritance of male sterility in flax[3] and maize,[4] and the rather similar inheritance of the female factor through the cytoplasm in Lymantria (p. 216).

---

[1] *Rev.* East 1934, Goldschmidt 1934*b*, Pellew 1929, Wettstein 1937*a*.
[2] Cf. Michaelis 1937.   [3] Gairdner 1929.   [4] Rhoades 1935.

In lower organisms, Wettstein[1] has described some very clear instances of cytoplasmic inheritance in mosses. Crosses within the species *Funaria hygrometrica* show ordinary Mendelian behaviour of the type to be expected in organisms with an important haplophase; that is to say, the gametes unite to give hybrid diploid sporophytes, whose appearance is governed by the dominance relations of the genes involved, and these sporophytes give haploid gametophytes which show

Fig. 124. **Cytoplasmic Inheritance of Male Sterility in Flax.**—The male sterile plants have much reduced petals and aborted anthers. They arise in crosses between normal tall and low-growing "procumbent" plants, but only when a factor or factors from the tall nucleus are present homozygous in procumbent cytoplasm. Typical crosses are as follows: $\boxed{T}$ = tall cytoplasm, $\boxed{P}$ = procumbent cytoplasm, *t* factor or factors in tall nucleus concerned with male sterility, *p* procumbent factors. (After Pellew.)

1. Tall $\boxed{T}$ *tt* × Procumbent $\boxed{P}$ *pp*
   F1 all male fertile,
   F2, etc., all male fertile.

2. $\boxed{P}$ *pp* × $\boxed{T}$ *tt*
   F1 $\boxed{P}$ *pt*
   F2 $\boxed{P}$ *pp* : 2 $\boxed{P}$ *pt* : 1 $\boxed{P}$ *tt* (male sterile)

3. $\boxed{P}$ *tt* × $\boxed{T}$ *tp* (from 1)
   F1 1 $\boxed{P}$ *tt* (sterile) : 1 $\boxed{T}$ *tp*
   $\downarrow$ selfed
   1 $\boxed{P}$ *pp* : 2 $\boxed{P}$ *tp* : 1 $\boxed{P}$ *tt* (sterile)

segregation. When crosses are made between two species of Funaria, *F. hygrometrica* and *F. mediterranea*, the same purely genetic inheritance is found for some characters (e.g. shape of paraphyses), but in respect of other characters the hybrids differ according to which species was used as female parent and supplied the cytoplasm. Thus if we symbolize the haploid gene complement of the two species by *Hy* and *Me*, and the two cytoplasms as $\boxed{Hy}$ and $\boxed{Me}$, we find that the shape of the capsules differs in the reciprocal hybrids $\boxed{Hy}$ *HyMe* and $\boxed{Me}$ *HyMe*. From the diploid sporophytes, diploid gametophytes can be obtained by regeneration (cf. p. 213) and these show more numerous distinguishable characters, many of which show the same dependence on

[1] Wettstein 1926, 1927, 1928*a*.

the nature of the cytoplasm. But, as we have said, the degree of importance of the cytoplasm is very different in different characters, and

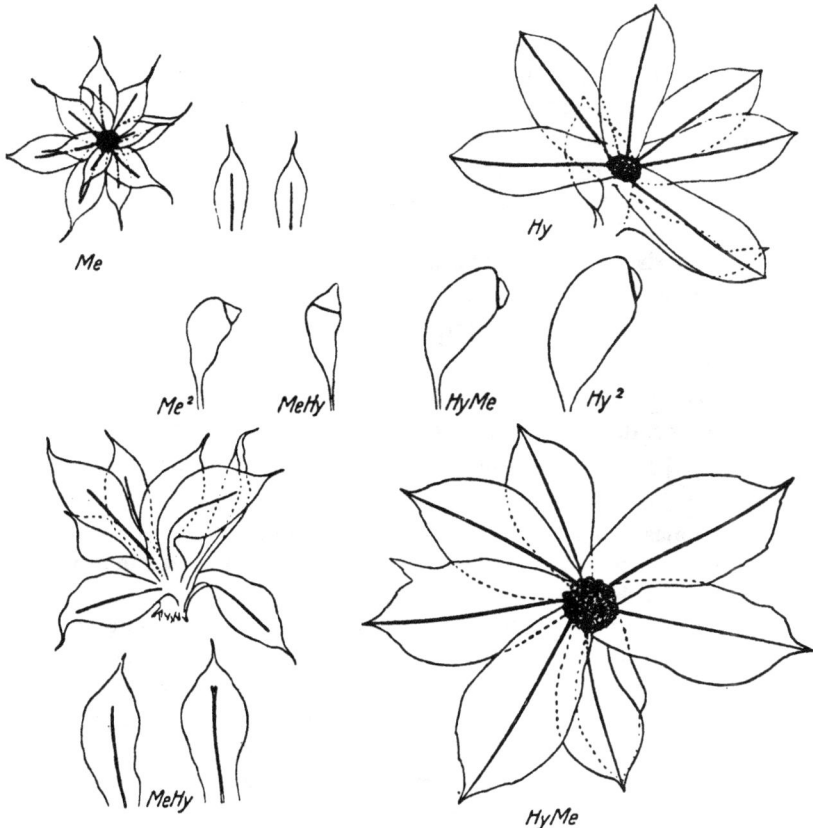

Fig. 125. Differences in Reciprocal Hybrids in Mosses.—The species used were *Funaria mediterranea* and *F. hygrometrica*; in the hybrids the maternal parent is written first, e.g. in *HyMe* the maternal parent was *hygrometrica*. Above are shown the pure species (haploid gametophytes). In the next row are the capsules of the pure and hybrid diploid sporophytes; note the resemblance to the maternal parent. In the third row are the diploid hybrid gametophytes obtained by regeneration from the sporophytes; note resemblance to mother in leaf points and ribs.

(From Wettstein.)

Wettstein concludes that development must always be regarded as the result of the interaction between the genes and the cytoplasm.

The nature of the cytoplasmic factors is obscure[1]; we can point to no

[1] Cf. Correns 1937, Renner 1934, Imai 1937.

particular structure which carries them, except in plants in which chlorophyll diseases or deficiencies may be inherited through the plastids, and even here it is not entirely clear whether the plastids themselves are deficient in some way or whether the abnormalities in the production of chlorophyll are caused by interaction between normal plastids and a modified cytoplasm. Even the unlocalized cytoplasmic factors seem to be quite self-perpetuating and constant. Wettstein has made a cross between *Physcomitrium piriforme* (*Pi*) ♀ × *Funaria hygrometrica* (*Hy*) ♂. In this, an intergeneric cross, the cytoplasmic influences are even more effective than before; the *Hy* chromosomes cannot survive when isolated in the [Pi] cytoplasm, so that no [Pi] *Hy* gametophytes are formed from the F1. But, by various regenerations, etc., polyploid types can be made including [Pi] *PiHy*, [Pi] *PiHyHy*, [Pi] *PiHyHyHy*, which were viable. Wettstein showed that even after six generations in which the [Pi] cytoplasm was exposed to one, two, or three doses of the *Hy* genotype, the characteristics of the [Pi] cytoplasm were still present and the [Pi] *Hy* type was still not able to survive. Thus the cytoplasmic factors, in this case at least, are only very slowly, if at all, affected by the chromosomes. This cannot be an entirely general rule, however, since in Limnea we know a case in which the cytoplasm is altered in a single generation by the genes which it contains (p. 143).

In some pairs of species, the cytoplasm of one is poisonous to the chromosomes of another. Thus a whole linkage group of factors from *Vicia faba major* is eliminated when in *V. faba minor* cytoplasm.[1] In crosses between different species of echinoderms, many of the paternal chromosomes may be eliminated during cleavage or later, and in some cases the reciprocal effect, a morphological influence of the chromosomes on the cytoplasm, has been observed.[2]

2. Persistent Modifications[3]

In some organisms, belonging to very different groups, experimental treatment has produced changes which are maternally inherited for some generations and then gradually lost. They seem to be caused by persistent, but not unalterable, changes in the cytoplasm. They were discovered in Protozoa,[4] where they were produced by heat and chemicals; and it was found that repeated treatments increased the effects

[1] Sirks 1932.   [2] Cf. Schleip 1929.   [3] *Rev.* Jollos 1938.
[4] Jollos 1921. For an interpretation in terms of gene mutations, see Raffel 1932.

produced. Similar persistent modifications have also been obtained in Drosophila by heat.[1] A culture heated for some hours to just below the lethal temperature (i.e. to about 37° C.) shows three types of effects: (1) a rise of the gene-mutation rate (p. 381); (2) the production of non-hereditable abnormalities which often closely parallel gene effects (p. 191); and (3) persistent modifications which are inherited through the female for a few generations and then gradually return to normal. In Drosophila further treatment does not increase the degree of modification.

Some authors have been tempted to suggest that persistent modifications may play an important part in evolution by bringing about an adaptation to the environment. The experimentally produced modifications are, however, not adaptive in nature; they seem to be caused by a slight alteration to the cytoplasm but there is no reason why this alteration should lead to the production of organisms which are adapted to the inducing conditions. Moreover, a persistent modification of the cytoplasm could only be important for evolution if it caused the genes to become altered in conformity with it, and, at least over the fairly short periods for which experimental evidence is available, there is no evidence that this is possible.

Certain phenomena of variation as it occurs in nature have been discussed in terms of persistent modifications. Woltereck[2] has shown that in Cladocera some species have many local forms in different regions, while others are much more constant. The variable species are those inhabiting lakes in northern Europe and America, which they can only have reached comparatively recently, after the last Ice Age, say ten thousand years ago. Their variations are strongly inherited and look as though they were adaptive; the pelagic types, for instance, occur only in lakes with deep water with a thermocline (a sudden change of temperature at a certain depth). *Daphnia cucullata* from L. Frederiksborg in Denmark (3–4 metres deep with no thermocline) were transferred to L. Nemi in Italy (34 metres deep, thermocline). They acquired a pelagic form. After fourteen years they were taken back into the laboratory and in about forty generations they had been brought back to the original Frederiksborg type. Thus the modification which they acquired in the L. Nemi conditions was persistent but not fixed. Another race, from L. Esrom in Denmark, which was in appearance very like the L. Nemi form, could not be converted in the laboratory

---

[1] Jollos 1934. For criticism, see Timofeeff-Ressovsky 1937.
[2] Woltereck 1932, 1934.

into a Frederiksborg type. Woltereck suggests that this difference in fixity of type is because the Nemi race was only exposed to the Nemi conditions for something like five hundred generations, while the Esrom race had probably been inhabiting that lake for a period covering some sixty thousand generations. But it is still doubtful whether the difference in the time of exposure has acted by fixing an originally labile cytoplasmic modification rather than in the more conventional way of giving time for selection to build up in L. Esrom a race with its characteristics controlled by genes.

3. *Maternal Effect*

Several cases are known in which adult characters clearly depend on the nature of the cytoplasm of the egg, and this again is under the

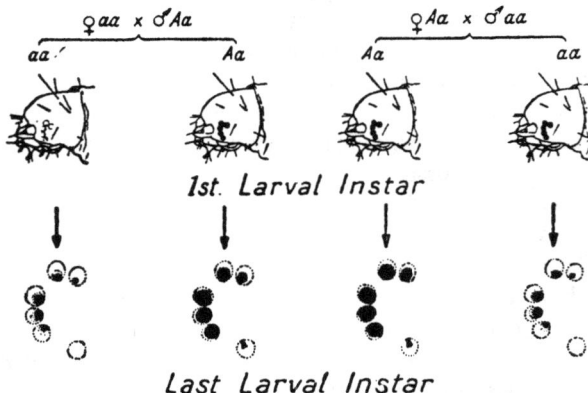

Fig. 126. Maternal Effect in Ephestia.—Diagram of the ocelli in the first and last larval instars of the F1 from the cross *aa* × *Aa*. Note that if the cross is made with *Aa* as the female parent the F1 are at first all alike, but that the ocelli of the *aa* animals gradually become pale as the initial store of pigment becomes exhausted. The upper row of figures shows the whole head of the larva, the lower the ocelli only.

(From Kühn.)

direct control of the genes in the mother. A well-known example is the inheritance of direction of coiling in Limnea, which is discussed in another connection on p. 191. Another very beautiful example has been described by Kühn[1] in Ephestia. The gene *A* causes a more intensive pigmentation of the larval ocelli (and also greater pigmentation in many other regions). In a cross *aa* × *Aa* segregation is quite normal if the mother is *aa*, but if the cross is made the other way, the progeny of the

[1] Kühn 1936.

Aa mother are at first all alike, with dark ocelli. In later stages, the ocelli become lighter in half the larvae and in the last larval instar the two classes *aa* and *Aa* can be easily distinguished. One must assume that in the second cross the *aa* animals at first develop pigment because some precursor has been carried over into the egg from the *Aa* mother, but that further pigment cannot be produced when once this store is used up. Kühn put this interpretation beyond doubt by showing that if an *A* testis is implanted into an *aa* female, it releases a substance into the blood which causes the darkening of the host's eyes; and further, the eggs laid by this *aa* female now possess an initial store of pigment exactly as do the eggs of an unoperated *Aa* female. Maternal effects are also known in silkworms,[1] Gammarus,[2] and Drosophila,[3] for example.

[1] Toyama 1909.
[2] Sexton and Pantin 1927.　　[3] Dobzhansky and Sturtevant 1935.

CHAPTER 13

Evolutionary Mechanisms

1. The Mechanisms of Evolution

(a) *Species Initiation.*—From a consideration of the nature of taxonomic differences discussed in Chapter 12 and the evidence of the actual course of evolution discussed in the first chapter of this part, we can form some idea of the kinds of mechanism which have to be taken into account in any complete theory of evolution.

In the first place, taxonomic groups usually differ in large numbers of small factors each with comparatively unimportant effects; these factors may be genes or small inversions, translocations, etc. Examples of groups which differ in this way are the geographical races of animals, but most related species probably also possess the same sort of differences, though if the species will not give fertile hybrids it is not easy to demonstrate this conclusively.

Secondly, species may differ in a few changes on a coarser scale. Examples are polyploid series, or related species which show a few large translocations, inversions, etc. Groups which differ by such crude chromosome changes are usually only ranked as separate species if they also differ in a large number of genes. Both the large-scale differences and the numerous small-scale ones may occur in a single step if chromosome rearrangements take place in a hybrid, and this sudden origin of fully fledged new species is clearly very important in plants. It is probably less important in animals, where hybrids usually cannot perpetuate themselves. In considering the origin of an animal species which differs from the old in large-scale chromosome changes, two processes are therefore involved: the origin of the chromosome change and the accumulation of the small-scale changes.

We see, then, that the accumulation of numerous small differences is important in every type of evolution except in the formation of new species by chromosome changes in a hybrid, and even in this case, the differences between the two parents may have arisen in this way. The process by which the gradual accumulation occurs is usually taken to be by the action of natural selection working on the random gene and chromosome mutations which are continually occurring. The detailed theory of how natural selection acts will be discussed in a later section.

Here we must discuss how a single population can become separated into two groups which follow divergent evolutionary paths.

If we have a homogeneous population of a species living under uniform conditions in a certain area, natural selection will be working equally on all parts of the species, and although it may cause a gradual transformation of the species so as to adapt it better to its environment, there is no reason why the population should break up into two or more separate specific groups. We should expect to find the gradual transformation of the entire population described in the Gryphaeas and Micrasters. Distinct species can only be formed if the hereditary material of the population is split into two or more parts which do not mingle. This splitting will occur if the population is divided, by geographical or ecological barriers, into groups of organisms which do not interbreed.

If we examine pairs of nearly related species which have probably diverged from each other only recently it is rather commonly found that some of the gene differences which accumulate in the two races are such as to lead to complete infertility of the hybrids, even if this was not present originally. For example, in *Drosophila pseudoobscura* there are two local races known as race *A* and race *B*. These differ in several specific inversions, but the chromosomes are all able to pair with their homologues of the other race; the hybrid males are nevertheless sterile, and this is apparently due to gene differences. Dobzhansky[1] has shown that numerous genes are involved, located in all the chromosomes. It is interesting to find that within each race there exist different strains with slightly different complexes of these "sterility" genes; it seems that the interracial sterility is built up from a basis of the minor differences occurring between the strains. Similarly, the sterility of hybrids between *D. melanogaster* and *D. simulans* is not caused by a failure of meiosis conditioned by chromosome changes, but by gene differences which prevent the germ cells even beginning to undergo the reduction divisions.[2] Other pairs of species may be prevented from interbreeding by ecological factors, such as different dates of maturing, or lack of sexual attraction, etc.

Now the genes responsible for interspecific infertility or failure to breed could not, in general, spread within an interbreeding population and cause it to split into two groups, since they would lower the average fertility, and thus the selective advantage of the organisms which

[1] Dobzhansky 1936*b*. Cf. Dobzhansky and Koller 1938.
[2] Kerkis 1933.

carried them. If, however, a population is already split into two groups by some extraneous agency, such as a geographical barrier, it is easy to imagine that each group may accumulate factors which are not deleterious within the group but which lead to infertility or lack of breeding between the groups. It is probable, then, that most species are initiated in this way[1] by a preliminary geographical isolation which allows the different groups to diverge in evolution before they have a chance to mingle. We shall find (p. 295) that at least some degree of geographical isolation also provides the conditions under which the modification of the genotype proceeds most rapidly under the influence of natural selection. Geographical isolation therefore appears to play a very important part, both in the original separation of a population into two parts, which we may call species initiation, and in the subsequent divergence of these two parts into two different species.

Some types of chromosome change, such as polyploidy and translocation, may immediately lead to sterility of the new type with the old. It is clear that in Nature the occurrence of polyploidy takes place and may initiate new species in plants. The position as regards translocations is more obscure; it is known that many species differ by translocations (e.g. in Drosophila) but no translocations have yet been found in natural populations, except for the segmental interchanges discovered in plants (p. 110), and in fact it is difficult to see how a translocation could ever be perpetuated.

(b) *Species Divergence.*—We have seen that evolution includes both the changing of the genetic constitution of a population and often its splitting up into separate groups which become distinct species. Although Darwin named his great work *The Origin of Species*, he actually dealt mainly with the first of these processes, and the theory of species initiation has ever since lagged far behind that of species divergence.

The greater part of evolutionary theory deals with species divergence, or the alteration of the genetic constitution of populations, and this, since it is the part dealing with the origin of new characters, is the more important part of the theory. The main theory which requires consideration is that of natural selection, though we shall also have to discuss the hypothesis of the inheritance of acquired characters. The theory of natural selection is as follows. If a species contains two hereditary varieties, and if one of the varieties produces proportionately more offspring than the other, then the rapidly breeding variety will become relatively more frequent and will eventually entirely replace the

---

[1] Sturtevant 1938.

other. In general terms the theory is obviously true. The points which require discussion are, firstly, whether species contain hereditary varieties which leave different numbers of offspring, and if so how are such varieties inherited; and, secondly, exactly what are the quantitative results of natural selection, and how far can they provide a theory of evolution adequate to explain the distribution of animal and plant species. We can give a fairly complete qualitative answer to the first question, and a discussion of the consequences of natural selection in exact mathematical terms is at present being worked out by several authors; but our quantitative knowledge, about the amount of variation found in natural species, the degree of selective advantage involved, and the rate of evolutionary change, is still extremely small.

A great deal of the discussion of evolution has been concerned with the discovery of particular cases which are either easy to explain on the basis of the theories held by the author, or, more often, seem difficult to explain on the theories held by his opponents. We shall not have space to discuss many of these problems here. The realm of Nature is so manifold, that we must expect to, and do, find the most extraordinary things in it, such as small bugs with an undeniable facial resemblance to crocodiles. Many of these peculiar phenomena can still be given no very plausible explanation on any hypothesis. It is more important to try to form a theory which will deal adequately with the normal and straightforward evolution such as we see it in the majority of species and in the palaeontological record.

2. *Variation Within a Population*

We have seen that environmental factors may cause the splitting of a species into more or less separate populations whose hereditary material does not mix. Natural selection can cause one of these races to replace the other, but it cannot transform one of the populations unless there are other hereditable variations within that population for it to work on. We are here concerned with the nature of the variation within an interbreeding population, which can be selected for or against in such a way as to change the average appearance of the group.

If we draw a large random sample from an interbreeding population and measure some character, we shall always find some variation between the different individuals. If we plot, against any particular value x of the measurement, the number of individuals in which that value is found, we obtain a curve giving the frequency distribution of the population with respect to the character. The distribution is usually represented by a bell-shaped curve; there are a majority of individuals with a more or less average value of the measurement, and fewer with either very high or very low values. The normal frequency curve is symmetrical about the mean value, but we may get asymmetrical or "skew" curves. The technique of comparing such curves with one another is part of the subject of statistics; full treatments on the matter will be found in books on that subject.

The variation expressed in such a frequency curve may not all be hereditable, since different individuals in the population probably developed in different environments which may have affected the character measured. Before we can get an idea of the hereditable variation, we must eliminate this environmental factor as far as possible. The variation which is left usually still shows a fairly normal frequency distribution, but it may now have a much smaller spread. It is, however, often still continuous over its whole range, and does not usually consist of a few distinct groups of individuals, though this may be true in a few populations which happen to be heterozygous for one or two genes with marked effects. The common form of continuous variation is nevertheless inherited by normal Mendelian genes, but it is controlled by many genes each of which has only a slight effect. Crosses between the extreme forms will therefore give intermediate F1's while in F2 segregation will cause very wide variation (cf. p. 159).

A beginning has been made with identifying the variability of natural populations due to mutant genes with clear-cut effects or to chromosome aberrations. The fullest analysis is of wild populations of Drosophila.[1] Any given gene is, of course, pretty rare, and exists in the heterozygous form (p. 289), but each individual fly may be heterozygous for at least one mutant. Nearly all the mutants which have been isolated from wild populations are deleterious; any useful ones would be selected for and would comparatively soon spread through the species. Autosomal recessives seem to be commoner than sex-linked recessives or autosomal dominants. This may be explained partly because there are more autosomal genes than sex-linked ones, and partly because an autosomal gene more often occurs in a heterozygous state in which it is hidden from the action of natural selection than do either of the more exposed types such as sex-linked genes or dominants, and will therefore be more widespread, for a given mutation rate and selective disadvantage.

There can be no doubt that many of the variations due either to

[1] Dobzhansky and Queal 1938, Dubinin *et al.*, 1934, 1936, Gordon 1936, Sturtevant 1937, Timofeeff-Ressovsky and Timofeeff-Ressovsky 1927.

genes with large or small effects affect the fitness of the organism as measured by the number of offspring it leaves, since we can often see that the varying character is one which plays an essential part in life. Differences in fitness caused by genes with strong effects have been occasionally measured in laboratory conditions (see p. 301), but we know very little about even the order of magnitude of the variations in fitness found in Nature. The laboratory measurements give quite a high value for the selective disadvantage of the genes concerned, and a similar value is found for the few genes which seem to be better fitted than the wild type. If these measurements can be taken to give a fairly accurate picture of the state of affairs in Nature, the differences found are quite adequate for the rate at which evolution normally proceeds.

3. *The Measurement of Selective Advantage*

The "advantage" which one variety can have over another must, if it is to be effective, be expressed in the production of offspring. The simplest way of giving a measure of selective advantage is probably that adopted by Haldane.[1] If an organism of variety A produces 1 offspring and one of variety B produces $1 - k$, then k is the coefficient of selection in favour of A.

Fisher[2] has given a fuller discussion based on considerations used in life assurance statistics, and this shows more completely what is at issue; the coefficient to which it leads is perhaps theoretically sounder and is easier to apply when generations overlap, as they do in any real case.

Suppose that the chance of an individual surviving to age x is l_x, and that its chance of reproducing itself between ages x and $x + dx$ is $b_x dx$, then its chance of surviving and reproducing is $l_x b_x dx$. The expectation of offspring from a newly-born individual is clearly the sum of this over his whole life from $x = 0$ to $x = \infty$, i.e. $\int_0^\infty l_x b_x dx$.

This is known as the net reproduction rate, and for bisexual species in which the sex ratio is not unity it is better to calculate it for females only.

If the population is constant in numbers, the net reproduction rate must be unity. If the population is increasing, the expectation of offspring is more than 1. Fisher puts this the other way round. For each child born now in an increasing population, less than one parent was

---

[1] Cf. Haldane 1932*b*. [2] Fisher 1930, cf. Charles 1934.

born earlier. Let the smaller number of parents born x years ago be represented as e^{-mx}; the contribution of these parents to the present birth is $e^{-mx}l_x b_x dx$, so that a single birth now is the result of the sum of all these contributions and we have $\int_0^\infty e^{-mx}l_x b_x dx = 1$. This is sufficient to define the value of m, which is called the Malthusian parameter of population increase. It is positive for an increasing population and negative for a decreasing one. The relative fitness of two varieties A and B can be measured by comparing their Malthusian parameters m_a and m_b. If A is stationary in numbers ($m_a = 0$) and B is falling, the ratio of A to B in the next generation is $1 : e^{mb}$, where m_b is negative. Haldane represents this as $1 : 1 - k$, so that k is nearly equal to $- m_b$ when both are small.

4. Selection of Single Genes in Infinite Populations[1]

The simplest case of selection is that of a single gene in an infinite population. The theoretical result depends on how the gene is inherited and the type of mating which occurs. For instance, if the population is one of self-fertilizing or apogamous plants, heterozygotes will be eliminated by the system of mating and we shall have only types AA and aa. Let the ratio of these in the nth generation be $u_n AA : 1 aa$. If the coefficient of selection (in Haldane's sense) in favour of AA is k, then in the next generation there will be u_n of AA and $1 - k$ of aa. Thus $u_{n+1} = \dfrac{u_n}{1 - k}$ which is nearly $e^k u_n$ when k is small. Then clearly $u_n = e^{kn}u_0$, where u_0 is the initial proportion of AA.

A more important case is that of a population mating at random. Suppose we have dominants, heterozygotes and recessives in the proportions $p_n AA : q_n Aa : r_n aa$. If mating occurs at random, the proportions of the different types of matings will be as shown below and the offspring they produce can be calculated.

Mating	Proportion	Offspring
$AA \times AA$	p^2	$p^2\,AA$
$AA \times Aa$	$2pq$	$pq\,AA : pq\,Aa$
$AA \times aa$	$2pr$	$2pr\,Aa$
$Aa \times Aa$	q^2	$\tfrac{1}{4}q^2\,AA : \tfrac{1}{2}q^2 Aa : \tfrac{1}{4}q^2 aa$
$Aa \times aa$	$2qr$	$qr\,Aa : qr\,aa$
$aa \times aa$	r^2	$r^2\,aa$

[1] Cf. Haldane 1932b.

Total offspring:

$$(p^2 + pq + \tfrac{1}{4}q^2)\, AA : (pq + 2pr + qr + \tfrac{1}{2}q^2)\, Aa$$
$$: (r^2 + qr + \tfrac{1}{4}q^2)\, aa$$
$$= u^2\, AA : 2u\, Aa : 1\, aa, \text{ where } u = \frac{p + \tfrac{1}{2}q}{r + \tfrac{1}{2}q}$$

If we repeat the calculation to obtain the next generation, we have simply to substitute u^2 for p, $2u$ for q, and 1 for r in the above result, when we obtain

$$(u^2 + u)^2\, AA : 2(u^2 + u)(1 + u)\, Aa : (1 + u)^2\, aa$$
$$\text{i.e.} \quad (u + 1)^2\, [u^2\, AA : 2u\, Aa : 1\, aa]$$

Thus in the next and subsequent generations the proportions of the different genotypes remain the same. The expression $u^2\, AA : 2u\, Aa : 1$ aa represents the equilibrium for a population mating at random, and this equilibrium is attained after the first generation of random mating.[1] It is clear that if a is rare, nearly all the a genes will be found in heterozygotes, and homozygous recessives will be extremely rare.

In this expression $u = \dfrac{[AA] + \tfrac{1}{2}[Aa]}{\tfrac{1}{2}[Aa] + [aa]}$ and thus measures the ratio of A to a genes. It is more convenient to work with this as a variable rather than the proportion of dominant or recessive zygotes.

If the nth generation of a randomly mating population is $u_n^2\, AA : 2u_n\, Aa : 1\, aa$, the next generation, after selection against the recessives, is $u_n^2\, AA : 2u_n\, Aa : (1 - k)\, aa$, so that

$$u_n + 1 = \frac{[AA] + \tfrac{1}{2}[Aa]}{\tfrac{1}{2}[Aa] + [aa]} = \frac{u_n^2 + u_n}{u_n + 1 - k}$$

$$u_{n+1} - u_n = \Delta u_n = \frac{ku_n}{u_n + 1 - k}$$

This is called a finite difference equation, i.e. an equation which expresses the relation between two members of a series of discrete values. But if selection is slow, we shall be concerned with a series of many generations, and can regard the difference between successive generations as infinitesimal. That is, we can treat the equation as an ordinary differential equation, and we can also, if k is small, neglect it in comparison with 1, and write

$$\frac{du}{dn} = \frac{ku}{1 + u}$$

[1] Hardy 1908.

which can be solved to give

$$kn = u_n - u_0 + \log_e\left(\frac{u_n}{u_0}\right)$$

If we consider the population to have originally consisted entirely of AA, $u_0 = 1$ and $kn = u_n + \log_e u_n - 1$. If this equation is plotted, it will be found that selection is very slow for, or of course against, rare recessives.

Haldane has worked out similar equations for many different types of selection and different types of inheritance and breeding. In nearly all these investigations the method is in principle the same as that shown

Fig. 127. The Rate of Natural Selection.—The table gives the number of generations required for a given change in proportions of the dominant, it being supposed that the dominants have 1,000 offspring for every 999 of the recessives ($k = 0.001$). Notice that the change is very slow when recessives are rare. Exactly the same figures are obtained if selection is against the dominant, but the table must then be read from the right towards the left; e.g. it will take 309,780 generations to decrease the dominants from 99,999 to 99 per cent.

Proportion of dominants:			0,001–1 %	1–50 %	50–99 %	99–99,999 %
Autosomal gene	—		6,920	4,819	11,664	309,780
Sex-linked	—	—	6,916	4,668	5,593	10,106
				(in homogametic sex)		
			6,928	5,164	11,070	20,693
				(in heterogametic sex)		

above; a finite difference equation is derived relating the value of u in two successive generations, and then some method is found of solving these equations, usually with the help of approximations which only hold when k is small. The only other example we have space to mention here concerns the effect of mutation.

Natural selection is usually balanced by natural mutation; in fact, perhaps the most common function of natural selection is to eliminate harmful genes which otherwise would quickly spread throughout a species. If a recessive gene a has a selective disadvantage k, and the mutation rates $A \rightarrow a = p$ and $a \rightarrow A = q$, we have in the $n + 1$th generation

$$u_{n+1} = \frac{A \text{ genes}}{a \text{ genes}} = \frac{(u_n^2 + u_n)(1 - p) + (u_n + 1 - k)q}{(u_n + 1 - k)(1 - q) + p(u_n^2 + u_u)}$$

$$u_{n+1} - u_n = \Delta u_n = \frac{ku_n}{u_n + 1} - pu_n(u_n + 1) + q(u_n + 1)$$

At equilibrium this $= 0$ and we get nearly $u + 1 = \sqrt{\dfrac{k}{p}}$. If the gene is lethal and k nearly $= 1$, the frequency of recessive zygotes at equilibrium is nearly equal to the mutation frequency.

5. Fisher's Fundamental Theorem of Natural Selection

Any natural population contains many different genes, and the fitness or selective advantage of an individual depends on the whole assemblage of genes in its genotype. It is clear, in a general way, that the more varieties there are in a population, the faster will natural selection be able to pull it along the evolutionary path by eliminating the unfit and causing the relative increase of the fit. Fisher[1] has made this idea precise in his *Fundamental Theorem of Natural Selection*, which states that the rate of increase in fitness in a population is proportional to the genetic variance.

Suppose, in a population whose variability depends on many genes, one gene is present with the proportions $pB : qb$. Then if a substitution of B for b in the kind of genotypes met with in the population has an average effect on fitness of α, a change dp in the proportion of B will have an effect αdp on the fitness of the population. Now suppose the individuals with the gene B have an average advantage of a in fitness (measured as Malthusian parameters) over the individuals with b; a need not be the same as α, depending on the system of mating, etc.

Then the rate of increase of the proportion $\dfrac{p}{q}$ would be $\dfrac{e^{k+a}}{e^k} = e^a$

and we have $\dfrac{p}{q} = e^{at}$ or $\log \dfrac{p}{q} = at$. Differentiating this with respect to t, and remembering that $\dfrac{dp}{dt} = -\dfrac{dq}{dt}$ we get

$$\left(\frac{1}{p} + \frac{1}{q} \right) dp = a\, dt.$$

Whence $dp = pqa\, dt$, since $p + q = 1$.

and the rate of increase in fitness $\alpha \dfrac{dp}{dt} = pqa\alpha$

Now to find the genetic variance we consider a measure of fitness X which measures the increments which must, for any genotype, be added

[1] Fisher 1930.

to the average value of the fitness (x_{av}) to give the "expected" fitness of that genotype. The increments for the genes B and b must clearly be qa and $- pa$, so that they will cancel out for the population as a whole. The genetic variance is defined as the mean value of X^2, and Fisher shows that this is the same as the mean value of Xx, i.e. the product of the expected value and the actual value. Now we have expressed the average difference between individuals with B and b as a, so that to find the value of Xx we have to consider

p individuals with $X = qa$ and which together have a mean $x = x_{av} + \frac{1}{2} a$

and

q individuals with $X = - pa$ and which together have a mean $x = x_{av} - \frac{1}{2} a$

So if we measure both X and x from the average value of x as origin, we get $Xx = p . qa . \frac{1}{2} a + q . (- pa) . (- \frac{1}{2} a) = pqaa$, which is the same as the rate of increase of fitness derived above.

This theorem, of course, can only be regarded as an abstract statement of one of the elements in a normal situation. It takes no account of new mutations or migration. It is a statement of the relation between natural selection and variance, the other factors being disregarded in a population which is not in equilibrium with its environment.

6. *Selection in Finite Populations*

According to the Haldane equations, all populations would fairly rapidly get into an equilibrium state balanced between selection and mutation. Thus no evolution will occur unless the environmental conditions change or new genes appear. Completely new genes probably occur very seldom, and the Haldane equations probably find their main application in cases where an advantageous gene has been kept rare by some mechanism, and is then released for selection to start working on. This might happen, for instance, if the gene was originally harmful but became advantageous through some change in environmental conditions.

There is another mechanism, perhaps more important, which can hold back a favourable gene; that is random extinction in a finite population. A certain number of the genes are formed continually by mutation, but if the population is fairly small and is subject to a high death-rate, there is a considerable chance that, after one mutation to the advantageous allelomorph has occurred, the whole stock of organisms bearing the allelomorph is wiped out, and the general spreading of the gene has to wait till its next occurrence. The theory of this random extinction has been worked out by Fisher and Wright.[1]

Wright's investigations, in fact, have led him to attach considerable importance to chance survival as a principle in evolution. His method is to study the frequencies, in a population, of different gene ratios $\left(\dfrac{p}{q} \text{ or } u \text{ above} \right)$. For instance, if selection is very severe, we tend to find the whole population homozygous and uniform, i.e. all the gene ratios are either ∞ or 0. The rate of evolution is the rate at which genes become "fixed," i.e. uniform, throughout the population. Wright considers mutation and migration as well as selection and random survival as factors in evolution.

He concludes that the rate of evolution is very dependent on the size of the population. In very small populations it is slow because of the small amount of variation available for selection to work on, and what evolution there is is mainly due to random survival. In large populations, again, nearly all genes are present in fixed ratios depending on the equilibrium between mutation and selection; the only evolution is of the Haldane type depending on the occurrence of new favourable mutations. An intermediate population is not so small that evolution is held up by random extinction of useful genes, but is small enough for the death-rate to produce a random fluctuation of gene ratios round their equilibrium values.

Wright points out that in assessing the fitness of a population we must consider all the genes simultaneously. For instance, in a population containing mainly the allelomorphs A and b, it might be useful to change A to a, but only if at the same time B is substituted for b. Wright discusses these changes in terms of probability surface rather like the "valley model" of development which was proposed in Part 2. Here each point represents a population of a certain genetic constitution; if we are talking in terms of a three-dimensional model, each point really only pictures the proportions of the two genes, while the third dimension represents fitness. In Wright's model, the fittest populations lie at the top of hills, and selection will continually be trying to keep the population there, unless a new mutation occurs which makes some new constitution become fitter; this would be equivalent to raising a new hill nearby, and the population would

[1] Wright 1931, 1932, 1935.

promptly move up to the top of it, by a Haldane process. The model is particularly valuable, however, in picturing what happens when no new genes occur. The frequency surface then remains unaltered, and evolution can only occur if the population can be caused to leave its original hill, cross a valley, and ascend to the top of another hill. Wright comes to the conclusion that in a middle-sized population, random

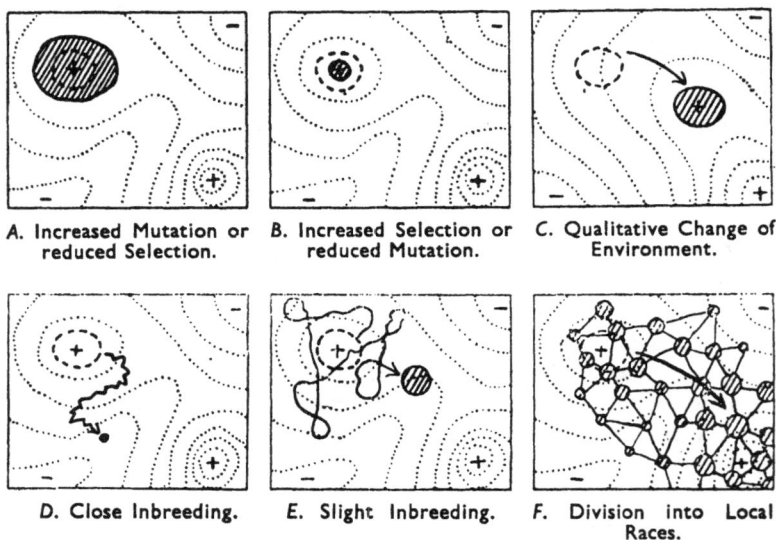

A. Increased Mutation or reduced Selection.

B. Increased Selection or reduced Mutation.

C. Qualitative Change of Environment.

D. Close Inbreeding.

E. Slight Inbreeding.

F. Division into Local Races.

Fig. 128. **Diagrams of Evolutionary Processes.**—Each diagram represents part of the field of possible gene combinations in a population; the fitness of the combination represented by any point is supposed to be plotted vertically and the resulting surface is indicated by the contour lines. In each diagram the population initially occupies the area enclosed by the dotted line, and the diagrams show how various processes shift the population about the field, i.e. change its gene composition.

(From Wright.)

fluctuation of gene ratios may be sufficient to carry the population down to the bottom of a valley and on to the slope of another hill, which it then ascends under the influence of selection. In this case the direction in which evolutionary change goes on, i.e. the choice of which hill is ascended, is under the control of chance, though in a long enough period the broadest, and therefore probably the highest, hill (representing the fittest population) is likely to be reached. Over a long enough period, then, the chance element in evolution is less important.

These populations of intermediate size, however, only move about the frequency surface rather slowly. Wright shows that if a large population becomes split up into localized groups, which usually inbreed but between which there is a certain amount of migration and cross-breeding, the conditions for evolutionary change are much better. The individual groups may be small enough for random fluctuation of gene frequencies to carry them rapidly about the frequency surface, but the cross-breeding saves them from degenerating into a condition when all their genes become fixed; they thus have both the rapidity of movement of the small groups and the store of variation of the larger groups, which enables the movement to go on longer. The occurrence of mutations has the same effect as cross-breeding in replenishing the store of variation in a group small enough to undergo rapid changes in gene frequencies. The best conditions for movement from one hill on the frequency surface to another, that is, for evolutionary change, therefore comes with a suitable balance between all the evolutionary factors; the group can be large, but if so it should be split into smaller groups, which have a considerable amount of inbreeding but some outcrossing, the mutation rate should be moderately large, and the intensity of selection should be enough to keep random fluctuation within bounds but not so severe as to eliminate it altogether.

It is clear that the rapidity of evolution is very largely influenced by the size of the population. However, the concept of population size in this connection is not at all simple. What is important is not the total number of existing individuals but is the size of the effective breeding population. For instance, in many species the population is almost wiped out every winter, and the only supply of variation on which evolution can be based is that contained in the remnant which survives to begin breeding in the spring. Similarly, inbreeding lowers the available variability. Thus the effective breeding population is a magnitude which can scarcely be measured directly, and we still have very little knowledge about it. It has been shown, however, that in some natural populations it is small enough to permit quite large random fluctuations in gene ratios. Dobzhansky[1] found that in *Drosophila pseudo-obscura* a given lethal factor is often more concentrated in a small isolated and inbreeding race than it is in the whole population of which the isolated community forms a part; this local concentration of the lethal can only be due to a chance fluctuation.

These conclusions seem to be in general agreement with those of Fisher and Haldane, though these two authors allow rather less im-

[1] Dobzhansky and Koller 1938*a*.

K*

portance to random fluctuations in gene ratios. The whole set of theoretical investigations provides a satisfactory picture of many of the most important characteristics of evolution. We have an account of the gradual transformation of widespread species, by Haldane selection of new mutations; of the production of many apparently non-adaptive varieties in small groups or groups with localized inbreeding; and of the comparatively rapid emergence of a new species, which may found an important new genus or family, when a new type of adaptation becomes possible through the occurrence of a new gene or the production of a new ecological niche.

The main phenomenon for which some explanation in terms of natural selection is required but is not yet provided is trend-evolution of the type shown by the Gryphaeas (p. 242). The difficulty is not so much the evolution in one direction over a long period of time. That might be brought about by selection in a gradually changing environment or it might be due to a sort of inertia which we should expect to find in evolution; granted that a character is dependent on the interaction of many genes, it will be easier to continue a line of evolutionary change, for which many of the modifiers are already present, than to start off on a completely new line. Thus the uni-directional nature of trend evolution is not particularly surprising. What is remarkable is that the trend continues so far as to lead to the extinction of the race.

Fisher suggested that in certain circumstances the genotypic inertia may be manifested in another way and if the gene ratios are changed by selection, they may overshoot the equilibrium point at which the population would be best adapted. It is conceivable that it was something of this kind which caused orthogenetic trends to go so far as to lead to the extinction of the species. Haldane has suggested another possibility, based on the supposition that there was stringent selection in young stages; selection for strength in the young might, for instance, favour individuals with a heterogonic growth mechanism which could not be controlled in later development and produced adults with some parts excessively developed. But it is not clear whether this intra-specific selection could lead to the extinction of a species in competition with the other species in its habitat.

It has sometimes been suggested that trends occur, not directly under the control of natural selection, but by progressive mutation in a single direction. Such a directed series of mutations would be as inexplicable, on our present ideas, as the trends themselves; and further, it is easy to see, from the discussion on p. 290, that mutation pressure cannot be effective against even quite small selective advantages. It is quite impossible for evolution to be directed by mutation unless the characters in question have very little adaptive significance indeed, or unless we postulate mutation rates of a totally different order from anything we know at present.

Probably the main connection in which mutation pressure has any influence on the course of evolution is in causing the reduction of organs which are no longer of any adaptive significance; in total absence of selection, random mutations can accumulate. Now it is usual to find that the highest mutation rate at any locus is to a hypomorphic allelomorph determining a reduction from the normal size. Thus in time the population is filled with a hypomorph of low efficiency and the organ is reduced or even disappears. A similar argument applies to parts of the Y chromosome in which crossing-over is suppressed. Genes in this region being always masked by their wild allelomorphs in the X will tend to become "inert" by the pressure of mutation towards hypomorph and eventually amorph allelomorphs.

7. Selection of the Genotypic Milieu

There is considerable evidence that the genotypic milieu of any species is adjusted to the particular genes which occur; this phenomenon has been discussed in relation to dosage compensation and dominance (Chap. 7). It was in connection with the latter problem that Fisher[1] first drew attention to the evolutionary aspects of the matter. He argued that there was no reason, on *a priori* grounds, why new mutations should not be partially dominant, with some expression of the mutant character in the heterozygote. It is therefore necessary to find some explanation of the fact that most rare genes are recessive to the wild type. Now these genes will usually be present as heterozygotes, and the fact that they have remained rare indicates that they are deleterious. The main influence of selection on a partially dominant gene will therefore be on the heterozygote, and these will be the more stringently selected against the more markedly they show the mutant character. If there are any modifiers which reduce the expression of the mutant gene in the heterozygote, these modifiers will confer some protection on heterozygotes containing them and will therefore have a positive selective value. Fisher supposes that this is sufficient to cause an accumulation of such modifiers, which will eventually suppress the mutant character entirely in heterozygotes, so that the mutant gene will

[1] Fisher 1928, 1931, Ford 1931.

behave as a recessive. He has named the phenomenon the evolution of dominance.

The mathematical basis of the theory has met with criticism from Wright,[1] who maintains that the process is essentially a second order one, and could not proceed fast enough to have any practical consequences, and points out, further, that the modifiers would probably have first order effects of their own on fitness, which would quite outweigh the slight advantages they would confer in the rare cases when they were in heterozygotes of the original mutant. Haldane[2] made rather similar criticisms of Fisher's original theory and put forward the suggestion that natural selection of heterozygous harmful genes picks out, not a set of modifiers which suppresses the expression of the gene, but a wild-type allelomorph which has a considerable margin of safety and thus has the same effect. We may suppose that the wild-type expression is often the upper limit of possible gene effect; if the original wild-type gene AA reaches this limit, but the heterozygote Aa does not, selection will pick out another allelomorph A', in which $A'A'$ again reaches the maximum and so does $A'a$ (p. 185).

Even if these criticisms are accepted, there is considerable evidence that the genotypic milieu is subject to some sort of evolutionary control. Harland[3] has described a mutant Crinkled Dwarf in cotton. This is completely recessive in sea island cotton (*Gossypium barbadense*) in which it was first found, but gave intermediate heterozygotes in the American species *G. hirsutum* where the mutation was not known to occur naturally. This looked like a clear case in which the modifiers conferring dominance on the wild type had not been accumulated in hirsutum where the gene did not occur and selection therefore had had no chance to act. But further work showed that the situation is more complicated. Harland discovered three normal allelomorphs of Crinkled and three different genotypic milieus. One of the normal allelomorphs in *G. hirsutum* shows complete dominance in its own milieu, although Crinkled does not occur. On the other hand, the three normal allelomorphs have different dominance relationships when got into the same milieu, so there is some evidence for the type of phenomenon postulated by Haldane.[4] (Cf. p. 167).

Fisher[5] has made a special investigation to test his hypothesis, based

---

[1] Wright 1929.   [2] Haldane 1930.   [3] Harland 1936.
[4] For a gene whose normal effect, the production of black pigment, is exaggerated into the production of melanotic tumours when it is in the genotypic milieu of a foreign species, cf. Kosswig 1929.   [5] Fisher 1935.

on the argument that selection *for* a character, such as is practised by man in building up domestic varieties, should act in exactly the opposite way and should favour modifiers tending to increase the expression of the gene in heterozygotes. In a domestic animal like the fowl, one therefore finds that genes from domestic races tend to be dominant over the wild allelomorphs. By continued crossing, however, it should be possible to free the genes from their domestic genotypic milieu and get them into an otherwise wild-type background, when they should show the usual recessiveness. A long series of crosses was made, and the prediction confirmed. This proves that selection of the milieu is possible, and that it occurs in artificial selection; it is not a complete demonstration that it also happens in selection in Nature.

Modification of the genetic background, if it is considered a possibility, could be invoked to explain several difficulties in evolutionary theory. For instance, we find cases of mimicry in which the model and mimic species resemble one another in many details, and yet differ only by a single gene. It is very difficult to believe that a random gene mutation has produced such a complicated resemblance. But we can suggest that the mutation at first produced only a crude likeness which has been refined and perfected by the gradual modification of the milieu. Similarly the insensible gradations found in trend-evolution may have been produced not by selecting a long series of slightly different allelomorphs, but by a gradual modification of the background against which only a few major genes were working.

8. *Evidence for the Occurrence of Natural Selection*

Natural selection is a necessary consequence of hereditary variation in fitness; the inevitability of the process is so obvious that Darwin was able to use it to convince the world not only of the mechanism but, more important, of the fact of evolution. Competition and natural selection between species is also a common fact of observation, although rather little exact quantitative work has been done on it. Gause[1] is studying the matter in the simple case of competition between two protozoon species in culture, and Timofeeff-Ressovsky[2] has described the competition between *D. melanogaster* and *D. funebris*.

Natural selection between varieties of a species,[3] which is the kind which is important for evolution, has been even less studied, either in Nature or in the laboratory. Harrison[4] has described an interesting case.

[1] Gause 1934.
[3] *Rev.* Robson and Richards 1936.
[2] Timofeeff-Ressovsky 1933.
[4] Harrison 1920.

A wood in Yorkshire was divided in two in about 1800 by a wide avenue, and in 1885 one of the two portions was planted with birches, while the other part remained mainly a pine wood. In 1907, 85 per cent of the moths *Oporabia autumnata* in the birch wood belonged to the light variety, while in the dark part of the wood there were only 4 per cent of the light forms, and the other 96 per cent were of a dark variety. Presumably this divergence in genes ratios had been brought about since the wood was replanted in 1885, and certainly it must have been produced since the wood was divided in 1800. Moreover, the change is clearly related to the environment and is therefore probably due to natural selection. In fact, evidence was found that in the dark wood the light forms were at a selective disadvantage, since the proportion of lights among the dead moth wings left by predatory animals (bats, birds, etc.) was much higher than 4 per cent.

Another well-known example of natural selection is that described by Sukatschev,[1] who cultivated together various clones of apomictic dandelions, and studied their fertility and viability. There were considerable differences in fitness between the varieties, but these differences depended on the environmental conditions; the forms which were best adapted under conditions of crowding did not always do better in less crowded cultures. Similar evidence is provided by Turesson's[2] studies of genetical varieties of plants from different ecological situations; when they were all cultivated under the same conditions, their fitnesses were not identical.

In recent years several rather complete investigations have been made on the selective value of "protective colouration." Sumner[3] reported experiments in which fish (*Gambusia*) were allowed to adapt to dark and light backgrounds, and the pale and dark specimens mixed and exposed to predators (herons or penguins). With a mixed population in a light-coloured tank, about 60 per cent of the fish eaten during the experimental period were dark coloured, while if the experiment was made in a dark tank, only about 40 per cent of those eaten were dark. Similarly, Isely[4] allowed birds to attack different coloured grasshoppers placed on backgrounds on which they were either concealed or conspicuous. Within a given period, about 85–95 per cent of the conspicuous grasshoppers were eaten, but only 35–45 per cent of the concealed. Although these experiments deal with colour variations which are not hereditary, they demonstrate that apparently protective

[1] Sukatschev 1928.
[2] Turesson 1925, 1930, 1931.
[3] Sumner 1935.
[4] Isely 1938.

colouration has a real value in nature, and are thus evidence that natural selection is a real phenomenon.

Fig. 129. Selective Advantages of Some Genes in *Drosophila funebris*.—The influence of selection was measured by making up a cross in which a 1 : 1 ratio was expected, and determining the deviation from this ratio. The selection is influenced by temperature, crowding and the genotypic milieu. Note that *ev* has some positive advantage over the wild type. The selective advantages are given as percentages compared to wild type. (Data of Timofeeff-Ressovsky.)

Gene.	Normal Conditions.	Temperature.			Crowding.		
		15–16 °C.	24–25 °C.	28–30 °C.	Sparse.	Crowded.	Very Crowded.
eversae *ev* ..	104±0·4	98	104	98	102	104	96
singed *sn*	♂79±1·0 ♀88±0·8						
Venae abnormes *Va'*	89±0·7	96	89	81	99	89	77
miniature *m* ..	70±0·9	91	69	64	93	69	47
lozenge *lz* ..	74±1·2						
Bobbed *bb* ..	85±0·8	75	85	94	77	85	92
Gene combinations *ev.sn*	103±0·5						
ev.Va'	84±0·8						
ev.bb	85±0·8						
sn.Va'	77±0·9						
sn.m	67±1·3						
Va'.m	83±0·8						
Va'.lz	59±1·2						
Va'.bb	79±0·9						
m.bb	97±0·3						
lz.bb	69±1·0						

It is common to find that different genes have different selective values under laboratory conditions. Timofeeff-Ressovsky[1] in particular has studied the relative viabilities of different genes in Drosophila, by

[1] Timofeeff-Ressovsky 1934*a*.

setting up a cross in which a 1 : 1 segregation would be expected, and then counting the divergence from this expectation which has been brought about by differences in selective values. In these experiments, however, the selection was usually only followed for one generation, and only sufficed to measure the relative viabilities of two genotypes under rather unnatural conditions. L'Héritier and Teissier[1] have carried on such cultures through many generations and followed the gradual elimination of certain genes. They have also made some experiments under natural conditions. Thus the common occurrence of wingless species of insects near the sea-shore has led to the suggestion that in such situations the winged forms are at a selective disadvantage owing to the danger of being carried out to sea by winds. L'Héritier, Neefs and Teissier[2] therefore released mixed cultures of winged and *vestigial* (wingless) Drosophilae. They found, as predicted, that after some time the winged forms were rarer than the *vestigials*. It is not perfectly clear from their account that they had eliminated the possibility that the winged forms had merely flown away elsewhere but the experiment is mentioned here as an example of a type of work which is of considerable importance for the experimental study of evolution but which has been surprisingly little taken up.

9. *The Inheritance of Acquired Characters*[3]

The theory of the inheritance of acquired characters is primarily a theory of the origin of hereditable characters, rather than a theory of their preservation, as is the theory of natural selection. In fact, a belief in the inheritance of acquired characters can easily be combined with a belief that all evolutionary advance is limited by natural selection. Darwin at one time held both views.

The inheritance of acquired characters may be taken to imply either of two things: (1) that if an environmental agency affects the body of an organism and produces a somatic change *A*, in the course of generations *A* will become hereditarily fixed and inherited independently of the environmental stimulus; or (2) it may imply in addition that *A* is a change adapting the organism to the environmental conditions to which it is subject. It is of course the second of these possibilities which gives the theory its attractiveness and causes it to be tenaciously held in the almost complete absence of evidence in its favour. This second form of the theory is usually associated with the name of Lamarck.

[1] L'Héritier and Teissier 1937. [2] L'Héritier, Neefs, and Teissier 1930.
[3] *Rev.* Detlefsen 1925, Haldane 1932, Robson and Richards 1936.

The evidence against the theory of the inheritance of acquired characters is, in particular cases, very strong. Thus the docking of sheeps' tails and circumcision of man have been carried on for hundreds and perhaps thousands of years without producing any hereditable effects. On the other hand, we can find plausible cases on the other side. For example, the thickened skin on the sole of the human foot, and the sternal and alar callosities of the ostrich[1], seem to be directly related to the pressure arising from the habitual positions of these animals; the simplest hypothesis is that, originally, the thickenings were reactions to pressure. But now the thickenings arise in the embryo, before any pressure is exerted. They are therefore hereditarily fixed, and it is argued that natural selection could not explain this fixation since the callosities can be of no service to the embryos. The conditions can, perhaps, be compared with the much wider class of phenomena known as double assurance in embryonic development. These are cases in which an organ is normally induced by an organizer from competent tissue, but in which the organ can develop by self-differentiation even if the organizer is removed. An example is the lens in the frog, normally induced by the eye cup, but able to develop in its absence. There is no obvious advantage in this, since it profits a frog little if it retains its lens but loses its eye. But we know too little about the genetic control of competence to assert that this exaggeration of competence is useless. It may be that the evolution of a double assurance mechanism is akin to the selection of genes with a margin of safety adequate to deal with minor variations in developmental conditions.

Another phenomenon which is brought forward as evidence of Lamarckism is the production by particular environments of forms similar to varieties which are known to be hereditable. This is taken to show that the inherited varieties originated as reactions to similar conditions, perhaps very long continued. The data have recently been summarized by Rensch[2] from a faunistic point of view. There is however nothing particularly mysterious in the fact that hereditable varieties can be imitated by the action of environmental agents during development. The phenomenon has been discussed (p. 191) and we saw that the environmental action can be explained in terms of the genotypic constitution of the organism, while there are no grounds for suggesting a relation in the reverse direction and trying to explain the genetic control as a result of the environmental effect.

[1] Duerden 1920.　　　　　　　　　　[2] Rensch 1929, 1936.

Finally, we come to experiments especially undertaken to prove or disprove the inheritance of acquired characters. The experiments are too many to be fully discussed here. Very many of the experiments must be dismissed for lack of adequate controls. It is essential to show that any effects which may be produced cannot have been due to selection, and this can best be done by using only genetically homozygous stocks in which there is no basis for selection to act on. As an example of an experiment which fails in this respect, we may take work of Harrison[1] on gallflies (*Pontania salicis*). These normally lay eggs on the willow *Salix Andersoniana* and allied species. Harrison liberated some in a district where the only available willow was *S. rubra*, and found that after five years the flies had become acclimatized to this species and refused *S. Andersoniana* when it was offered them. In the first two years of the experiment, however, most of the galls laid on *S. rubra* aborted, and it is therefore likely that in the initial stages there was stringent selection for individuals with a taste for this plant. It is also possible that in this experiment the kind of food plant selected by a female for oviposition is determined not by any direct genetic mechanisms but by the adult's memory of its own larval food. In some other experiments the selection is even more obvious. For instance, Dürken[2] found that if pupae of *Pieris brassicae* are exposed to orange light some of them become green. If these green pupae are bred from, a still higher proportion becomes green, and the high proportion of greens persists in later generations even without the orange light. This is clearly an experiment in selection rather than in the inheritance of acquired characters.

There are two outstanding experiments which require more serious consideration. Harrison and Garrett[3] claimed that if certain moths are fed on foliage impregnated with lead or manganese salts, such as occurs in many industrial districts, melanic forms are produced by gene mutation and breed true as ordinary recessives. This could provide an explanation for the melanism so commonly found in moths in industrial areas. It should be noted, however, that the phenomenon is not a straightforward Lamarckian one, since there is no suggestion that the melanism is adaptive. The effect postulated is a direct chemical attack on the germ plasm, producing specific mutations. Other attempts to alter the genes by chemical means have usually been unsuccessful.[4] Thermal and X-ray effects on the mutation rate are of course well

---

[1] Harrison 1927.　　[2] Dürken 1923.　　[3] Harrison and Garrett 1926.
[4] Sacharoff 1935.

substantiated phenomena (pp. 381, 382), but these agents do not produce specific changes in the way suggested by Harrison. The actual facts of the case, however, are still not entirely beyond doubt, since other workers[1] have failed to repeat the results and have suggested that the original stock may have contained the melanic gene; Harrison has criticized their work as involving too high a mortality, so that if any of the weaker melanics were induced they would not have survived. Perhaps it is necessary that the work should be repeated once more.

McDougall's[2] experiments on learning in rats are the most careful and complete of any purporting to give positive evidence of the inheritance of acquired characters. He found that if rats are taught a task (to choose a dark platform when presented with both a dark and a light one) their offspring tend to inherit the capacity to learn and can be more easily trained to choose a given one of the two alternatives. The experiments were made with an inbred strain, and great care was taken to avoid selection. So long as these experiments stand alone, as reasonably convincing evidence of the inheritance of an acquired character, they seem scarcely strong enough to support a theory of evolution of such enormous scope. Moreover, they have been repeated without success by Crew,[3] who, while not able to offer any complete explanation of McDougall's findings, failed to obtain a similar result and drew attention to certain disturbing facts, such as definite evidence of genetic heterogeneity for learning ability in the inbred stock he used, which would make it possible for selection to be effective.

[1] Hughes 1932.　　　　[2] McDougall 1927, 1930.　　　　[3] Crew 1936.

PART FOUR

Genetics and Human Affairs

Genetics, like all sciences, grew originally out of man's desire to control his environment. Its original exponents were stockmen, who tried to improve the strains of domesticated animals, and seedsmen, who needed to produce uniform seed of good quality. Once the theoretical science has begun, it takes on a life of its own, and its development is not directly dependent on the solution of practical problems. The application of genetics to plant and animal breeding forms a special branch of the subject, requiring something more than a knowledge of theoretical genetics; in particular, it is concerned with husbandry and plant and animal pathology. The first chapter of this Part therefore gives only a very summary account of it. The application of genetics to man himself is a less specialized subject, and can therefore be treated at somewhat greater length.

CHAPTER 14

Animal and Plant Breeding[1]

Genetics, as we saw in the Introduction, grew out of the practical problem of the improvement of agricultural crops and stock. The development of the subject which followed Mendel's discoveries has not, however, been primarily concerned with practice, but has been towards the formulation of a comprehensive theory and the elucidation of the fundamental biological problems which were raised. This scientific advance has only fairly recently begun to have much effect on breeding practice, and we are still a long way from being able to apply the whole of the theoretical knowledge at our disposal. In the peculiar economic situation in which man has found himself during the last few decades, production, even with little help from science, has been so far ahead of consumption that the application of genetics to this field has, in most countries, not been investigated whole-heartedly and on a large scale. The results which have been obtained are, however, already quite considerable, and we can envisage the possibility of quite startling changes in the world's agricultural economy, particularly in connection with crop plants.

1. The Methods of Breeding

The purpose of breeding is to produce animals and plants with useful characteristics. The methods of genetics are only useful in so far as these characteristics are hereditable. Crop and stock improvement has an aspect concerned with methods of husbandry as well as one concerned with genetics. The two methods of approach continually overlap in practice; the results which are obtained with a particular genotype depend on the conditions in which the zygote develops, and we often find that one genetic type is most suitable in one set of conditions, but that in other circumstances some other genotype gives a better result.

We shall be mainly concerned here with genetic improvement, and in so far as the characters involved can be treated from this point of

[1] *General references:* Babcock and Clausen 1927, Hudson 1937, Hunter and Leake 1933, Rice 1934.

view, the production of satisfactory animals and plants becomes a question of the production of satisfactory genotypes. Since no way exists at present of manufacturing particular genes to order, the problem reduces to that of selecting and bringing together the genes which chance has offered. The methods which are relied on for this purpose are selection, crossing, and inbreeding. These methods have probably been used to some extent ever since plants and animals were first domesticated, but we now understand more clearly what are their particular uses. By selection, we aim to breed from an organism containing useful genes, which are thus perpetuated. Crossing has several functions. By permitting segregation in the F2 and subsequent generations it allows us to break up a genotype into its constituent genes, and thus to select those which we require and reject the others. At the same time, it enables us to combine in a single individual the genes which previously had existed in different lines. Finally, by inbreeding, we perpetuate useful lines already in existence, and, combining inbreeding with selection, we can purify them by reducing them to a homozygous condition.

2. Selection of one Parent only

The most elementary form of selection practised by breeders is the selection of one parent only, the choice of the other being left to chance. This so-called mass selection has probably been practised since the earliest times; a careful farmer used for planting the seed taken from the healthiest and most productive of his plants. If a plant is self-fertilized, selection of one parent is in effect selection of both, and it would be possible to obtain a uniform and valuable variety quite rapidly in this way. In a cross-fertilized organism, in which only half the genetic factors of the offspring are contributed by the selected parent, one can only expect a slow accumulation of beneficial genes. In practice, the steady improvement may continue over very long periods, since the characters selected for, such as yield, are often such as to be the resultants of complex physiological processes and are therefore likely to be affected by many factors, which take a long time to collect. In a well-known case,[1] maize was selected for high and low protein content and high and low oil content; after more than twenty years each of the four selected lines was still changing, though slowly, in the required direction.

This laborious method of improvement has been of the greatest

[1] Winter 1929.

importance in the past. Several important early varieties of maize, such as Leaming, and Delta Farm White Dent, were obtained in this way, while probably all our agricultural crops have been improved by rather haphazard mass selection since agriculture first began. A recent example of successful selection is the production of a sweet lupin.[1] Most wild races produce an alkaloid in the seeds which renders them unsuitable for fodder. Forms containing very little alkaloid were found by German breeders after an intensive search, and pure lines bred by selection. Details of the work were not published and the seeds were not allowed to be exported. The entire work was therefore repeated, equally successfully, in the U.S.S.R. The plants provide a valuable fodder with high protein content.

Selection of one parent only is still one of the most important methods in animal breeding.[2] The world possesses enormous numbers of low-grade stock animals, and since the reproductive rate of animals is comparatively low, they cannot be simply destroyed and the next generation produced entirely from high-grade animals, as could be done with plants. The stock can, however, be improved by "grading-up," that is, by breeding from most of the females, but using a small number of selected males to father the whole of the next generation. The effects of this method of breeding may be very beneficial; at the Iowa Agricultural Experiment Station, the first generation from a poor scrub herd crossed with pure-bred bulls gave 55 per cent more milk and 44 per cent more butter-fat than their dams, the second generation 116 and 106 per cent more. That there is plenty of room for improvement of this kind is shown by the fact that the average yield of milk in the United States is about 4,000 lb. per dairy cow per year, with 160 lb. of fat, whereas the average production by pure-bred cattle is well over 10,000 lb. with 450 lb. of fat, going up to the record performances (probably only realizable under artificial and non-commercial conditions) of about 25,000 lb. of milk a year.

Equally valuable results have been obtained by grading up the enormous beef-producing herds of western United States by using suitable sires. Recently attempts have been made to give greater force to the method by increasing the number of females which can be served by a single superior male. A single ejaculation of semen contains very many times more sperm than are necessary to ensure fertilization, and by performing artificial insemination with diluted semen, a desirable male can be caused to produce offspring with at least a hundred

[1] Cf. Hudson 1937.   [2] Rice 1934.

times as many females as he could serve naturally. The sperm can, moreover, be shipped considerable distances before use.[1]

The best method of selecting males for use in grading up has been the subject of considerable investigation. It may be possible to select a male who himself shows the characters desired, as would be the case, for instance, with bulls of beef-producing types. But it is clear that what we really wish to select is a certain genotype, and this may be very imperfectly expressed in the phenotype of its bearer; in particular, bulls for breeding dairy cattle can in the nature of the case show no direct sign of their genotypic capacity for milk production. Attempts to correlate external characters with the nature of the hidden genotype have been made empirically by breeders frequenting the sale-ring, and scientifically by correlation studies, but in neither case have led to reliable or easily communicable results, though some breeders undoubtedly develop a flair for judging good breeding stock.

It should be possible to tell something about an animal's genotype from an inspection of his pedigree; and the recording of all pure-bred animals in herd books, etc., is an attempt to provide a basis for prediction of this kind. In the past, however, too much attention has been paid to the presence of a famous animal somewhere, perhaps quite far back, in an animal's ancestry, and it has been forgotten that the other animals involved, particularly those in recent generations, have made important contributions to the genotype under investigation. Again, indices of inbreeding,[2] designed to give a hint as to the degree to which an animal is likely to be homozygous, have usually been too largely influenced by inbreeding some generations back, whereas it is the recent generations which have most effect on homozygosis.

The only satisfactory way, in fact, of determining what an animal's genotype is like is to test it. We must allow the male to sire offspring and determine from the offspring themselves whether their father is capable of transmitting useful characteristics. Various technical methods have been proposed for disentangling the contribution of the male from those of the females in such tests, but we shall not here[3] go into these different methods of arriving at a "bull-index"; the essential point is to use the bull with several females and study the resemblances between his daughters. The effects of selection guided by such progeny performance tests, as they have been called, have been very strikingly shown in, for example, an experiment on egg-production at the Maine Agricultural Experiment Station. From 1899 to 1907 selection was

[1] Landauer 1933a, Walton 1933.　[2] Cf. Wright 1922.　[3] Cf. Rice 1934.

practised by breeding only from hens which laid at least 160 eggs in their first year and from cocks whose mothers had laid at least 200 eggs in their first year. Egg-production sank steadily. From 1908 onwards, the selected hens were those which had already had high-producing daughters, while the cocks were sons of hens which had had high-producing daughters. By this system of using breeding stock whose usefulness had already been tested, egg production in the selected flock went up steadily, at least till 1920.

The great difficulty in the use of progeny performance tests is, of course, the time they take; from the time of the male's maturity, a sufficient period must elapse for several of his offspring to be born and grow old enough for their characteristics to be tested. The test therefore requires that a considerable number of males shall be kept for rather a long period while they are being tested and their breeding life will be correspondingly shortened. The testing of dairy bulls would not be complete till they were about nine years old, by which time at least three-quarters of them would have been eliminated by death or for reasons of selection. The progeny test, therefore, although much more accurate than the pedigree test, does not in practice supplant it. Efforts have been made to work out, from an animal's pedigree, an index based not on the actual characters of the ancestors but on their progeny performances. These indices, like all those based on pedigrees, can only have a statistical justification, since they can take no cognizance of segregation of important factors. But for characters such as milk production, which are obviously influenced by many genes of more or less equal importance, the basis for a statistical treatment is present and the indices may be useful when large numbers of animals are bred from, though unreliable in particular cases.

3. Selection of Both Parents

Selection, either by phenotypic characters, pedigree or progeny performance tests, may be applied to both parents. This is common practice in animal breeding, where matings are usually controlled in any case, but is less common in plants where controlled mating requires the rather laborious process of artificial pollination. A classical example of this type of selection is the investigation by Castle[1] of the hooded pattern in rats, in which two lines, one of increasing and one of decreasing pigmentation, were built up.

Castle originally suggested that the selection had actually altered the

[1] Castle 1916.

genes concerned, but he later demonstrated the correctness of the more orthodox view that the selection had merely accumulated minor factors affecting the extent of the pattern. He crossed rats with extremely small pigmented areas with rats from a normal unselected stock, and recrossed the hooded segregates into the same normal line; after several

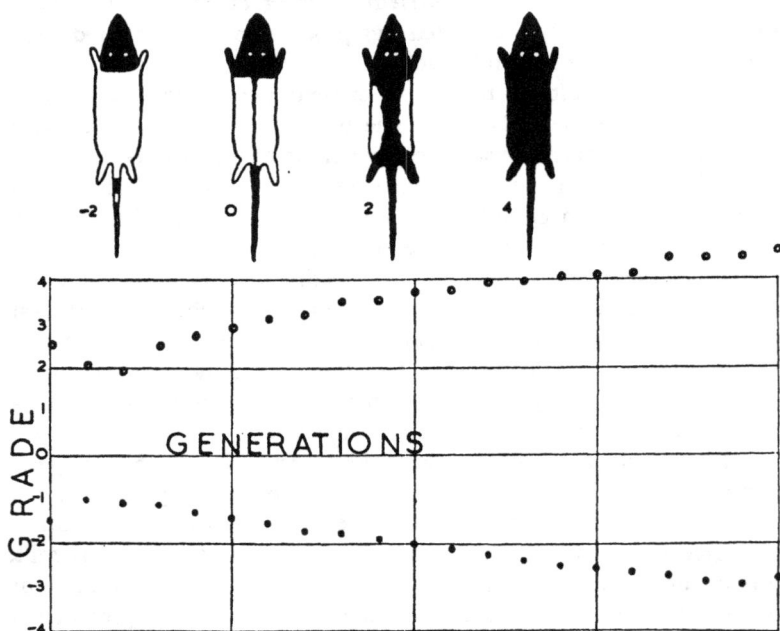

Fig. 130. **Selection in Rats.**—The hooded rat has very variable pigmentation. Four of the arbitrary grades used for measuring the pigmentation are shown above. Below are the results of selection for twenty generations of inbred lines; hollow circles selection for increased pigmentation, full circles selection for decreased pigmentation.   (Data of Castle.)

generations of such crossing, the hooded gene from the low pigmented line had been got into a normal genotypic milieu and again produced a normal, and thus higher, degree of pigmentation. Thus the low pigmentation in the selected line was due only to the accumulation of modifiers tending to reduce pigment.

4. *Inbreeding and the Production of Pure Lines*[1]

Inbreeding is the breeding together of related organisms. Their

----
[1] *General references:* East and Jones 1919.

relationship means that some one individual ancestor has contributed to both their genotypes. Thus inbreeding is to some extent the breeding of like with like, and will clearly have a tendency to bring together like genes and produce homozygosis. In the absence of selection, the rate of approach to homozygosis is dependent mainly on the type of inbreeding practised, though it is modified by mutation, linkage, etc. The most rapid approach is with the closest type of inbreeding, namely self-fertilization, in which the number of heterozygous genes is halved in each generation. Thus a population of self-fertilized plants may be expected to consist of an assemblage each of which is nearly or quite homozygous, and a very short period of selection of individual plants will yield a set of completely homozygous types.

The first realization of this fact was by Johannsen,[1] who showed that a population of beans (self-fertilizing), which showed a certain variability in the weights of the individual seeds, was really composed of a mixture of pure lines, or homozygous forms, in each of which the variability was very much less. Moreover, within each pure line the variability was entirely due to random environmental fluctuations and was not inherited, so that selection within a pure line remained ineffective. Indeed it must do so until a mutation in the desired direction occurs, since in a completely pure line all factors are homozygous and no hereditable variation can exist. (Fig. 131.)

Since mutation and occasional chromosome rearrangements are bound to occur, at least rarely, some selection is necessary to keep even a pure line absolutely constant;[2] but granted a minimum of selection, further rigour has no effect. The de Vilmorin wheats, originating from single plants and self-fertilized, have remained substantially constant for over fifty years.

The formation of a pure line is usually an essential in the production of a new variety of a sexually reproducing plant, since the variety must breed true and yield uniform progeny. In plants reproducing vegetatively or by female apospory, it is not necessary to take any steps to see that the variety is homozygous, since it must breed true in any case; and in fact we find that many, if not most, vegetatively reproduced crop or ornamental plants are extremely heterozygous and often hybrid.

The production of a uniform variety is simplest, as we saw, in self-fertilizing plants, where individual plants are usually already nearly homozygous. Thus the offspring of a single plant may constitute a true breeding variety, and several very valuable types have been produced

[1] Johannsen 1903.   [2] Haldane 1936.

in this way, particularly in wheat.[1] The variety Red Fife, which spread through the whole wheat-growing area of Canada in the latter part of the last century, originated from a single plant. A farmer named Fife obtained seed, which originally came from Russia, and planted it in the

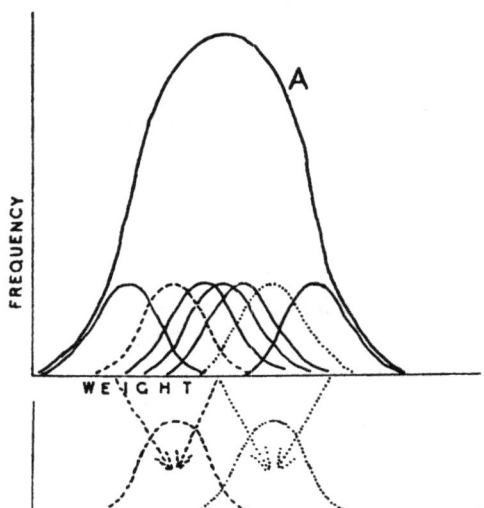

Fig. 131. **Pure Lines.**—If one plots the frequency of different weights of, e.g. beans, in a large population, a curve is obtained like that marked A. This is the sum of the frequency curves of the different individual families of beans. Since beans are self-fertilizing, any heterozygote (Bb) will have half its offspring homozygous (BB or bb), and thus after some generations nearly all the individuals in a self-fertilizing population become homozygous. All the members of a family are therefore genetically identical, and the variation in weight within a family is due only to variations in environmental conditions. The lower part of the figure shows that if we take the lightest and the heaviest members of a family (e.g. that giving the dashed frequency curve) their progeny will, under the same conditions, give the same frequency curve, which proves that there is no genetic difference between them. The same is true for any other family, e.g. the dotted one. Each family is a pure line.

spring only to find that it was a winter sort. Only one plant ripened, and from this plant the whole Red Fife variety was derived. Another and more recent example is the variety Kanred, produced by the Kansas experiment station as the best of 554 pure lines started in 1908 from single plants of the Russian variety Crimean. The preliminary testing took eight years, but in the seven years after being put on the market Kanred proved so successful that it was grown on approximately two million acres.

[1] *Rev.* Hunter and Leake 1933.

In normally cross-fertilized plants, pure lines are not obtained so easily. They can be prepared by artificial selfing. The efficiency of this process in bringing about homozygosis is best seen if we compare the results of selection in cross-fertilized and self-fertilized lines. Thus five years of selection for high protein content in self-fertilized maize gave a greater rise in average protein content than twenty-three years of mass selection, while plants from the sixteenth generation of mass selection could still be rapidly improved by self-fertilization and selection.[1]

Comparatively few varieties which are themselves valuable have been isolated by self-fertilization of normally cross-bred plants. This is because it is commonly found that the inbreeding of such plants leads to a general loss of vigour. In an outbreeding population, rare recessive genes with adverse effects on vigour can spread fairly widely through the species in a hidden, heterozygous condition, and many, if not most, individuals may contain one or more such factors (p. 286). Inbreeding, and particularly selfing, will tend to produce individuals homozygous for such factors, which will then be expressed in a lowering of the vitality of the plant, since any dominant factors there may be for normal health will produce no more effect when homozygous that they did when heterozygous. The loss of vigour on inbreeding is found to be fairly general in practice, though it may be absent in particular cases; for instance, Collins[2] has described an inbred strain of Saxton June maize which was extremely vigorous and presumably originated in a plant which happened not to contain any deleterious genes.

On crossing two different varieties or inbred strains, the effects of the recessive genes of one plant are hidden by the dominant allelomorphs of the other, and perfectly healthy plants result. The phenomenon is known as heterosis,[3] or hybrid vigour. The effect usually disappears in generations later than the F1; the reason for this is supposed to be that the beneficial factors in one strain are often linked to deleterious factors, so that the segregates in F2 and later will be homozygous both for beneficial and deleterious genes and will therefore have only a mediocre appearance.

Ashby[4] has attempted to explain the physiological processes underlying heterosis. He suggests, from observation on maize, that the growth rate of the hybrid is actually always that of the faster growing of its inbred parents and that its greater vigour is entirely due to an

---

[1] East and Jones 1920.
[2] Cf. Babcock and Clausen 1927.
[3] *Rev.* East 1936.
[4] Cf. Ashby 1937.

initial advantage conferred on it by the greater size of seed formed on outcrossing, that is to say, by an increased growth rate of the hybrid embryo between fertilization and seed formation. This, however, seems unlikely to be a general explanation. Sinnott and Houghtaling[1] have shown that although the different sizes of squash fruits are often a consequence of differences which can be found at a very early stage of development, in other cases they are due to different amounts of growth of primordia which were originally of the same size. And since there undoubtedly are some hereditary differences of the growth rates in later stages of plant development, it is difficult to see why they should be denied any part in producing heterosis. Finally, the theory cannot be applied to animals, in which there is no sudden check in development comparable to seed formation. Castle and Gregory[2] have shown that the eggs of rabbits of a large race cleave faster than those from a small race from the very earliest stages, and this higher growth rate continues throughout the life of the animals.

Heterosis has already found practical application, particularly in maize breeding. Inbred lines are prepared, which although homozygous for some deleterious factors and therefore weakly in constitution, are nevertheless cleared of the most damaging factors which are completely lethal in homozygous condition. On crossing two such lines, extremely sturdy and productive plants are raised in F1. A further increase in yield can be obtained by crossing two hybrid F1's, giving a "double-cross" F2, but beyond that it does not seem profitable to go.

Hybrid vigour can be perpetuated in plants which propagate vegetatively, and valuable hybrid trees (e.g. poplars) have been prepared in this way.

In crosses between animal varieties, hybrid vigour is usually found, and advantage is often taken of it in raising beef cattle, for instance. A similar enhancement of vigour is sometimes found in wider crosses, between different species, as in the well-known case of the mare-ass cross, which produces mules, and in various crosses between bovine species. But in such cases it is by no means clear that the hybrid vigour can be explained entirely as a result of bringing into the genotype dominant allelomorphs of deleterious factors, since it is probable that the genotypes differ more profoundly than in the possession of different allelomorphs at the same loci. One must suppose that in those cases in which hybrid vigour is found in species crosses, it is a more or less fortuitous result of a happy combination of the two gene complexes;

---

[1] Cf. Sinnott and Dunn 1935.   [2] Castle and Gregory 1929.

certainly there are very many cases in which the combination is not successful and the hybrids are weakly or even inviable, while in the best cases they are often sterile or nearly so, even though sturdy.

An increase of vigour on crossing, though often found, is perhaps not so general among animals as it is in plants, and this is probably correlated with the fact that the results of inbreeding are not so uniformly unfortunate. Animal populations do not seem to be so permeated with harmful recessives as are cultivated plants, and it is often feasible

Fig. 132. The Pedigree of a Famous American Bull, Comet.—Notice the large amount of inbreeding in the ancestry of this superlatively fine animal.

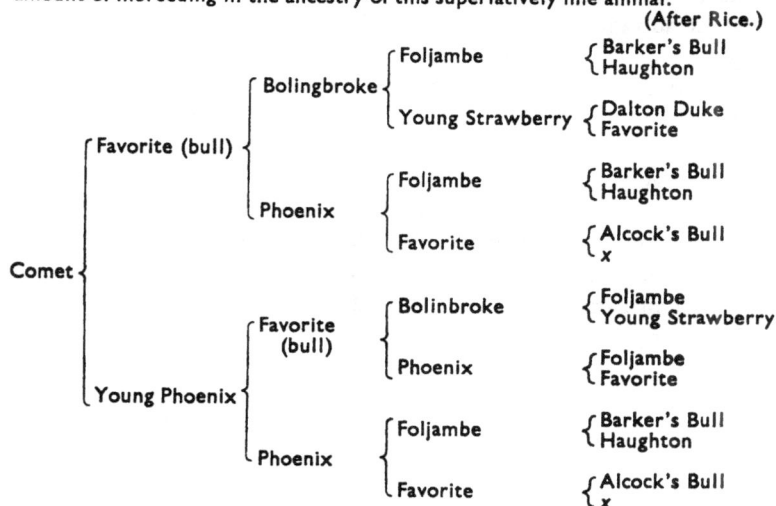

(After Rice.)

Comet			
Favorite (bull)	Bolingbroke	Foljambe	Barker's Bull / Haughton
		Young Strawberry	Dalton Duke / Favorite
	Phoenix	Foljambe	Barker's Bull / Haughton
		Favorite	Alcock's Bull / x
Young Phoenix	Favorite (bull)	Bolinbroke	Foljambe / Young Strawberry
		Phoenix	Foljambe / Favorite
	Phoenix	Foljambe	Barker's Bull / Haughton
		Favorite	Alcock's Bull / x

to practice quite close inbreeding without harmful results. In fact, close inbreeding is an extremely valuable method of stock improvement and has been involved in the building up of nearly all the best animal stocks. It is the only method available for creating a homogeneous population, and if combined with selection gives a population homozygous for favourable genes. We owe the development of longhorn cattle, Leicester sheep, and shire horses largely to Bakewell, who, in the middle of the eighteenth century, dared to go against the prevailing prejudice against "incest" and practice inbreeding. Even full brother and sister matings, which is as close inbreeding as is possible in a bisexual organism with separate sexes, has been shown to have no bad effects for four generations in a certain line of pigs.[1]

[1] For a detailed study of inbreeding in rats, see King 1918, 1919.

L

5. *Variation*

The process on which animal or plant improvement immediately depends is selection, but the results which can be obtained depend on the sort of variations which are available. One of the main difficulties of the breeder is to get into his material all the variants which he desires; only then can he begin to select and combine together the favourable factors.

The variations on which useful breeds have been built up depend mainly on gene mutations; chromosome changes, the other main source of variation, usually lead to infertility and therefore are only important in vegetatively propagated plants (but see p. 322). Not very many favourable gene mutations affecting vigour, yield, etc., have actually been observed at the time of their occurrence. An exception is perhaps the mutation to a many-leaved, unbranched growth habit in tobacco, which occurred in the Connecticut Cuban variety and may prove to have economic importance.[1] Somatic mutations may cause the development of shoots with mutant characters, and these "bud-mutations" have been extensively tested in vegetatively propagated plants such as Citrus fruits, and several of commercial value have been found both in Citrus fruits and apples. The nectarine is a bud mutation from the hairy peach, to which it is recessive.

The majority of spontaneous mutations lead to a decrease in vigour, but they may still be valuable in horticultural plants if they produce larger or more brightly coloured flowers, etc. The date of origin of the factors affecting habit, flower colour and flower shape in plants such as sweet-peas can be fairly accurately determined from old catalogues.[2] The enormous variety of such plants which are now available has been formed by hybridization between the stocks in which the mutation originally occurred, and a conscious application of Mendel's laws makes it possible to produce any combination at will.

The breeder seeking genes of a particular kind may succeed in finding them in the mixed progeny of commercial seed. We have already mentioned examples of the production of valuable new pure lines from such material. Another striking example is provided by the velvet bean Stizolobium, which was originally limited to the South-East United States until early maturing varieties, probably dependent on factor mutation, were found, which enabled it to be grown over a much wider area.

[1] Cf. Babcock and Clausen 1927.
[2] Cf. Babcock and Clausen 1927, Crane and Lawrence 1934.

Very often, however, the desired character cannot be found in the crop plant itself, or not in the best varieties of it, and the factor must then be brought in from another species or variety. For instance, the northern range of the orange in California was greatly extended by combining, as far as possible, the good qualities of the normal orange with the cold resistance of the Chinese trifoliate orange, whose fruits are themselves valueless. Another example of very valuable advances obtained by wide hybridization is the work of the Russian breeder Michurin[1] with fruit trees. Michurin collected wild forms from the enormous forests of plums, cherries, peaches, etc., which are found in Persia and adjacent regions (their centre of origin, cf. p. 250), as well as local races from the Far East. Crosses between these and European forms have given some exceptionally cold-resisting types, which seem likely to be of the greatest possible value in the north of the U.S.S.R. One of his most spectacular successes is the production of a mountain-ash (one of the hardiest of trees) with palatable berries. Michurin developed a considerable number of practical aids to the carrying out of such wide crosses, which are often difficult to bring off. He used a technique of "vegetative rapprochement," attempting to overcome interspecific sterility by bringing the plants vegetatively in contact, either by grafting a whole branch of one on to the other, or by tying a piece of the style of the pollen parent on to the style of the ovule parent. The physiological basis of many of these methods, if they are in fact successful, is not understood, but in the example given above it is perhaps not impossible to see how the procedure might work. More doubt attaches to Michurin's belief that a young hybrid is in a very unstable condition, and can be altered, even in its genotype, by external agents, e.g. by being attached, either as a stock or as a scion, to a graft of some other species which acts as its "mentor." Very much more evidence than is yet available would be necessary before this pre-Mendelian type of concept can be accepted. Unfortunately Michurin also believed that better results were obtained by the use of mixtures of pollen, so that it is impossible to be certain of the nature of many of his alleged hybrids.

In recent years, Russian scientists in particular have been very active in collecting from all over the world wild races or species which seem to contain valuable mutations.[2] It may be expected that the utilization of this material will lead to great advances, though it is still too early for much to have come of the collections as yet. An example of the richness

[1] Michurin, cf. Hudson 1937. [2] *Rev*. Hudson 1937.

of variation made available for breeding is provided by the potato. The European varieties are derived from a single species, *Solanum tuberosum*, but in South America there are at least thirty cultivated species, forming a polyploid series with $2n = 24, 36, 48, 60$ and 70. Among these are frost-resisting forms, which can perhaps be cultivated in the far north of Europe, and also types which form tubers under conditions of short periods of daylight, which may perhaps be useful and adaptable to tropical conditions. The economic and social possibilities of any large extension of the potato belt are bound to be enormous.

Artificial polyploids (p. 256) made from species hybrids have already been introduced into horticulture, e.g. *Primula kewensis*. None have yet been made which are satisfactory from an agricultural point of view, but great hopes are held out for the future of the doubled wheat-rye hybrid described by several Russian authors;[1] it may prove a very valuable cereal on light dry soils. The method may be expected to find considerably greater application in future.

Perhaps the most important achievements to date of the method of wide crossing are those obtained by wheat breeders.[2] English wheat, at the beginning of the century, was extremely prolific but lacked the hardness of grain required to give good baking flour. This hardness was present in the lower-yielding Canadian wheats, Red Fife (p. 316), and its derivative Marquis. Biffen crossed Red Fife with an English variety, Rough Chaff, and showed that hardness depended on a dominant gene; and by selection in the F2 and subsequent generations he isolated a segregate, Burgoyne's Fife, which combined hardness with some of the high-yielding qualities of the English wheats. Subsequently he developed Yeoman I and later Yeoman II from a Red Fife-Browick cross, and in these two varieties the hardness is transferred to the English wheats with little, if any, loss of yield.

Biffen also investigated the inheritance of resistance to the important wheat disease known as Rust, and found that it was determined by a single recessive factor in crosses between varieties of ordinary vulgare 42 chromosome wheats (cf. p. 258). The picture soon became more complicated, however, and it was discovered that there are very many different physiologic forms of the rust fungus, and that resistance to one form did not necessarily imply resistance to another. An attempt was therefore made to get into vulgare wheats the factors determining the general resistance to all forms of rust which can be found in the 14 chromosome Emmer and 28 chromosome Durum wheats. The F1 of a

[1] Meister 1930.   [2] *Rev.* Hunter and Leake 1933.

durum-vulgare cross is highly infertile, as a result of irregular segregation at meiosis, but the viable gametes tend to have chromosome numbers near to the haploid numbers of either their durum parent (14) or their vulgare parent (21). By inbreeding or back-crossing, some 28 and 42 chromosome plants can be raised, and among the latter are some (e.g. Hope)[1] which combine general rust resistance with the high yield and good quality of the vulgare varieties used.

Some of the new wheat varieties discovered by the Russian collecting expeditions may simplify this task considerably.[2] For instance, they describe a species *Triticum macha* which is like the usual 28 chromosome wheats but actually has 42 chromosomes, while there is a 28 chromosome form, *T. persicum*, resembling the usual 42 chromosome group. This *T. persicum* is claimed to be a tetraploid hybrid between *T. dicoccoides* $n = 14$ and the grass *Aegilops triuncialis* $n = 14$.

The breeding of disease-resistant varieties is now one of the most important phases of plant breeding. In the past, entire crops have been wiped out by epidemics. Coffee growing in Ceylon, for instance, had to be given up after the attack by Hemileia on the *Coffea arabica* then under cultivation; attempts are being made to introduce into the valuable arabica types the resistance of *C. canephora*, which itself gives a low-quality product. The destruction of French vines by Phylloxera attacking their roots is another well-known example of the devastation wrought by disease, and in this instance it has hitherto been impossible to combine resistance with quality, so that growers have had to graft fine varieties on to the roots of the low grade but resistant American species.

The existence of different physiologic forms gives trouble in coping with diseases in other plants as well as in wheat. Thus Salaman[3] attempted to make Phytophthera-resistant varieties of potato by bringing in factors from the species *S. demissum*, which is highly resistant (although Phytophthera does not occur in its native regions; an example of adaptation to non-existent conditions!). This attempt appeared successful until a new strain of Phytophthera arose, against which the demissum resistance was inoperative, when the work had to be begun again; it is now once more believed that a resistant variety has been produced.

In some parts of the world, as in Russia, the production of cold-resistant forms is one of the most important problems facing breeders. In this connection a theory of breeding has been developed by Lyssenko[4]

[1] McFadden 1930.   [2] Cf. Hudson 1937.
[3] Salaman 1934.   [4] Cf. Anon. 1935.

for which fundamental importance has been claimed. Lyssenko first studied the technical device of vernalization, which enables one to alter the date of ripening of a crop by suitable treatment (with heat, moisture, etc.) of the seeds. This led him to the theory that the vegetative period in the growth of a plant consists of a series of phases, during each of which the plant demands certain particular external conditions. In consequence, one variety may, in a certain region, be late because the demands of an early phase are slow in being fulfilled, while another variety may also be late but because of a slowing down of a later phase; a suitable hybrid between the two complementary types may then be earlier than either. It is probably too soon for a considered judgment to be given of the value of this method of approach, but *a priori* it seems likely that an analysis of the developmental effects of the genes will prove valuable. Unfortunately, Lyssenko combines this apparently sound theory with the wildest hypotheses, such as the denial of segregation or of the existence of genes, and the rejection, on philosophical grounds, of the whole edifice of genetics as developed since the rediscovery of Mendel's work.

Breeding for disease resistance has not been practised with any great success in animals, though it is known that hereditary differences in susceptibility occur.[1] Thus native African cattle are often markedly more resistant to local tick-borne diseases than are imported European cattle, and Brahman cattle seem to be resistant to Texas fever; hybrids with European cattle have been bred, but no definite breeds have been started.[2] Resistance, however, often seems to be polyfactorial and to depend largely on environmental factors which are difficult to control, and in practice selection would not be easy by any method less drastic than allowing the disease to kill off all the animals it could.

Even in animals, however, the search for favourable variations has led outside the confines of the species. Not many of the species hybrids which have been made have yet proved their worth in practice, except the mule, which is an old-established type for certain purposes. The cattalo, a cross between the cow and the American bison, may prove valuable in the inhospitable plains of Texas; the F1 males are infertile, but the females could be back-crossed and fertile hybrids were obtained in later generations.

[1] Cf. Gowen 1937, Crew and Roberts 1933, Mohr 1934.   [2] Cf. Rice 1934.

CHAPTER 15

Human Genetics

A. THE METHODS OF HUMAN GENETICS

The inheritance of hereditary characters in man does not differ in principle from heredity in other organisms, but the study of human genetics is so important and differs so much from animal or plant genetics in the methods which it employs that it must be given separate consideration. Since controlled breeding is impossible, genetic analysis has to be carried out by special and rather roundabout methods, which will be discussed in this chapter. We shall first consider the study of rare mutant genes with relatively clear-cut effects and then pass on to the analysis of the genetic differences which exist within a population of what we should call "normal" men and women.

1. *The Identification of Genes*[1]

The first task of human genetics is to find the genes. We can expect that some genes will have clear-cut, noticeable and constant effects on development; they have good expressivity. Two well-known examples are the genes which determine the different blood groups; the two dominant allelomorphs *A* and *B* specifically cause the presence of the isoagglutinins *A* and *B*, the compound *AB* having both and the homozygous recessive neither. Another similar example is the dominant gene which confers the capacity to taste thiophenylurea as a bitter substance (to homozygous recessives it is nearly tasteless).[2] Rarer examples are the genes for haemophilia, albinism, etc.

In other cases we find, and indeed must expect, complications. These are of two sorts. In the first place, there may be two or more different genes which produce identical effects. Thus there seem to be at least two different forms of retinitis pigmentosa, one determined by an autosomal dominant, and one by an autosomal recessive, while there may be a third due to a sex-linked factor. In some cases the discrimination of two different methods of hereditary transmission of apparently similar pathological conditions has led to a more complete examination

---

[1] *General references:* Baur, Fisher, and Lenz 1931, Blacker 1934, Gates 1929, Hogben 1931, 1933, Holmes 1934.   [2] Blakeslee and Fox 1932.

which demonstrates that the two conditions are not quite identical. In this way genetics may be able to render valuable service to medicine by indicating a more fundamental classification of diseases. It seems clear, for instance, that in the psychological sphere many genetically and medically different entities are at present grouped together under such terms as feeblemindedness and schizophrenia.

The second sort of complication which must be expected is in a way the reciprocal of this. The same gene may have different effects in different genotypes or different environments. There are two causes of variability involved. On the one hand, it is quite likely that some of the human genes have a penetrance less than 100 per cent. For instance, it is very probable that diabetes is inherited by a recessive gene,[1] but one

Fig. 133. Blood Group Inheritance.—The most important groups are those determined by the genes A, B, and O, which are all allelomorphs. When bloods of the different groups are mixed, the reaction between the agglutinins and agglutinogene causes the blood corpuscles to adhere, as shown in the tables.

Blood Group	Genotype	Agglutinogen in Cells	Agglutinin in Serum	Agglutinates	Agglutinated by
I	O	O	ab	II, III, IV	None
II	A	A	b	III, IV	I, III
III	B	B	a	II, IV	I, II
IV	AB	AB	o	None	I, II, III

tends to find a smaller proportion of diabetics than would be expected from the various types of matings. The penetrance of the diabetes gene is not complete; but in the homozygous recessives in which it has not produced diabetes it may have a milder effect in causing abnormal blood-sugar values and reduced sugar-tolerance, so that such individuals can be identified by suitable tests and steps taken to control the disease.[2]

On the other hand, the expression of a gene is often considerably influenced by the external environment in which the zygote develops. We must never lose sight of the fact that a certain gene in a certain genotype determines only the potentialities of the zygote, and that in development only some of these potentialities will be realized. Genetics requires a concept akin to the old idea of "diathesis." A man with "a gene for a character A" has in reality a capacity, determined by the whole genotype as well as the particular A gene, to develop the character A under certain external conditions. If the conditions are not

[1] Pincus and White 1934.
[2] For a review of hereditary metabolic disorders in man, see Macklin 1933, 1934.

normal, A may fail to occur, or occur in a modified form. We shall discuss later in a particular case how far Nature (the genotype) or Nurture (the environment) can be said to be responsible for the manifestation of A. In the present context, the point is that in order to identify human genes, genetics must learn to classify together all the

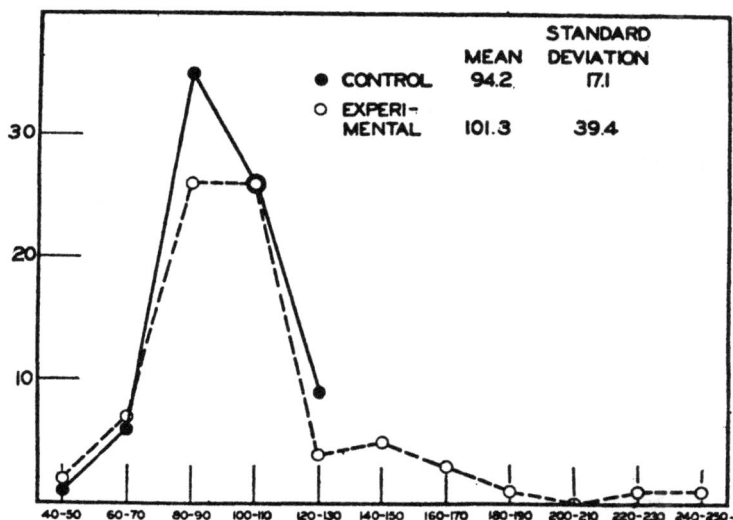

Fig. 134. **Blood Sugar in Relatives of Diabetics.**—The frequency diagram of blood sugar concentration in normal individuals is shown by the full circles, that of relatives of diabetics by hollow circles. Note how some of the latter, without developing definite diabetic symptoms, have an abnormally high blood-sugar content. There is some doubt whether diabetes mellitus is due to a recessive or a dominant gene, but whichever it is, the gene is incompletely penetrant, and we are here discovering cases in which it is very weakly manifested.

(From Pincus and White.)

different manifestations of a single genetic "diathesis." This task is still largely unachieved.

2. *Autosomal Dominants*[1]

Probably the easiest genes to identify are rare autosomal dominants with good penetrance. In a randomly mating population, the frequencies of the different genotypes for a rare dominant gene A are $p^2\, aa : 2pq$ $Aa : q^2\, AA$, where $p + q = 1$ and p is much larger than q (cf. p. 289).

[1] For mathematical methods of the following three sections, see Hogben 1931, 1933.

L*

Thus there are very few homozygous dominants and nearly all the individuals showing the character will be heterozygotes. If such an individual marries a normal, his offspring will show a 1 : 1 ratio of affecteds (showing the character) to normals. This requirement is sufficient to identify some characters, e.g. diabetes insipidus, brachydactyly, as due to autosomal dominants.

With rather commoner genes, the homozygous dominants become more important and cannot be neglected. If we know the frequency of the character in the population, i.e. $\dfrac{2pq + q^2}{p^2}$ we can calculate p and q, and from them the proportions of different kinds of matings and finally the proportion of affecteds which should be expected among offspring of matings between affecteds and normals. The agreement between the theory and the actual data is quite good for some common genes, such as those for the blood groups or taste-capacity, but is sometimes not so good for other semi-rare genes, when an excess of affecteds is found to occur. This is probably because the assumption of random mating throughout the entire population does not really apply. Rare genes do not immediately spread throughout the entire population but tend to be concentrated in certain localities, owing to some degree of inbreeding within local groups. In such a group, the concentration q of the gene is higher than it is in the population as a whole, and there are therefore relatively more homozygotes and fewer heterozygotes among the affecteds than would be suggested by a calculation based on the frequency of the gene in the whole population.

Characters dependent on dominant genes will clearly run in families, recurring generation after generation; they will be "hereditary" in the usual medical sense, whereas characters due to recessive genes, as we will see, tend to occur in several of a group of brothers and sisters and then to disappear, a type of behaviour which is called "familial." Quite often, a "hereditary" trait occurs sporadically, the proportion of affecteds among offspring of affecteds being much less than a half, and the trait sometimes skipping a generation altogether. The suspicion may arise that we are dealing with a rare recessive, and this may sometimes be the case. But the gene may also be a dominant of low penetrance. In this case we can identify it by another test. Consider a family tree containing a rare dominant. Very nearly all matings of affecteds with normals will be $Aa \times aa$ (since AA is very rare), and it is easy to see that, if the penetrance is constant, the proportion of affecteds to normal will be the same among the parents, sibs or offspring of an affected individual, and there will be half this proportion among his nephews, nieces, uncles and aunts, and a quarter among his first cousins, and so on. On the other hand, if the gene was a recessive, most of the affected × normal matings will be *aa* × *AA* and there will be no affecteds among the offspring, or among the parents who would usually be *Aa* × *Aa*. If the character depended on two complementary factors, the affected: normal ratio would fall off very rapidly in passing to more distant relatives of an affected individual. Thus this test may suffice to identify a dominant even of low penetrance.[1]

A true understanding of the heredity of a disease is of course more

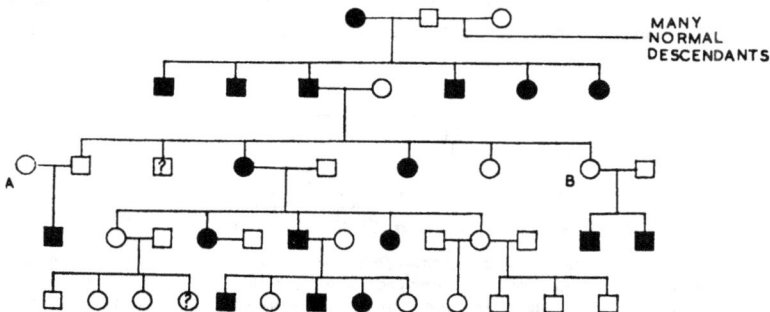

Fig. 135. **A Pedigree of Glaucoma.**—The character behaves as a dominant; it is usually handed on only by individuals who themselves show it. But at *A* and *B* it is transmitted by non-affected individuals, so that its penetrance is not complete. Males squares, females circles, affected individuals black.

(After Holmes, data of Courtney and Hill.)

difficult to obtain when penetrance is low, but it possesses peculiar importance in such cases. Some non-affected relatives of an affected individual will in reality carry the gene, and if they can be detected it may be possible to help them to adjust their environment so as to minimize the chance of their gene being effective.

3. *Autosomal Recessives*

The equilibrium proportions of a random mating population can be written $1 AA : 2u Aa : u^2 aa$, and if *a* is rare, *u* is small, and u^2 very small. Thus most recessives will be produced by matings of the type $Aa \times Aa$, which will occur in a relative frequency $4u^2$, whereas matings $Aa \times aa$ will only occur with frequency $2u^3$. Recessives therefore rarely have affected parents, and usually marry the com-

[1] Levit 1936.

moner type of homozygous normal and have no affected children. On
the other hand, a quarter of their sibs are likely to be affected so that
the character appears as a "familial" one.

A character determined by a recessive gene cannot be identified by
following it through several generations, since, as we have just seen, it
usually occurs suddenly in a lineage and then disappears again. If the
presence of a recessive gene is suspected, we should expect the ratio of
affecteds to normals, in families showing any affecteds, to be 1 : 3,
since these families will be produced by $Aa \times Aa$ matings. The
difficulty in testing this is that human families are so small that some
$Aa \times Aa$ matings will give only one or even no affecteds. In casually
collected statistics as to the incidence of familial diseases, families with

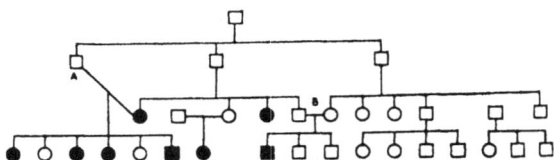

Fig. 136. A Pedigree of Albinism.—(Males squares, females circles, affected
individuals black.) The trait is inherited as a recessive. Note the inbreeding,
between an uncle and his albino niece at A and between cousins at B. It is only
because of this unusual amount of inbreeding that the character appears in two
successive generations.

(After Tertsch.)

only 1 affected are often omitted, since they do not seem to show any
"familial" incidence. Even if the statistics have been carefully collected,
$Aa \times Aa$ families which happen to contain only normals and no
affecteds will not be recognizable. The proportion of affecteds among
the families which are recognized as involving the gene therefore tends
to be too high. Compensation can, however, be made for this. Agreement with the theoretical 1 : 3 ratio is then often very good.

The most conclusive test for a recessive gene is derived from a
consideration of inbreeding. Consider a heterozygous Aa individual.
He will only have affected offspring if he marries another heterozygote
(or a homozygote, which is unlikely). Now since a is rare, Aa mates will
be fairly rare; but since the Aa individual has received this gene from
his parents and grandparents, it will also have been handed down to
some of his collaterals, such as cousins. Thus if the individual marries
his cousin, there is a considerable chance that the mating will be
$Aa \times Aa$ and that affected offspring will appear. Applying this argu-

ment backwards, we can see that affected individuals tend to be the offspring of related parents. The calculation required to put this argument on a quantitative basis is rather complicated, and so many factors enter in that exact predictions can hardly be made. But qualitatively we may say that if the parents of affecteds show an excess of inbreeding over that characteristic of the population as a whole, that is strong evidence that we are dealing with a recessive gene. This test is particularly useful in distinguishing between recessives and characters due to two complementary genes, which otherwise show rather similar behaviour.

4. Sex-linked Genes

In man, the male is the heterogametic sex. Rare sex-linked genes can therefore show in males, but in females tend to be covered by their

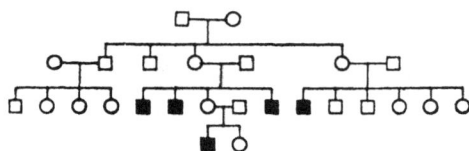

Fig. 137. A Pedigree of Red-Green Colour Blindness.—(Males squares, females circles, affected individuals black.) The trait is inherited as a sex-linked character. Note that it is transmitted by females, but is only shown by males.
(After Holmes, from Groenouw.)

normal allelomorphs. It is often concluded without more ado that a character which is found in males but is also transmitted through normal females must be due to a sex-linked factor. But clearly this condition is also fulfilled if the gene is autosomal but its expression is limited to the male sex. Additional criteria of sex-linked inheritance, sufficient to distinguish it from sex-limited inheritance, are (1) that the trait is not handed on from father to son (except in the rare case where the female is heterozygous for the gene), and (2) that at least one female showing the character is known. For many of the characters usually listed as sex-linked neither of these conditions is fulfilled by the data as yet available.

Still further information is required before it can be decided whether a sex-linked factor is dominant or recessive. Clearly the criterion should be the appearance or non-appearance of the character in a heterozygous female. In a random mating population, the frequencies of the various genotypes should be $pA : qa$ in males and $p^2 AA : 2pq Aa : q^2 aa$ in females. The ratio of q affected males to q^2 affected females, appropriate for a recessive gene, can be verified in the case of colour blindness, but in very few others. Many of the sex-linked genes usually considered to be recessive cannot strictly be proved to be so. For instance, hæmophilia (failure of the blood to clot) is found only in the male sex. It is not handed on from father to son and therefore is probably due to a sex-linked gene though its failure to show in females may indicate that it is also sex-limited in expression (there is some evidence that the presence of the female sex hormone prevents the expression of hæmophilia).[1] In this case the question of recessiveness and dominance does not arise, and the gene may be classed as indeterminate. In other cases,

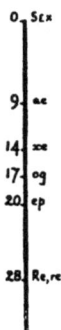

Fig. 138. **Provisional Map of the Human X Chromosome** (and of the homologous part of the Y).—The genes indicated are: *ac* achromatopsia, *xe* xeroderma pigmentosa, *og* Oguchi's disease, *ep* epidermolysis bullosa, *Re*, *re* dominant and recessive retinitis pigmentosa. The map was derived from investigation of partial sex linkage.

(From Haldane.)

it is known that a gene has some degree of dominance, since some effect is shown in the heterozygous female, but it is not known whether the dominance is complete or whether the homozygous female would show even more marked effects. Levit[2] has proposed calling such genes "conditionally dominant."

The quantitative requirement that, from a mating involving a sex-linked recessive $Aa \times AY$, the affected males should form a quarter of the offspring, can be tested in some cases (e.g. hæmophilia, some forms of night blindness, etc.), correction being made as before for the possibility of missing some families in which by chance no affected offspring occur.

Genes inherited in the Y chromosome have also been described in man. If they are confined to the Y, they pass from father to son and do not appear in females, but the linkage is usually incomplete and crossing-over takes place between the X and Y (p. 80). The principle of the

[1] Pratt 1932.   [2] Levit 1936.

method of searching for such genes in man is as follows for the simplest case, that of a dominant. If a man receives an incompletely sex-linked dominant from his mother, his constitution will be $AX.aY$ and on mating with a normal female $aX.aX$, he produces mainly affected daughters and normal sons, with a few affected sons and unaffected daughters due to crossing-over. But if he received this gene from his father, it will be in the Y and will go mainly to his sons. Thus in both cases the majority of affected children of affected fathers should be of the same sex as their grandparents. The conditions for incompletely sex-linked recessives are more complex. Haldane[1] has presented evidence that five or six genes are incompletely sex-linked in man, and the degree of linkage made it possible to draw a provisional linkage map of the homologous parts of the X and Y chromosomes (Fig. 138).

5. Continuous Variation

The greater part of the variation in normal human populations appears to be continuous; we can at best draw arbitrary lines between the tall and the short, the clever and the stupid, etc. The methods of spotting genes which we have hitherto discussed are quite inapplicable to such a situation, and other methods of attack have to be found.

We may expect that the differences between human beings are partly due to genetic differences and partly due to differences in the environments in which the genotypes have developed. The problem which is often posed is to determine the relative importance of heredity and environment in producing the differences which occur. But this is, strictly speaking, a meaningless question. We have seen (p. 189) that if two different genotypes are reared in two different environments, the difference between the resulting phenotypes is a function of all four variables, the two genotypes and the two environments; it is therefore impossible to get any true insight into what is happening by means of observations on only two objects, namely, the two phenotypes. The problem must be analysed into two separate questions. Let us consider two environments and let the same genotype develop in each of them; are the environments such that the phenotypes differ, and if so by how much? Again, what differences are produced when two different genotypes develop in the same environment?

The technique which is employed in the study of continuous variation is that of statistics.[2] The two main notions involved in the discussion which follows are variance and correlation. Suppose we select

---

[1] Haldane 1936a.   [2] Cf. Fisher 1928a, Pearl 1930, Yule 1929.

a group of individuals; they may be a group all possessing some characteristic, such as blue eyes, or they may be all the individuals we can take the trouble to collect. Then let measurements be made of some character, such as height. It will in general be found that intermediate values of the measurement are the more frequent, and if we plot against a measurement X the frequency with which it occurs, we obtain a typical bell-shaped frequency curve with its highest point in the middle of the range and falling off towards the extremes at each end. The variance is a measure of the amount of scatter in measurements of this kind; it is related to the area enclosed under the curve and it is calculated from the squares of the deviations of individual measurements from the average.

Correlation coefficients are measures of the degree of relationship between pairs. Suppose we measure the heights of a set of brothers; then a high correlation (i.e. nearly 1) means that there is a strong tendency for tall boys to have tall brothers, and short ones short brothers; a correlation of 0 means there is no connection between the height of a boy and his brother, while a negative correlation coefficient means that tall boys tend to have short brothers.

Perhaps the simplest use of correlation coefficients in genetics is to calculate the correlation between different relatives, e.g. between fathers and sons, parents and offspring, etc. This procedure has been carried out in great detail by the biometric school, led by Galton and Karl Pearson. They found that the correlation between offspring and parent, for many characters, was about a half, while between offspring and grandparent it was about a quarter. At first the conclusion was drawn that characters which show continuous variation are inherited by blending inheritance, that is, that discrete hereditary units were not involved and that no segregation and recombination occurred.[1] It was supposed that each parent contributed half the genetic constitution of the child, which was a sort of average of its father and mother. A considerable controversy developed on this question between Pearson, who argued on the basis of statistical averages, and Bateson, who upheld Mendelism on the basis of observations of individual matings. The issue was finally settled when it was shown, partly by Pearson himself and finally by Fisher,[2] that the data of the biometricians could be entirely harmonized with the requirements of Mendelian theory. It is possible to calculate the magnitude of the correlation between various relatives from Mendelian principles,[3] but only if certain assumptions

[1] Cf. Pearson and Lee 1903.   [2] Fisher 1918.   [3] Cf. Hogben 1933a.

are made as to the degree of dominance shown by the genes in question, and as to the frequency of the genes in the population. It turns out that the assumptions which are necessary to derive, from Mendelian principles, correlation coefficients similar to those actually found, are quite reasonable ones. The sort of correlations which are found, therefore, give no reason for denying the Mendelian analysis.

But actually very little can be deduced from correlation coefficients of this kind, because they are affected by too many environmental factors.[1] The biometrical assumptions assume that all genotypes develop in the same environment, and this is certainly not true. If we consider a correlation between brothers, there will be differences between the environments of the two brothers of a pair, and the greater these differences are, the less alike will the brothers be and the lower the correlation between them. On the other hand, the environment of one pair of brothers will differ from that of another pair, and the greater these differences, the more boys of one family will resemble each other rather than boys of another family, and the greater the correlation between brothers will be. The actual correlation found will therefore be the resultant of the true genetical correlation in a uniform environment, with the differences within a family tending to lower it and differences between families tending to raise it. There is no way of assessing how important these environmental factors have been in a given case, and even if we find exactly the correlation which would be expected on certain genetical assumptions, we cannot jump to the conclusion that those assumptions are justified.

6. The Genetic Analysis of a Continuously Varying Character: Intelligence

We cannot derive any insight into the actual genetical situation unless we can find some way of simplifying the usual complicated state of affairs. As an example of what can be done in this way, we shall take the analysis of human intelligence.[2]

The range of variation is apparently quite continuous between idiots and geniuses. The first necessity is to attempt some method of classification. In the lower grades, some more or less specific types of idiocy or insanity can be distinguished. Thus Huntington's chorea is a disease involving progressive mental degeneration; it has been clearly shown to be due to a simple autosomal dominant. Amaurotic family idiocy is

---

[1] Cf. Hogben 1933.
[2] *General references:* Freeman 1934, Hogben 1931, 1933, Holmes 1934, Penrose 1936.

another disease involving strong mental symptoms, and is probably caused by a recessive. In other cases of more or less defined pathological conditions, such as microcephaly, the hereditary determination is still obscure; and there are still other kinds, such as Mongolian idiocy, where there is evidence that environmental conditions are very important (Mongolian idiots tend to be more frequent among children born late in their mother's life). Many cases of very low-grade mentality belong to less well-defined types, and evidence as to their hereditary nature is still rather vague. But it is suggestive that in an overwhelming

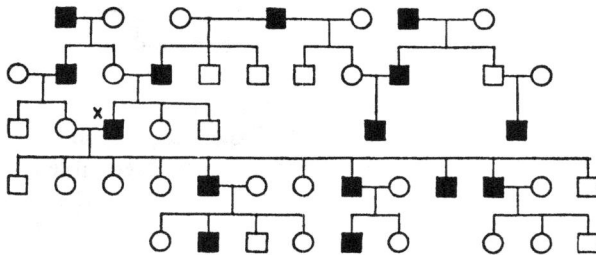

Fig. 139. **A Pedigree of the Darwin Family** as an example of how intelligence "runs in families."—Males squares, females circles. The males marked in black attained intellectual eminence, most of them having been Fellows of the Royal Society; no attempt has been made to assess the intellectual achievements of the women. Charles Darwin is marked with an X; note that he married his cousin, and that the inbred family was extremely successful.

(After Holmes.)

proportion of cases in which one of a pair of identical twins is mentally defective, the other, which must have exactly the same genotype, shows a similar type of defect. Moreover, matings of two low-grade defectives tend to give a great majority of low-grade deficients. Probably multiple factors are involved.

When we consider the sub-normals, normals, and geniuses the unravelling of the genetical situation becomes more important and more difficult. There are several well-known pedigrees which exhibit an apparent inheritance of either superior or inferior mentality. Famous cases of the former are the Darwin family, of the latter the Jukeses or the Hill-folk.

But we may ask how far this inheritance is genetic or how far it is due to the persistence of a favourable or unfavourable social environment. We require to know how much difference can be made by the range of environments which actually occur in human society, and how this compares with the range of mentality which is found. For anything like an exact analysis, some measure of intelligence is necessary. It is usual to use one or other of the "intelligence tests" devised by psychologists starting with Binet.[1] No one knows exactly what these tests actually measure, but the measures are fairly reproducible when the same individual, or better, the same group of individuals, is measured at different times, so that the thing which is measured seems to be something fairly definite, and is presumably one of the factors referred to when the word intelligence is used in the ordinary unanalysed way. To allow for age, the measure is usually expressed as the Intelligence Quotient or I.Q., which is the actual measured intelligence expressed as a percentage of the average intelligence at that age.

Taking the I.Q. as the variable, we can discover the correlations between different pairs of relations. Between pairs of sibs (brothers and sisters) of the same sex, it is usually about $0 \cdot 4$[2] and gives no indication of the influence of sex-linked genes, which would tend to cause a higher correlation between female pairs than between males. But that is all that can be deduced from the data on ordinary sibs. To progress any farther we must make uniform either the environments or the genotypes. A moderately uniform environment is provided when children are, at a very early age, adopted into an orphanage or other institution, though even then the uterine environments of successive children in a family are probably by no means the same. The facts about orphanage children, however, are not very certain. Some investigators have found a high correlation between sibs reared in an institution (e.g. $0 \cdot 53$ for Gordon's data analysed by Elderton),[3] others have found a correlation no higher than that between normal children (e.g. Davis).[4] The results of the reverse process, increasing the environmental differences between sibs by having them adopted into different homes, seem to indicate an effect of the environment, since the correlations are usually lowered.[5] But this method hardly leads to quantitative estimates and the data are still scanty.

We can obtain rather fuller information from cases in which the genotypes are the same, i.e. from identical twins. As is well known, there are two sorts of twins, fraternals, which are ordinary sibs which happen to be conceived at the same time, and identicals, both of which proceed from the same fertilized egg and therefore have identical

[1] Cf. Dearborn 1928.   [2] E.g. Herrman and Hogben 1933.
[3] Elderton 1922.   [4] Davis 1928.
[5] Freeman, Holtzinger, and Mitchell 1928.

genetical constitutions.[1] Two fraternal twins have the same average genetic difference as two ordinary sibs, but since their environments are generally more similar than those of sibs, they tend to be more highly correlated. Identical twins are in much the same environmental situation as fraternals but in addition their genotypes are exactly the same, and their correlation is correspondingly higher still. The similarity between them extends to the most unexpected traits. Lange,[2] for instance, has shown that if one of a pair of identical twins has been to prison, there is a very high likelihood that the other will commit some crime and be caught.

It is perhaps easiest to consider the resemblances between twins in

Fig. 140. Criminality in Twins.—Notice that if one of a pair of identical twins shows criminal tendencies, so does the other in a high proportion of cases. If the twins are fraternals, it is more common to find only one of a pair affected.

Type of Twins				Adult Criminals	Juvenile Delinquents	Behaviour Problems
Like sexed pairs, probably identical						
Both affected	..	..	..	25	39	41
One affected	..	..	..	12	3	6
Like sexed, probably fraternal						
Both affected	..	..	..	5	20	26
One affected	..	..	..	23	5	34
Unlike sexed, fraternal						
Both affected	..	..	..	1	8	8
One affected	..	..	..	31	32	21

(After Holmes, from Rosanoff, Handy, and Rosanoff.)

terms of average differences. The average difference in I.Q. between fraternal twins in two well-known investigations was about 16–17 per cent (Holtzinger,[3] and Herrman and Hogben),[4] while that between identicals was only 8–9 per cent. It seems that the variability between twins is only cut in half when their genotypes are made exactly the same, so that roughly half the variability of the fraternals must be due to the slight environmental differences between them. Such an estimate of the magnitude of the effect should, however, be accepted with some caution. If the tests are not quite reliable, we would find some variation in score even if we measured the same individual on different days, and a part of the variability of the identicals may be due to slight unreliability of the tests. If this is so, the influence of the environment will have been exaggerated in the estimate given above. But the facts

[1] For a review of the biology of twins, see Verschuer 1932. [2] Lange 1930.
[3] Holtzinger 1929. [4] Herrman and Hogben 1933.

that variability is certainly not abolished in identical twins, and that fraternal twins have a higher correlation than normal sibs, make it certain that environmental differences, even the slight ones between members of the same family, have a recognizable effect in determining scores in intelligence tests.

The environmental differences between identical twins brought up in the same family are very small. We might expect to find out more about the possible effects of the environment from a study of identicals which have been reared apart. Comparatively few such cases have been studied yet, but in those which have the average difference is greater than that of identicals reared together and less than that of fraternals reared together. Newman, Freeman, and Holtzinger[1] have made a detailed study of nineteen pairs. Differences in intelligence, when they

Fig. 141. Intelligence Tests on Twins and Sibs.

(From Herrman and Hogben.)

	Correlation coeff. of I.Q. scores	Mean difference in I.Q.
Identical twins	0·84 ± 0·04	9·2 ± 1·0
Fraternals, both same sex	0·47 ± 0·08	17·7 ± 1·5
Fraternals, of unlike sex	0·51 ± 0·06	17·9 ± 1·5
Sibs, pairs of like or unlike sex ..	0·32 ± 0·09	16·8 ± 2·3

occurred, could nearly always be correlated with differences in the amount of schooling or other cultural opportunities which the pairs had received (11 pairs). Moreover, in five out of the six cases in which there had been marked differences in opportunity, there were corresponding differences in ability; in the sixth case, the twin who had had the greater opportunities had also suffered a severe illness which may well have annulled the cultural advantages she had enjoyed. The authors conclude that their studies provide convincing evidence of the marked effect of education on ability. Comparisons were also made of the temperaments of the twins. These are of course much more difficult to study, even with the aid of a whole battery of tests. The interesting point emerged that, while twins reared in different environments might show considerable differences in superficial behaviour, such as conformity to canons of polite society, they were often very similar in more fundamental traits. An extreme example was the handwriting of two brothers; there were differences in most of the usual features relied on for describing handwriting, such as form of letters and pressure, but both writers suffered from a slight muscular tremor.

[1] Newman, Freeman, and Holtzinger 1937.

The parts played by heredity and environment in the development of several other characters of man have been studied in the same way as has just been described for I.Q. In particular the study of twins has proved extremely revealing in several other connections.[1] A very important part in this study has been played by the Medico-Genetical Institute in Moscow,[2] where a large team of doctors collaborated with scientists and statisticians in the study of "normal" or medically important characteristics. They found, for instance, that identicals show greater similarity than fraternals in age at teething, age at learning to sit or to walk, in susceptibility for certain diseases (e.g. scarlet fever) but not for others (e.g. tuberculosis).[3] Again, twins are the ideal human experimental material, since it is possible to treat one twin and keep the other as a control which is known to be exactly comparable. This method has as yet been very little employed. One of the most promising experiments dealt with different methods of teaching. One twin is allowed to build models with wooden blocks, imitating a structure which is in front of him. The other does the same, but paper is glued over the structure so that he cannot see how it was made. It is reported that after two months (half an hour's play a day) the twins of the second group were considerably superior to those of the first, both in model making and in other activities whose connection with model making was not at all obvious.

B. THE GENETIC STRUCTURE OF HUMAN POPULATIONS

1. Race[4]

Man is a very variable animal. An Australian aborigine, a Chinaman and a West European differ as much from each other as do many related species of monkeys. But all the living types of man are mutually fertile, and it is usual to classify them all in a single species *Homo sapiens*.

The local varieties of man are spoken of as races. But, because we know more, or want to know more, about man than about, say, Lymantria, and because there are very many more specimens available for examination, the concept of race breaks down when we try to apply it

---

[1] *Rev.* Verschuer 1931.   [2] Cf. Muller 1935*b*.

[3] Evidence for inherited susceptibility to this disease has been found in other cases.

[4] *General references:* Castle 1930, Fisher 1930, Freeman 1934, Haldane 1934, 1938, Hogben 1938, Holmes 1934, Huxley and Haddon 1936.

in any strictly defined sense. In Lymantria we are only interested in the sex genes and a few others. If two groups of Lymantria collected in different places differ in respect of some or all of these genes we call them different races. But in man we are interested in so many characters that we can hardly ever find two individuals which do not differ in some respects; we can never deal with groups which are uniform for all the characters we can examine. The only hope of applying a race concept is to limit the sphere of reference, and to do that we shall only take account of certain sorts of characters.

If we limit the number of characters severely enough, we can distinguish a few large groups of mankind, such that the different groups are pretty sharply distinguished from one another and we do not find many intermediates except obvious hybrids. Thus we can discriminate the "races" of whites, negroes, American Indians, yellows, Australians, etc. But these groups are themselves not at all uniform. For instance, the genetic basis of colour seems to be different in different groups of negroes; West Africans differ from whites in many polymeric colour genes, so that in hybrids such as American mulattoes there is a segregation into a continuous series of colour grades, whereas South Africans seem to differ in only a few major genes, and hybrids show a fairly clean segregation into dark and light types.[1] However, when we try to bring such differences into the classification, and discriminate minor races within the great groups of whites, blacks, etc., we find that it is extremely difficult to draw any dividing lines within the general chaos of variations.

We can set about the attempt in two ways. In the first place, we may define a set of characters which we take to be characteristic of a certain race. For instance, we might say that the Nordic race has long heads, blond hair, and blue eyes. But this only has any sense if we believe that a group of men having these characteristics exists now or existed at one time. In the absence of any detailed knowledge about the past, the justification for such a definition reduces to whether or not we can find such a group at present. We must therefore tackle the problem in the second way, which is to examine human populations and see how they can be grouped. If we do this, we can in fact find groups which are homogeneous in the sense that, although they show variation, this variation is continuous and the group cannot be split up into smaller groups which are significantly different from one another.[2] But these homogeneous groups are extremely small and insignificant. They are

[1] Gates 1929.   [2] Cf. Morant 1928.

usually isolated communities living under fairly primitive conditions and have a high degree of inbreeding. We might attach a definite meaning to the term race if we defined it with reference to such groups, but the concept would not be very useful. We should in the first place have to make sure that the characters concerned were hereditarily and not environmentally determined. This is not always certain; it is probable that even the cranial index, that sheet-anchor of the physical anthropologist, can be modified by the environment, since it is less extreme in the American-born children of immigrants than it was in their parents.[1] Even when this point had been disposed of, the races would only be definitely characterized in respect of the points mentioned in their definition; there is no reason to suppose that because a population is homogeneous for long-headedness it is therefore homogeneous for other characters such as intelligence. Finally, such a concept of race would be a purely statistical one, defined in terms of populations, and could not be applied to individuals. In the first place, different races overlap, so that the longest-headed members of a short-headed group have longer heads than the shortest variants of a long-headed group. Even if we knew an individual belonged to one of the pure races, we could not tell, looking at him alone, which of the races it was. And, further, as we have said, only a few small populations do consist of pure races. The vast majority of the population is made up of individuals who would have to be considered as very complex hybrids showing all sorts of segregation and recombination.

Thus the attempt to classify mankind into genetically homogeneous groups becomes progressively more difficult as we take account of more genes. If we try to derive a concept of race which will be relevant to all characters of a man, the attempt is a complete failure. Segregation and recombination has gone so far in most sections of the human population that it is impossible to summarize an individual by any description less complete than a specification of his whole genotype.

2. *Gene Differences Between Races*

Nearly all attempts to define races are based on examination of phenotypic characters, since we know much too little about the genes. It is certain, however, that there are differences in the frequencies of some genes in different populations. For instance, the blood groups of many races have been tested.[2] Very few "races" are homogeneous, though some American Indian tribes possess no B and A is fairly rare

[1] Boas 1928.   [2] Cf. Davis 1938.

among them, the majority belonging to group O. In all other populations both A and B are present, though the proportions vary somewhat and these variations may be correlated with other genotypic differences. Thus Hungarian gipsies show a gene ratio more like that of Indians, from whom they are probably descended, than that of the Hungarian population by which they are surrounded. Bernstein[1] has suggested that primitively the human race contained neither A nor B, and that A arose in Europe, B in East Asia, both genes having since spread from their place of origin. But the evidence is not very convincing, and as homologous genes exist in the higher apes, it is perhaps unlikely that they have arisen anew in man.

Some of the rare pathological genes are more common in some racial groups than in others, e.g. family amaurotic idiocy (recessive?) occurs more frequently in Jews. But a recent investigation showed that most such genes have a very similar incidence in Japanese and Europeans.[2] There may also be differences between the major racial groups in the frequency of the extremely important genes determining susceptibility or resistance to diseases. Negroes are usually much more susceptible to tuberculosis than whites and less so to malaria and various parasites. Also the commonest sites of incidence for cancer are different in different races, even when the total rate of incidence is the same; but within the same racial group the total incidence of cancer may be the same in men and women, but usually the disease attacks the genital tract of the latter and some other organ system of the former.[3] Our exact knowledge of disease resistance is, however, very incomplete and in many cases we do not even know for certain that it is genetically determined.

3. Nations and Races

The concept of nationality has, in recent years, often been confused with that of race. It should be unnecessary to state that nationality is not a genetical concept at all. It can be defined either in political terms, as all the people living under a certain government, or in cultural terms, as all the people enjoying a certain cultural tradition. Neither of these entities has anything to do with the concept of race, even on the loosest definition of that troublesome word. The habit of referring to alleged national characteristics, such as "British hypocrisy" or sense of "cricket," as though they were racially determined, is also clearly

---

[1] Bernstein 1931.   [2] Komai 1934.   [3] Kennaway and Kennaway 1937.

unjustified. Not only are nations not necessarily races, but we know very little about the mental differences which presumably exist between different racial groups.

A few attempts have been made to investigate such differences and evaluate them in genetical terms. American negroes, for instance, nearly always give lower average I.Q. scores than whites, from whom they differ "racially" as well as culturally.[1] Typically, the average negro I.Q. is about 75–80 per cent of that of the whites. But there is considerable overlap when individuals are considered, and about 25 per cent of negroes equal or surpass the white average. The genetic significance of the results is very difficult to assess. Unfavourable environment certainly plays some part in lowering the negro attainment; thus negroes from northern states score better than those from the poorer southern states, and when the negroes and whites live under approximately the same conditions, as in the small island near Jamaica investigated by Davenport and Steggerda,[2] their average scores are more nearly the same (but in the former case there may have been some selection of better negroes in the northern states and in the latter of poorer whites to live on the island). Tests on negro infants of 2–11 months, on whom the external environment has not had much time to act, show that they also score slightly lower than white infants in the so-called Baby Tests, but the difference is not so large as when adults are compared. Even at this age, the negro children are somewhat inferior to the whites in physical development, so that it is probable that their environment, including pre-natal environment, had been inferior. There is as yet no way of telling whether these environmental effects can account for the whole difference between whites and negroes; it is perhaps rather unlikely that they can. It must be remembered, however, that the I.Q. tests do not measure all aspects of personality, and there may be other respects in which negroes surpass whites.

A large-scale investigation of natio-racial differences was made at the time conscription was introduced in the United States in 1917. Brigham[3] found that most immigrant peoples, except those from northern Europe, made lower average scores than native-born Americans. But the scores of Americans from different states were quite highly correlated with the expenditure of the state on education, and there was also a high correlation between the national scores and the rating of the educational facilities provided in the countries of origin. It

[1] *Rev.* Freeman 1934.
[2] Davenport and Steggerda 1929.   [3] Brigham 1922.

therefore became clear that the results of the tests were by no means independent of environmental factors, and their genetic significance is therefore doubtful.[1]

Rather similar results were obtained from tests of Massachusetts school children.[2] Children of North Europeans and Jews scored high, those of South Europeans, Mexicans, and Negroes low, even if the parents had lived some time in the United States before the children

Fig. 142. National and Racial Differences in I.Q.—The Pintner-Patterson performance tests were given to groups of one hundred boys from Paris, Hamburg, and Rome, and from rural districts in France, Germany, and Italy. The rural groups included only boys who could be definitely classified as Mediterranean, Alpine, or Nordic in racial type.

	Average Score	Range
Paris	219·0	100–302
Hamburg	216·4	105–322
Rome	211·8	109–313
German Nordic	198·2	69–289
French Mediterranean	197·4	71–271
German Alpine	193·6	80–211
Italian Alpine	188·8	69–306
French Alpine	180·2	72–296
French Nordic	178·8	63–314
Italian Mediterranean	173·0	69–308
All city	215·7	
All rural	187·1	
All Nordics	188·5	63–314
All Alpines	187·5	69–306
All Mediterraneans	185·2	69–308

(Data of Klineberg, from Freeman.)

were born. But though these results may perhaps give some indication of the relative merits of different immigrant stocks (provided environmental factors can be neglected) the immigrants are only a selected group of their nation and cannot be directly compared, as the way in which they were selected may have differed from nation to nation.

An attempt was made to classify the Massachusetts children into racial groups (Nordic, Alpine, and Mediterranean) on the rather rough criterion of eye colour. There was little difference between Nordics, Alpines, and Mediterraneans but considerable difference between different national groups belonging to the same race. The same phenomenon was found by Klineberg,[3] who examined Germans, French, and

[1] Brigham 1930. [2] Cf. Freeman 1934. [3] Klineberg 1931.

Italians in their own countries. His data also show differences between groups belonging to the same "race" but having different nationality, but in addition to this there were differences between groups of different "race" within the same nation. There was also a very significant fact that all city-dwelling groups, independent of nation or race, scored alike and higher than all the rural groups.

4. Race Crossing

It is often claimed that the crossing of widely different races of man inevitably leads to undesirable biological results. On the other hand, there is no doubt that most national groups have arisen through a fusion between at least fairly different peoples, and in some cases the hybridization may have been fairly wide; for instance, the Japanese are probably a mixture of Mongolian and Malay stocks.

The main disadvantage of crossing which would be suggested by biological theory is that it would lead to the production of a disharmonious assemblage of characters. If there are gross anatomical, physiological, or mental differences between the parents, the children may develop an assortment of characters which do not fit well together. Davenport[1] has drawn particular attention to this possibility and has stated that it is often found for anatomical characteristics, for instance, in white Bushman crosses. But it will occur the more often the fewer the genes on which the differences depend; if the parent genotypes differ in multiple genes, we should expect something more like blending inheritance, and this is also found. It is impossible to say *a priori* whether two races differ in many or only a few genes; we have already mentioned that the colour differences between whites and South Africans are due to a few genes, those between whites and West Africans to many. Only painstaking investigations of particular crosses, such as those made by Gates[2] and Lotsy,[3] can solve the question for each individual case.

It is extremely difficult to determine whether disharmonious mentalities are produced by crossing, as is often alleged. In some cases, e.g. in South America, crossbreds are not at all inferior in mentality to the pure races, while recently Germany has discovered, perhaps with some dismay, how many of her most famous sons have some Jewish genes in their chromosomes. But such equality or even superiority is only found when there is no social handicap in being a crossbred. In only too many parts of the world the social disadvantage of being a

[1] Davenport 1917.   [2] Gates 1929.   [3] Lotsy 1928.

half-caste is probably the greatest influence on half-caste mentality, and from such cases no genetical conclusions can be drawn. Castle[1] goes so far as to say "Human race problems are not biological problems any more than rabbit crosses are social problems." One cannot deny, however, that the human problem has a biological element, and perhaps in the future it will be possible to analyse this more adequately in countries like the U.S.S.R. where an attempt is being made to preserve strict race-equality in social matters.

We should expect the later generations from wide crosses to produce segregates more extreme than either of their parents. This does not often seem to happen. Muller[2] has discussed the question thoroughly, and suggests that the extremes of variation in a single "race" are probably due to rather rare genes, so that unless some of these happen to be included in the parents of the cross there will not be an adequate genetic basis for the cross-breds to produce extreme segregates.

5. Genetic Differences Between Classes[3]

Men are not only divided into racial and national groups, but, in all civilizations except some of the most primitive and the most advanced, they are also stratified into classes. In societies founded on the labour of slaves captured in war, we should expect to find considerable genetic differences between the ruler and the ruled. But few such societies now exist. In this section we shall discuss the genetic differences between classes in present-day capitalist civilizations.

If mental tests (I.Q. tests, performance tests, etc.) are made on individuals belonging to different classes, it is commonly found that the average score for professional men is higher than for shopkeepers, for those again higher than for skilled workers, and for unskilled workers lowest of all. There is of course an enormous overlap; and since the lower categories are much more numerous than the higher, the greatest number of individuals above normal may come from one of the intermediate groups such as the skilled workers. Even the average differences, though statistically significant, are not large, the limits of the range being about 15 to 30 per cent in different investigations.

Before any genetical significance can be attached to these results, we must discuss how large a part environmental agencies play in producing them. Unfortunately this question is nearly always confused by poli-

---

[1] Castle 1926.   [2] Muller 1936.
[3] *General references:* Cf. Freeman 1934, Gray 1935, Haldane 1938, Holmes 1934, Lorimer and Osborn 1934.

tical considerations. The Left usually wish to show that the differences are entirely accounted for by the educational disadvantages under which the poorer classes suffer, while the Right wish to regard them as solely genetic. It is difficult to see any necessity for these attitudes. If there is any injustice in the present distribution of the physical or mental amenities of life, that injustice is not increased by showing that the servant is as good as his master, nor lessened by demonstrating that we are only kicking the weaker brethren. If these considerations are

Fig. 143. Average I.Q. of Children according to the Father's Occupation.— Data are given from four different investigations.

Father's Occupation	Average I.Q. of Children			
Prof_ssional	116	125	114	116
Semi-professional	112	120		
Managers			113	
Clerical ..	108	113	112	
Business..				107
Tradesmen			109	
Foremen			106	
Skilled Workmen			102	98
Farmers			99	91
Semi-skilled	105	108		95
Slightly skilled ..	104	107		
Unskilled	96	96	94	89

(After Freeman.)

borne in mind, it should be possible to approach the question with no preconceived bias, and to obtain an objective scientific judgment on it.

It is extremely doubtful, however, if we yet have adequate data to make a final decision as to the relative importance of heredity and environment in producing class differences. The study of twins and of exceptional families makes it likely that hereditary differences such as occur in the population can easily produce effects of the right order of magnitude (p. 338). But so, clearly, can the environment, whose most profound effects are seen in studies of "wild boys" who have been abandoned far from the haunts of man and grown up entirely on their own resources.[1] James the Fourth of Scotland is said to have performed the experiment of isolating a young child on an island in the Firth of Forth; when visited some years later he was able to speak good Hebrew. Most parts of the world are less like the Garden of Eden; the wild boys

[1] Itard 1932, Squires 1927.

do not develop any mode of speech and their intellectual development is very meagre indeed.

But the environmental disadvantages of the lower classes are cer-

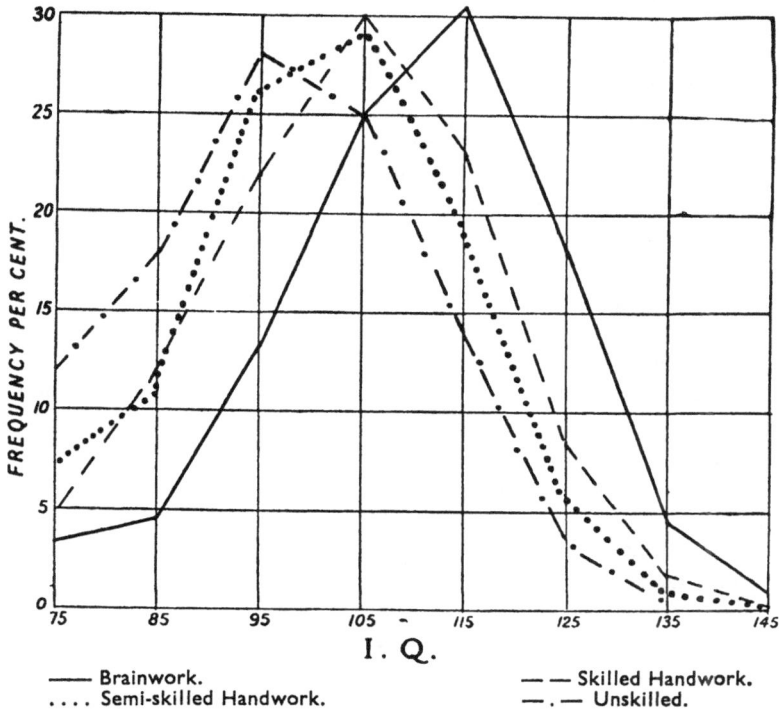

Fig. 144. **Intelligence Quotient and Social Class.**—The I.Q. was measured for children in elementary and secondary schools in Northumberland, and the occupation of the fathers of the children was noted. The I.Q.'s were grouped in classes centring round values of 75, 85, etc., to 145, and the occupations were grouped in four classes (Brainwork (lower ranks of professions), Skilled Handwork, including shopkeepers and policemen, Semi-skilled Handwork, Unskilled Handwork). The graphs show the percentage frequency of children with the different I.Q.'s and belonging to the different social classes determined by their father's occupation. Note that the average I.Q. increases towards the higher classes; but note also the enormous overlap between classes.

(Data of Duff and Thomson.)

tainly not of this order. We get rather nearer to them in investigations on children who have been prevented by physical illnesses from attending school for the full period; their intelligence test scores are adversely affected more or less in direct ratio to the length of time they have lost.[1] Again, children of bargees and gipsies only attend school very irregularly, and while the I.Q. of the youngest tested (four to six years) is only slightly below average (90 instead of 100) the average gradually falls with age and for the group between twelve and twenty-two years is only about 60 per cent of the normal. We have already mentioned (p. 344) the fact that the scores of different American conscripts was correlated with the expenditure of their state on education. Finally, the differences between the averages for the highest and lowest classes (15 to 30 per cent) is not much more than twice the average differences between identical twins (10 per cent) which have been produced by the very slight environmental differences acting on two individuals of the same sex and birth rank in the same family. It is possible, however, that the environmental differences which produce the variation between two identical twins reared together are largely intra-uterine, and it may not be justifiable to compare this with the environmental differences between the classes.

We can certainly conclude from this that the genetic differences between the classes are considerably less than their average performances would suggest. But they may not be negligible. Particularly the lowest class of unskilled labourers, who very rarely produce any individuals above the average, may really be genetically slightly inferior. But conditions of labour have so changed recently that many unskilled workers are now recruited from families which used to belong to the skilled class. Perhaps an investigation on children at the same school, some from families which have for generations been unskilled or casual labourers, and others from families which have recently been forced down into this class, would show whether there was a true genetic difference between the two groups. Perhaps also an investigation of the effect of prolonged unemployment of the parents on normal working-class children would reveal the effectiveness of the kind of environmental differences which are found in present-day society. But data on these two points are not yet available.

6. *Changes in Population Composition*

There is not space in this book to discuss the general trends of population change either in total numbers or in age composition,[2] and we shall confine ourselves to a summary of views on the changing genetic structure of human populations. With the enormous decrease in death-rates brought about in the last hundred years by advances in medical

[1] Gordon 1923.   [2] Cf. Hogben 1938.

services, differential mortality no longer plays the main part in altering the proportions of different genotypes from generation to generation in races fairly well adapted to their environments, such as the Western Europeans. Rare deleterious genes are, of course, still being eliminated, more rapidly if they are dominants, very slowly if they are recessives. In races recently brought into contact with new death-dealing agencies, for instance, in savage races newly introduced to tuberculosis and alcohol, natural selection for resistance probably proceeds with great vigour through the higher death-rate of susceptible individuals. Some savage races, such as the Polynesians and Maori, were reduced to a small fraction of their original numbers in the first years of contact with white culture, but now seem to be increasing again. But it is not clear whether this is because the more resistant stocks have been selected or whether it is a result of gradual amelioration of the culture contact; civilization which at first appeared in the guise of guns, traders, and alcohol, becomes in time a matter of medical stations and labour laws.

In European and American populations, the main evolutionary changes are brought about by the differential birth-rates, or better the different Malthusian parameters (p. 288) of the nations and classes. Unfortunately, the Malthusian parameters, or net reproductive rates (number of daughters per mother, corrected for death-rates and age composition), are only available for European nations and North America.[1] The most industrialized of these nations are failing to maintain their populations, having net reproductive rates of less than 1, while in others the rate is sinking and will probably soon be below unity. Special efforts have been made in Germany and Italy to encourage fertility, but it appears that the success which was at first obtained was only temporary. Of countries with a high reproductive rate, Russia is the main example in Europe; in 1927 the net reproductive rate in the European parts of the country was about 1·7, slightly more than double that of England and Wales. In other parts of the world, Japan is probably the country with the highest net reproductive rate, but only rough statistics are available; some students profess to discern the beginning of a reduction in Japanese fertility and foretell a fairly rapid approach to the conditions found in other industrialized countries.

The differential reproduction of different nations is a problem better treated from a political and social point of view than from a purely genetical one.[2]

[1] Charles 1934, Lorimer and Osborn 1934.
[2] For further discussion, see Duncan 1929, Thompson 1935.

M

Within Western European and American populations, important genetical effects have been ascribed to the differential fertility of different classes. Since age composition and mortality rates are much the same in all classes, these differences in fertility are probably not grossly distorted when judged on the basis of the birth-rates, which are usually the only statistics available. The corrections, if they could be applied, would probably increase rather than decrease the differences which are shown by the birth-rates.

Many investigators have shown that the birth-rate is highest in the lowest occupational class (unskilled labourers) and falls progressively

Fig. 145. **Fertility and class.**—The first column gives the number of children per wife, and the second the net reproductive rate in 1928, for different classes in populations in the United States.
(Data of Notestein, and Lorimer and Osborn, quoted from Holmes.)

	Birth rate	Net reproductive rate
Professional	1·51	0·76
Business and clerical	1·52	0·85
Skilled labour	1·78	1·06
Semi-skilled	..	1·03
Agriculture	..	1·32
Farm owners	2·33	..
Farm renters	2·58	..
Farm labourers	2·77	..
Unskilled labour	2·13	1·17

in higher and higher classes. As we have seen, the I.Q. is highest in the highest classes, so that the consequence which can apparently be immediately deduced is that the more intelligent are being gradually swamped by the higher fertility of the stupid. Fisher[1] has indeed produced an argument to show why this must always be so in any society of classes based on wealth. Social advancement is achieved by two sorts of persons, the intelligent and the infertile; by the latter because of the educational and other advantages enjoyed by children who do not have to share the family's resources with many brothers and sisters. In the upper classes, then, the intelligent mate with the infertile and do not maintain themselves. Fisher supports his hypothesis, which if it is true is of fundamental importance, with considerable evidence. Firstly, studies of genealogies of English peerages show that there is good correlation between the size of a woman's family and that of her mother

[1] Fisher 1930.

(correlation 0·21) and of her grandmother (correlation 0·1065). Fisher concludes that about 40 per cent of the variance in human fertility is due to hereditary causes. Secondly, he shows, again by data from noble families, that marriage with an heiress very frequently leads to the extinction of the family, heiresses being presumably highly infertile since they come from families in which no sons were produced. Fisher believes that this principle is responsible for the fall of past civilizations and is likely to bring down our own unless we learn to control it.

The argument has been criticized on two main grounds.[1] Firstly, it may be denied that the apparent differences in intelligence between the classes really represent genetic differences. We have discussed this question earlier in this chapter. It has been urged in this connection that the upper classes are primarily recruited from the infertile, intelligence having much less to do with social advancement than Fisher suggests, and in this case the differential infertility does not actually sift out and remove the intelligent. Secondly, it has been claimed that recent investigations, mainly in Germany, show that the differential fertility is disappearing, and that the birth-rate of the lower classes is sinking to the level of that of the upper classes. This has certainly occurred in some regions, but we require much more extensive data before we can judge how widespread it is. Germany in the post-war years is scarcely typical of European civilization as a whole. It is probable, however, that the first statistics on the matter, which dealt with a period when birth control was just coming into use among the upper classes, gave an exaggerated view of the differential fertility. Possibly, with the general adoption of contraception in wider circles of society, the birth-rates of different classes will converge to some extent; whether they will attain equality is open to question.

A second differential of the greatest social importance exists between rural communities with a typically high birth-rate and urban or industrial communities with a low one. The genetic effects, if any, of this are completely unknown.

The effects of modern warfare on population composition are difficult to assess. Some authors, carried away by the doctrine of the struggle for existence, have urged that warfare is the agency through which natural selection in man has its most important effects and that the abolition of war would be followed by the decline of human society into a state of sybaritic incapacity. This, however, turns mainly on a misinterpretation of the word struggle. The struggle for existence in

[1] Cf. Charles 1934. See also Haldane 1938.

Nature rarely takes the form of a combat between members of the same species; or at least this is only one of innumerable activities of life through which natural selection acts. Moreover, it must be pointed out that most wars are between national groups, and the selection which takes place is a selection, not of individuals but of nations. The qualities which make a nation best fitted to survive in such a struggle have, at the present day, more to do with its industrial organization and diplomatic finesse than with the biological quality of its members. It is certainly doubtful whether evolution guided by natural selection of societies on such a basis leads in the direction of a social order which the majority of men would consider desirable.

The effects of war on the populations taking part seems, *a priori*, likely to be genetically unfavourable. It is the sane and healthy who go to the front line and are killed. At least, this was the case till as recently as the American Civil War,[1] but the genetic effects may perhaps be ameliorated when the more indiscriminate modern weapons are used. If war appears unlikely to produce, as an agent of natural selection, results whose value is commensurate with the undoubted unpleasantness of the process, genetics suggest no reason why it should not be eliminated. There is no merit in natural selection for its own sake; it had led organic evolution into paths which seem degenerate and distasteful to us as well as into paths we consider progressive. People who wish to justify war on biological grounds must look to other fields than genetics for arguments.

7. *The Control of the Genetic Composition of the Human Population*

The process of evolution has produced the most highly developed object we know of—man. The power to control this process now passes, thanks to genetics, into the hands of man himself. It is inevitable that this power will sometime be used. The methods which must be employed are the province of the geneticist, and no one doubts that it is to him we must turn to obtain practical advice on how certain ends are to be attained. Many geneticists argue that this is the limit of their special responsibilities; in the framing of an evolutionary policy, they urge, their opinion should have no more weight than that of any other citizen. In this the geneticists are perhaps too diffident. The assessment of evolutionary value by the mass of mankind is often influenced by opinions as to the correctness of which the geneticist or social scientist has scientific evidence. As an example one may cite the Ger-

[1] Cf. Holmes 1934, Jordan and Jordan 1914.

man race theory, which allows a popular ideal to be stated in terms which can be shown to be without scientific foundation; and this basis could scarcely be removed without causing some alteration in the superstructure. Similarly, a geneticist would, surely, fail in his duty if he did not question any judgment based on the supposition that all men are born equal. A student of human genetics must perforce train himself to eliminate from just these judgments all the elements of irrationality which render most human opinions so fallible; about the subject of his own study, he, if anybody, should be able to give an objective and unbiassed assessment of value. If he observes the ordinary rules of scientific thought, he is likely to err if anything on the side of excessive caution rather than of wanton interference with human behaviour.

The genetic composition of the population could be improved either by lowering the fertility of the carriers of deleterious genes or raising that of the carriers of favourable ones. Of the methods of lowering fertility, sterilization is the most radical, and since, if breeding is to be discouraged, it might as well, in most cases, be prevented altogether, it is probably the most generally useful.[1] Naturally, such a procedure should not be applied except to individuals who are known to carry genes whose harmful effects are generally acknowledged; instances would be the hereditary absence of both arms and legs, amaurotic family idiocy, etc., which everyone would agree to be undesirable. The operation of sterilization has no effects other than rendering the operated individual sterile. The technique generally employed is severing the genital ducts, the vas deferens in the male, the fallopian tubes in the female. Sexual drive and potency remain normal.

The effect on the population is to eliminate a source from which the unwanted genes can be perpetuated. The phenotypic effect, of course, depends primarily on the mode of inheritance of the gene, since it is only possible to sterilize the individuals in which the gene is manifest. Dominants could therefore be eradicated in one generation, except in so far as they are renewed by mutation. Similarly, sex-linked recessives could be eliminated fairly rapidly since all the male carriers would show the presence of the gene. Haldane[2] has discussed the case of hæmophilia, a sex-linked recessive which causes a failure of the blood to clot, so that affected males usually die young from trivial wounds and fail to reproduce themselves. Since the frequency of the hæmo-

[1] For discussion, see Haldane 1938, Hogben 1931, Holmes 1934, Popenhoe and Johnson 1933. [2] Haldane 1935.

philia gene seems to remain constant, mutation must replace it as fast as it is lost by the failure of affected males to breed. This gives a mutation rate of about 1 in 50,000 life cycles, which, in terms of life cycles, is rather higher than is typical for Drosophila genes.

Matters are quite different for autosomal recessives. Only a very small proportion of carriers are homozygous and show the character (frequencies $p^2 AA : 2pq Aa : q^2 aa$). Sterilization of the recessive phenotypes can therefore only act very slowly to prevent the appearance of the character in future generations, and it will act the more slowly the rarer the gene is. It was at one time thought that, on account of this, sterilization could be of very little use, since most deleterious characters were thought to be inherited as recessives. It seems, however, that many apparent recessives are really dominants with low penetrance, in which case selection against them would be somewhat more effective.

Sterilization has been legalized and practised in some countries, the most important being the U.S.A. and Germany. In the U.S.A. the laws have been in force for some years and considerable experience has been gained in their effects. California is the state which has been most active in this respect, some 20,000 sterilizations having been performed up to 1935. Most mental defectives and cases of insanity submit voluntarily to the operation before leaving institutional care, and a fairly thorough supervision is kept over them after release. It is still uncertain exactly how mental defect is inherited, in so far as it is inherited at all and not environmentally determined, but it seems probable that among the many multiple genes which seem to be concerned (p. 336) some at least are partially dominant, so that selection against them may not be too slow. Moreover, there is a considerable tendency for the feeble-minded to marry each other, and the proportion of feeble-minded genes in the inbreeding groups may be quite high, so that even if they are all recessive selection can be expected to have appreciable effects. The method has, however, not been in use long enough to have produced any marked changes in the incidence of feeble-mindedness in the population. But it seems clear that the operation in itself produces no untoward symptoms in those who submit to it.

The social dangers of a procedure such as sterilization should be very obvious. It is, however, a subject on which opinions tend to be extreme, and some of its advocates, such as the authors of the model sterilization law,[1] suggest that it should be applied to some classes of

[1] Quoted in Haldane 1938.

individuals whose hereditary inferiority is by no means obvious, such as the blind, the deaf or the pauper. Clearly there is no genetic justification whatever for sterilizing an individual unless it can be shown that he carries defects which are certainly hereditary. The danger of a sterilization policy being influenced by political, religious, or other prejudices is very great. Moreover, it must be remembered that sterilization of a defective may not only remove a deleterious gene from the human stock, but may deprive it also of valuable hereditary material carried in the same individual. Mr. H. G. Wells, for instance, is a diabetic, but one would hardly on those grounds place him on the debit side of the human balance-sheet.

Sterilization seems to be almost the only method of reducing fertility which has been advocated. The project which is perhaps most characteristic of the eugenical programmes which have been put forward is the encouragement of fertility in the carriers of favourable genes. If this proposal is made in general terms, we are involved in the very difficult task of deciding which genes are unusually valuable and worthy of encouragement. Often, however, reformers content themselves with attempting to remedy ills which they claim are proceeding under our eyes; the most noticeable, of course, is the differential fertility between the upper, and apparently more intelligent, classes and the lower, putatively inferior, ones. This inequality, it is urged, should be evened out or perhaps reversed, the method proposed being usually a system of family allowances.[1] None of the systems of this kind which have been tried in various countries seem to have been effective; but most systems of family allowances have been designed rather to push up the birth-rate in general than to adjust relative fertilities. The levelling up of fertilities which has occurred in some places during recent years cannot be attributed to any conscious eugenical measures.

In the discussion, in a previous section, of the I.Q. of different classes, it became clear that the environment plays an important part in determining the differences which are found. No sensible eugenical proposals can neglect this important fact. It is clear, for instance, that it would theoretically have been possible to reduce the incidence of small-pox from that characteristic of mediæval Europe by breeding a resistant stock; but suitable hygienic measures such as vaccination are much easier to apply and much more efficient. Similarly, it is very probable that a much greater improvement in intelligence could be produced by measures of social amelioration than by any eugenical

[1] Cf. Fisher 1932.

steps which are within the bounds of probability. When it is remembered that we are still uncertain whether hereditary differences have any part at all in causing the class differences in intelligence, while it cannot be denied that the environment has some hand in this, the priority of environmental over genetical methods of raising the general intelligence becomes obvious. Moreover, the genetical methods involve us in something of a vicious circle; the self-control and conscious forethought required for eugenical progress is scarcely likely to appeal to a comparatively unintelligent populace, and it is not until people have attained a certain level of awareness, probably higher than that found in the majority to-day, that we may expect them to advance through their own eugenical efforts. Until the essential improvements in social hygiene have been made, and society reaps the very great advantages which can be confidently expected from such measures, proposals of "positive" eugenics for improving the human stock remain only of secondary importance.

The methods used in animal breeding could theoretically be used to improve mankind. In most cases, human customs and beliefs would be profoundly outraged if this were attempted. Thus few people would seriously urge a large-scale programme of selective mating or inbreeding of superior stocks. Muller[1] has pointed out that the technique of artificial insemination may, in time, remove the psychological objections which at present stand in the way of attempting "grading up," which, it will be remembered (p. 311), involves the use of all females but only selected sires. If men and women could be persuaded that a husband and father need not be the biological sire of his children, it would be possible to use only sperm from superior men, and to give every individual at least a haploid set of superior chromosomes.[2] Perhaps, if social progress continues, man may be educated up to the point of considering such possibilities seriously. Still further in the future we may envisage a tissue culture technique of preserving the gonads of highly superior men for perpetual use; and the artificial production of polyploidy and polyploid hybrids cannot be dismissed as fantastic, alarming though the prospect of such experiments may be.

[1] Muller 1935.   [2] Brewer 1935, Huxley 1936.

PART FIVE

The Nature of the Gene

One of the ultimate aims of genetics must be the elucidation of the nature of the gene, but science is still very far from attaining this goal. We can only approach the problem in several rather indirect ways. Firstly, we can investigate the nature of the chromosomes with which the genes are so intimately associated. Secondly, we can consider the developmental reactions determined by genes. Thirdly, there are some general genetical phenomena which have suggested hypotheses as to the nature of the gene. Fourthly, we can investigate gene mutation and particularly the induction of mutation by experimental means. In this part, these methods of approach are considered in order, and in the last two sections an attempt is made to formulate the chemical problems which are implicit in our knowledge of gene behaviour.

CHAPTER 16

The Nature of the Gene

A. CYTOLOGICAL CONSIDERATIONS

1. *Chromosome Movements*[1]

There is very little experimental work which gives us any detailed understanding of the nature of the forces which act on and between chromosomes during the process of nuclear division; the nucleus is too carefully guarded within the cell to be easily reached by our present experimental methods. But the process of division, while remaining fundamentally the same throughout the biological realm, suffers in the different groups various modifications which can be regarded as natural experiments and these are numerous enough to enable cytologists to make deductions about the mechanisms involved. Particularly the study of meiosis in polyploids, which has been very vigorously pursued in recent years, has been found to provide crucial tests for many of the possibilities suggested by the phenomena in simpler forms. There is no space here to summarize the detailed evidence, which belongs to cytology proper rather than to genetics. However, a short account of the general conclusions arrived at must be given, since a knowledge of the forces between chromosomes obviously has a bearing on our idea of chromosome constitution.

Two fundamental forces are involved throughout the whole of mitosis and meiosis; firstly, an attraction between similar genes in pairs, which leads to pairing in meiotic prophase, and secondly a repulsion between pairs of chromonemata, and presumably between individual pairs of genes, which leads to the diakinesis configuration and aids in terminalization. To these we must add repulsions firstly between centromeres and centrosomes, which cause the movement of chromosomes on to the metaphase plate, and secondly between centromeres, which aid in terminalization and cause the first part of anaphase separation. Finally, the later part of anaphase separation is due to the elongation of the region of the spindle between the two separating sets of chromosomes. These forces are operative partly within the fluid nuclear sap and partly within the semi-solid spindle which is formed after the nuclear membrane breaks down.

[1] Darlington 1937.

The identification of these forces with definite physical agents is still largely speculative, but the time is ripe for an attempt to see how far the theory of colloidal structures can account for them. In this attempt it will be advisable, following Darlington, to consider separately the forces concerned with the "internal mechanics" of the chromosomes, that is to say, forces which act over distances of the same order as the diameter of the chromonema, and those concerned with the "external mechanics," which act over distances of the order of the length of the

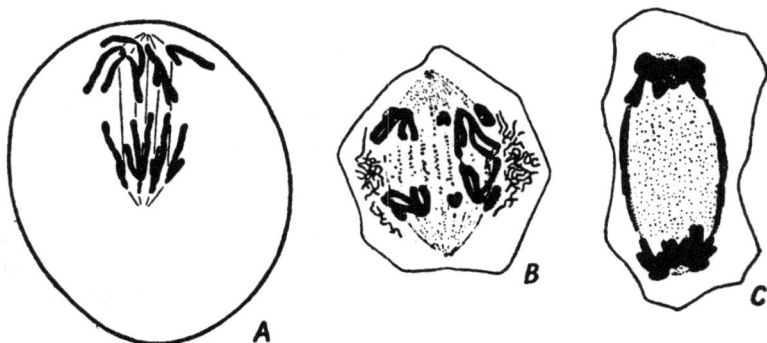

Fig. 146. **Anaphase Movement.**—A. Anaphase in a mitosis in a pollen grain of Podophyllum. One group of chromosomes is near the wall of the cell, and appears to have ceased moving, while in the other the chromosomes are still lying parallel as they are pushed into the centre of the cell. B, C. Early and late anaphses in the meiotic division of *Stenobothrus lineatus*, to show the elongation and narrowing of the spindle between the separating groups of chromosomes. Note the mitochrondria lying near the surface of the spindle (after Belar).

(From Darlington.)

chromonema; we shall postpone till the next section a consideration of the special forces which bring about the spiral coiling of chromosomes.

For the internal mechanics, colloid theory provides two forces:[1] a repulsion due to the formation of a double electron layer on the surface of neighbouring particles in a fluid medium; and an attraction due to the so-called London-van der Waal's forces. Both these forces are non-specific; that is to say they occur between any two particles and not only between similar ones. This is what is required of the repulsive forces between chromosomes and between centromeres and centrosomes. On the other hand, the attractions appear to be specifically

[1] The discussion largely follows unpublished suggestions of J. D. Bernal. For the basic conceptions, see Freundlich 1932, Hamaker 1937, 1938.

between similar sets of genes. It is possible, however, that the London-van der Waal's forces would act in such a way as to appear as specific attractions. The magnitude of these forces depends on the difference in refractive index between the particles and the medium. Now, the presence of chromomeres probably indicates that a chromosome at zygotene consists of a linear array of more and less refractive particles, and if this is true the attraction between two sections of chromosome will be greatest if the sections are homologous so that similar particles can fit side by side. A homologous pairing arrived at by chance in one region would therefore be expected to extend along the chromosome by a mechanism like that of a zip fastener. Some indication that these forces are really involved in pairing might be found if regions of chromosome which become precociously condensed in zygotene, and therefore have a higher refractive index than the rest of the chromosome, were to pair precociously. But the invocation of the London-van der Waal's forces to explain pairing can only be regarded as the simplest hypothesis, and is still in need of rigorous checking by specially designed observations and experiments.

For the forces of the external mechanics, operating over greater distances, it is probably necessary to look outside the chromosomes themselves to the spindle. The physical nature of the spindle is now becoming fairly clear. On fixation it coagulates as a "spindle-shaped" body like two cones united by their bases; internally it has a structure of long fibres. Micro-dissection of living spindles[1] shows that these fibres do not exist as such in life; the micro-needle can be moved through them without displacing the chromosomes to which they are apparently attached. On the other hand, the formation of fibres on fixation is undoubtedly an expression of some real factor in the constitution of the spindle. This factor can hardly be a magnetic, electrical or diffusion field, since spindles can be caused to bend into quite sharp U-shaped curves. The field can only be a material one, and one may probably assume that the spindle is a region in which elongated particles lie parallel to one another in an arrangement like that in a nematic liquid crystal.[2] Positive evidence for this suggestion is provided by the behaviour of the spindle in polarized light.[3] Moreover, the spindle is the equilibrium shape for a mass of orientated elongated particles floating in a liquid medium with which it is immiscible, while the radiating

[1] Chambers 1925, Wada 1935.
[2] For the physics of liquid crystals, cf. Bragg 1933.
[3] Schmidt 1936, 1937.

shape of the asters which are often found at the poles of the spindle round the centrosomes is the equilibrium for a similar substance arranged round a singular point or within a spherical wall.

The magnitude of repulsive and attractive forces between bodies within a liquid crystal will be expected to be different in different directions, and in general greater when parallel to the direction in which the particles are orientated. This seems to be the case with the repulsions between the centromeres,[1] which move the chromosomes along the axis of the spindle towards the poles and not centrifugally outwards from the metaphase plate. This mechanism cannot account for the whole of the anaphase separation unless we assume that the repulsion between centromeres and centrosomes becomes ineffective in the later stages. A more probable suggestion was made by Belar,[2] who showed that there is an active elongation and narrowing of the part of the spindle (the "Stemmkörper") which lies between the separating groups of chromosomes in mid-anaphase, and claimed that in the final movement the two groups are passively pushed apart by this elongating material.

2. Spiralization of Chromosomes[3]

Metaphase and anaphase chromosomes usually appear as fairly thick solid rods. Suitable methods of fixation and staining (particularly pretreatment with acids or ammonia vapour) reveal more structures. The chromosome consists of the chromonema thread which in mitosis is coiled in a single tight spiral, while in meiosis this spiral may be coiled again in a second spiral. The mitotic spiral is known as the minor spiral and has up to about thirty turns. The spiral which is superposed on this in meiosis is the major spiral and has about five or six turns; it has so far only been described in plants.

Cytologists have not as yet by any means reached agreement as to the phenomena of coiling. The debate is partly concerned with the relation between the spirals in two paired chromosomes, and this is also connected with the argument mentioned in Chap. 5 concerning the time of splitting of the chromosome thread. Another undecided point is the frequency of changes of direction of the spiral from right-handed to left-handed, and their possible relation to crossing-over. This question also has a direct bearing on the interpretation of coiling in terms of the internal structure of the chromosome. There are two

[1] Cf. Darlington 1936a.   [2] Belar 1927, 1929.
[3] General references: Darlington 1937, Geitler 1934, Kuwada 1927.

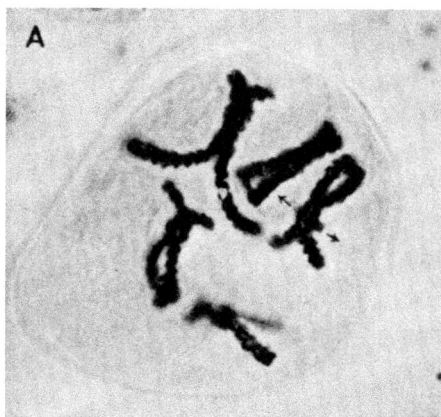

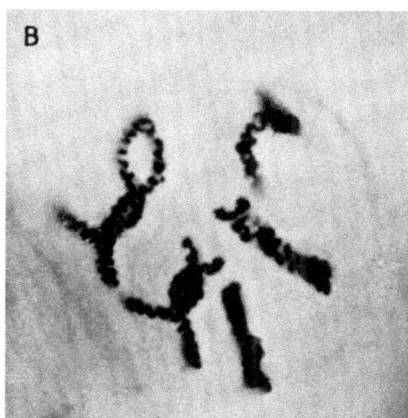

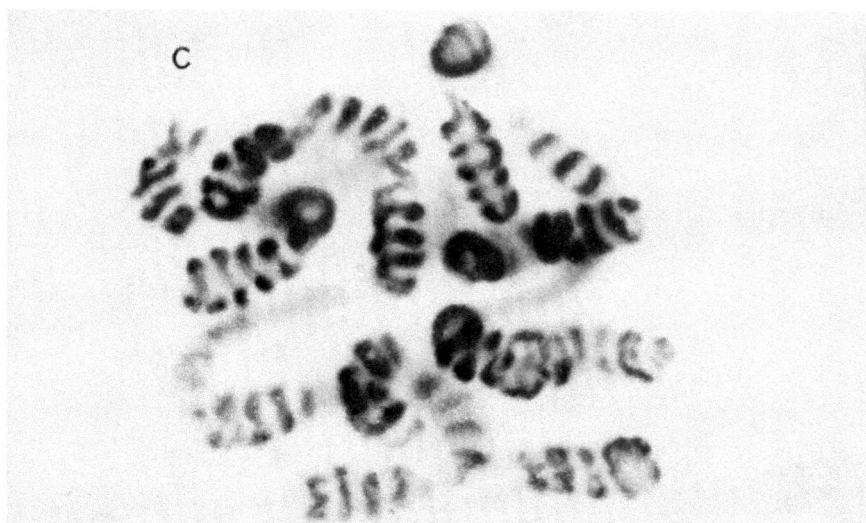

Plate 5.—**The Spiral Structure of Chromosomes.**

A. and B. Meiotic metaphases in pollen mother cells of *Trillium erectum.*
(From Huskins and Smith.)
C. Meiotic metaphase in *Fritillaria sp.* (Courtesy of C. D. Darlington.)

general lines which such an interpretation can take. It may, on the one hand, be suggested that the individual genes have some internal spiral arrangement (perhaps to be compared with the indications of spiral structure in the molecules of proteins such as insulin)[1] and that this determines the formation of the visible spirals; these would then be expected to be consistent for considerable lengths of the chromosome. Or, on the other hand, one may point out that any elongated body consisting of fibres orientated parallel to its length tends to become a spiral if its surface contracts; but in this case there are usually frequent

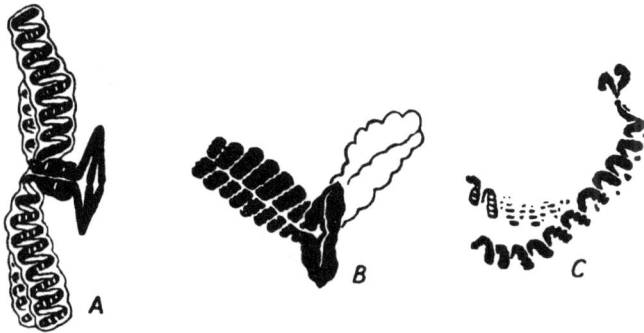

Fig. 147. **The Double Coil at Meiosis,** according to Darlington.—*A* and *B*. Lightly and heavily stained metaphase bivalents with chiasmata localized near the centromeres. *C*. A telophase bivalent, all in *Fritillaria* spp.

(From Darlington.)

reversals of the direction of coiling. In both cases the actual assumption of the coil must be regarded as a response to changed environmental conditions.

There are two particular coiling phenomena which seem to be generally agreed upon, and which seem likely to be of especial importance for the theory of gene structure; though in neither case is their interpretation easy. The first is the phenomenon which Darlington has spoken of as a "hysteresis," though this probably is a rather inappropriate name. At the end of a telophase, the chromosomes only partially unwind their metaphase coils (p. 43), and they appear again in the following prophase still with the remains of these coils. During the prophase, these so-called "relic" coils become finally unwound (and for a time the chromosomes, which have no room to lie stretched perfectly straight within the nuclear membrane, become thrown into

[1] Crowfoot *et al.*, 1938.

"super-spirals" which disappear as the chromosomes shorten). During the unwinding of the relic spirals, the chromosomes are already contracting and assuming the spiral which will be fully developed at the new metaphase. The remarkable fact emerges that these new spirals are not developed directly from the relic coils, but arise quite independently of them. Darlington states that the new coils always have a smaller amplitude than the relic coils, and gives it as a general rule that spiralization always proceeds from numerous coils with a small diameter of each turn to less numerous coils with a larger diameter.

The second phenomenon is that of relational coiling. When two

Fig. 148. Relational Coiling.—Three bivalents of *Chorthippus* in pachytene or early diplotene. Note the relational coiling of the chromatids and of the chromosomes. Each bivalent contains three chiasmata, indicated by dots (cf. Fig. 65).
(From Darlington.)

chromosomes pair in zygotene, they become coiled round one another; and when each chromosome splits in pachytene, it is seen that the sister chromatids are coiled round each other. This can be explained if we assume that each of the paired chromosomes is attempting to uncoil its relic coils, but that it is so closely associated with its partner that it is unable to rotate around its axis. The mechanism can very easily be demonstrated with two thick strands of wool. Twist each strand in the direction in which it has already been twisted during spinning, place the two side by side and let go. Each strand will now attempt to untwist; and this can be taken as analogous to the attempt to uncoil which we have assumed for the chromosomes. But it will be found that the small hairs on the two strands of wool become entangled and prevent the strands slipping over each other; the result will be that the strands become partially untwisted and also coiled round one another in a "relational spiral."

This coiling suggests one of two conclusions about the nature of the chromosome thread. Either the attractive forces in zygotene are much more intense than at later stages when each chromosome coils independently of its partner; or each chromosome has a single "sticky face" along which the adhesion takes place. The latter conclusion is perhaps unlikely; it would certainly raise considerable difficulties about the mechanism of reduplication of the chromosome (cf. p. 399).

It should be pointed out that a similar relational coiling, due to the assumption of a chromosome spiral, might occur between different regions of an unpaired part of a chromosome; the mechanism can be seen if a woollen thread is twisted tightly and the ends then brought nearer together, when a twisted loop known in textile circles as a "snarl" is produced. McClintock[1] has described associations of non-homologous regions in zygotene in maize which have been interpreted[2] as "snarl-pairings"; but until we have a physical picture of completely specific pairing forces, it is impossible to exclude the possibility that non-homologous chromosomes may sometimes pair in the normal way.

3. The Connections of Chromomeres

Chromomeres (by which we shall in this section mean the region of chromosome associated with one gene) are joined together in a linear array and must therefore have two ends which can join up with neighbouring chromomeres. There usually do not appear to be more than two such ends; if there were, we should be able to get branched threads in which one chromomere was joined to three others. Branched chromosomes have indeed been described, both on genetical and cytological evidence. But in some cases in Drosophila, Muller and Offermann[3] have shown by investigation of the salivary glands that the genetical evidence was at fault, and the cytological evidence in other forms seems to require further confirmation. At present it seems justifiable to conclude that the chromomere has only two ends which are capable of stable attachment. This may not be true of the centromere, at which branching of the chromosome seems to be possible (p. 95).

These two ends cannot be regarded as poles. That is to say, the chromomere is not in any sense like a little magnet with different north and south ends. This is clearly shown by the fact that they can join up in reverse order in inversions.

There is still considerable debate whether there are special one-ended chromomeres occupying the ends of the chromosomes. If this

[1] McClintock 1933.   [2] Darlington 1937.   [3] Kossikov and Muller 1935.

were true, one would have to consider the ends of the chromosomes as permanent organs just as are the centromeres. There seems, in fact, to be no case on record in which we can be certain that an end of a chromosome has been knocked off so that a new chromomere forms the tip; in all deficiencies which can be adequately tested, it is either proved, or at least still possible, that the deficient section does not reach quite to the end of the chromosome, and that the hypothetical end chromomere is preserved. However, it would in most cases be very difficult to prove rigidly that the end chromomere was missing, and the common occurrence of interstitial deficiencies may only indicate that they usually arise by the breakage and rejoining of loops (the second mechanism in Fig. 43) and not that the simple breakage (first mechanism) is totally impossible.

B. DEVELOPMENTAL CONSIDERATIONS

The "Mendelian factor" was at first a purely abstract idea, defined by the Mendelian laws which regulated its behaviour. The first suggestion as to a possible material basis for it was the Presence and Absence hypothesis, discussed and to some extent defended by Bateson. This was based on the observation that most mutant characters are inherited as recessives and can be considered as in some way defective when compared with the normal form. The hypothesis was therefore suggested that the dominant gene was a material entity responsible for the development of a character, while the recessive was a simple absence of this entity. The hypothesis soon had to be modified to account for multiple allelomorphism; it had to be admitted that some recessives involved only a partial absence (i.e. a quantitative diminution) of the dominant. A much more serious difficulty arose when it was discovered that not only may the dominant gene mutate to the recessive, but the recessive may mutate back to the dominant. It is fairly easy to conceive how a gene or part of one may become lost, so that a dominant becomes converted into a recessive, but much more difficult to see how a gene can be produced *ex nihilo* to give a "back-mutation." The final criticism of the theory, however, is that it is inadequate to deal with the facts of X-ray induced mutation; it limits speculation too narrowly to the consideration only of quantitative changes, whereas it is necessary to consider qualititative changes as well. The theory was essentially based on a direct translation of the developmental phenomena of hypo-hyper-morphism into terms of

gene-structure; we shall meet other cases in which a similar unsophisticated approach leads to an untenable position.

The discussion of genic action in Part Two emphasized the fact
that the system of gene-controlled processes concerned with the
development of each particular character often has two or more fairly
sharply distinct alternative modes of change. Similarly, the investigations on organizer phenomena have shown that embryonic tissue passes
through phases of competence when two or more alternative modes of
reaction are open to it. We cannot, however, conclude from these facts
that the individual genes have two or more sharply contrasted potentialities. The alternative reactions which we find in development may
well be functions of systems of genes, each of which continues, whichever alternative is actually followed by the whole system, to perform
the same fundamental reaction, though perhaps at different rates or
with different reactants in the two cases. There is at present no adequate
experimental evidence to enable us to decide whether this is true, or
whether we shall be forced to admit that the gene itself may be capable
of more than one primary reaction. The solution of this problem would
be the most important step which could be taken towards an understanding of the nature of genes as determinants of development.

It must not be forgotten that in the inert regions there may be genes
which have no developmental effects and the same may be true of
amorphs.

C. GENERAL GENETICAL CONSIDERATIONS

1. *Step Allelomorphism*

A group of Russian workers[1] have described a remarkable series of
multiple allelomorphs of the locus "scute" in *D. melanogaster*, from
whose developmental effects they derived a theory of the nature of the
gene. Each of the twenty-five or more allelomorphs which are known
removes some of the bristles normally appearing on the thorax of the
fly; there is considerable variation in the expression of each gene,
whose effect on each bristle can only be described statistically as a
reduction of so much per cent in the frequency of its occurrence.

The remarkable fact emerged, or appeared to emerge, that it was
possible to arrange the bristles in an empirically determined linear
order such that all the bristles affected by any one allelomorph formed
a set of neighbours. For instance, we might find that one allelomorph

[1] Cf. Dubinin 1929, 1932*a*, *b*.

reduced the frequency of the bristles *a*, *c*, and *f*; another affected *a*, *b*, *h* and *i*, while a third affected *e*, *f* and *g*. No regularity is apparent in this until we hit on the idea of writing the bristles in the order *e g f c a h i b d*, when we find that each allelomorph affects a compact group (i.e. *fca*, *ahib* and *egf*). Allelomorphs which behave in this way have been called step-allelomorphs.

From this it was argued that the empirical linear order of bristles corresponded to a real linear arrangement of sub-genes within the scute locus, so that each of the allelomorphs was produced by a change

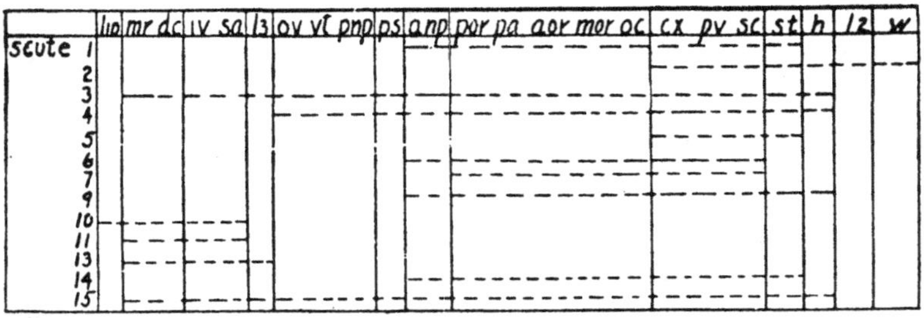

Fig. 149. **Step Allelomorphs of the Scute Locus** in *Drosophila melanogaster.*— The upper row of letters symbolize the names of the bristles on the back of the fly. When they are arranged in this particular order, it is found that each allelomorph, whose numbers are given on the left, affects a set of bristles which lie next to one another. The order in which the bristles have to be arranged has little obvious relation to their pattern on the body of the fly. l_{10}, l_3, and l_2 are lethals associated with some of the allelomorphs.

(From Child, after Dubinin and Friesen.)

in a contiguous set of sub-genes, and affected only the bristles controlled by those sub-genes. In confirmation of this, it was found that in compounds of two scute allelomorphs, only those bristles were affected which were common to both; which was interpreted to mean that only those mutated sub-genes were effective which were included in both allelomorphs and therefore homozygous.

This theory is obviously in a sense an extension of the theory of the linear order of the genes. But it is arrived at in a totally different way, by a direct reflecting back into the zygote of the order and system which can be found in the final product of development. It has all the simple logicality which is characteristic of such preformationist hypotheses; but it depends essentially on the statement that each scute allelomorph affects only a certain definite group of bristles, and it is just this most fundamental point which seems to be most doubtful.

Sturtevant and Schultz[1] showed that the effects of scutes are increased when they are combined with the IIIrd chromosome gene Hairless, which acts as a sensitizer by lowering the threshold for the bristle-suppressing action of scute. In such circumstances, a scute allelomorph is found to affect many more bristles than it normally does, and these may be scattered more or less at random over the sub-gene map. Similarly, Child[2] showed that the set of bristles affected by a scute allelomorph depends on the temperature during larval life. In both

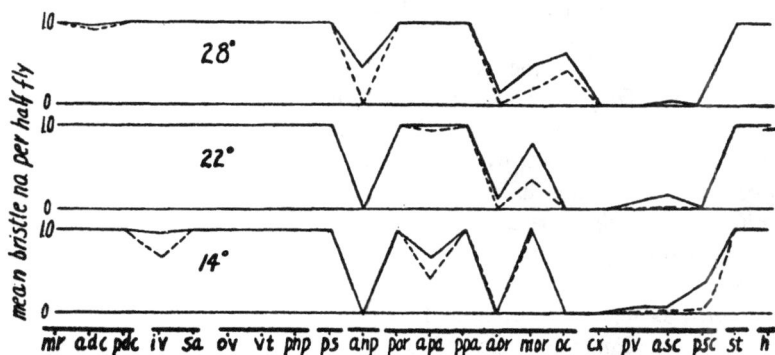

Fig. 150. **Effects of Scute-1 at Different Temperatures.**—In the bottom row, the bristles are arranged in the same order as in Fig. 149. The three curves above show the effect of scute-1 in removing bristles in flies reared at 14°, 22°, and 28° C. The effect on males, where it differs from that on females, is indicated by the dotted lines. Note that at the high and low temperatures, scute-1. may affect bristles which lie outside its normal region according to Fig. 149.

(From Child.)

these cases, the fundamental postulate that a scute allelomorph affects only a particular set of bristles is found to break down.

The remarkable data about the scutes still await a full solution. It is now known that scute-like effects can be the result of transpositions of genes in this region of the chromosome (position effects). It seems likely, however, that the explanation of the different patterns of effect of the different allelomorphs will have to be sought in a developmental theory rather than in the geometrical structure of the gene or the chromosome. Purely formal attempts to provide such developmental theories have been made by Goldschmidt[3] and Sturtevant and Schultz in terms of the diffusion of a "bristle forming substance." There is some difficulty, however, in formulating a theory which fits the facts, and still no experimental evidence that such a substance exists.

[1] Sturtevant and Schultz 1931.   [2] Child 1935.   [3] Goldschmidt 1931.

2. Mutable Genes

Some genes are known with abnormally high mutation rates, which may even reach 1 per cent of cells. Typically the mutations occur in somatic tissues as well as in germinal cells, so that mosaics or variegations are produced; the phenomenon is particularly well known in plants, where flecked or speckled colourations of leaves and flowers are sometimes due to this cause. One of the most fully studied cases is that described by Demerec[1] in *Drosophila virilis* for the locus miniature wing. The mutation relations of the different allelomorphs are very complicated; some are quite stable, while others mutate with considerable frequency, each individual mutation-step having a characteristic frequency, which may be different in different tissues, but which remains constant over many generations under the same conditions. The allelomorphs are classified in five main groups, known as miniature-1, miniature-2, etc., and each group may contain one or more members, e.g. mt-3α, mt-3β, etc. Members of different groups hardly ever, perhaps never, mutate to one another; they were originally obtained as separate mutations from wild type. Members of the same group have almost the same phenotypic effect, but differ in stability. Most of their mutations are back to the wild type allelomorph, but they also mutate, somewhat more rarely but still comparatively often, to each other. Thus the miniature-3 series has three allelomorphs, mt-3α, mt-3β and mt-3γ, which can be derived by mutation one from the other; mt-3α is unstable and mutates back to wild both in somatic and germinal tissues, while mt-3β is stable in somatic tissue but mutates germinally while mt-3γ mutates in somatic but not germinal tissue. Other genes are known which increase the mutation rates of these genes both in somatic and germinal tissues.

The explanation of these high mutation rates is obscure. In many cases we may really be dealing with ordinary genes whose stability is low. But in some cases the mutability is probably produced by special conditions. Eyster[2] studied the variegation produced by somatic mutation of colour genes affecting the pericarp in maize seeds. There are a series of multiple allelomorphs producing various shades of colour from red (top dominant) to white (bottom recessive). An intermediate gene, producing, say, an orange colour, is found to mutate somatically in both directions simultaneously, giving patches of red and white tissue. Eyster suggested that this might be due to the sorting out of

[1] Demerec 1933, 1935.   [2] Eyster 1925, 1928.

sub-units (genomeres) of which the gene was supposed to consist. The genomeres were assumed to be of two kinds, producing red pigment or no pigment; an intermediate gene contained a mixture of the two sorts, and the somatic mutation was supposed to result from a reassortment of the genomeres during the division of the gene in mitosis. Demerec pointed out that the hypothesis could not easily be reconciled with the constant mutation rates found by him in mutable genes which he studied; and there is evidence from X-ray work on mutation against the idea that the gene contains equivalent sub-units.

The segregation of genomeres would give rise to twin patches of coloured and non-coloured tissue. Similar patches have been found by Stern[1] in Drosophilas heterozygous for genes affecting the body surface, and he has proposed to explain them by a process of somatic crossing-over. When mitosis occurs in a heterozygote Aa, there are four threads $AAaa$ to be distributed to the two daughter cells, and the normal separation of A from A and a from a depends on the sister threads remaining attached to the same centromere until they get on to the metaphase plate and the centromere divides. If a somatic crossing-over occurs, and the two A's become attached to different centromeres, there is a chance that they will both pass into the same daughter cell; the division will then produce two homozygous cells AA and aa, from which twin patches of tissue with contrasted characters may arise. Somatic crossing-over seems to be a genuine phenomenon, at least in Diptera; it is not known whether it depends in any way on the close somatic pairing of homologous chromosomes found in this group.

In other somatic mosaics, still other mechanisms may be at work. Schultz[2] has pointed out that high mutation rates are often associated with translocations of part of the inert regions into the active sections of the chromosome, and has suggested an explanation for the mutation process based on the peculiar mechanical behaviour to be expected in such circumstances (but cf. p. 400).

3. *Position Effects*[3]

It was for a long time thought that the arrangement of the genes in the chromosome was completely indifferent and had no influence on their mode of expression. The first evidence against this was produced by Sturtevant,[4] who showed that two Bar genes adjoining one another

---

[1] Stern 1936.   [2] Schultz 1936.
[3] *General references:* Dobzhansky 1936a, Offermann 1935.
[4] Sturtevant 1925.

in the same chromosome have a greater effect in reducing the number of facets in Drosophila eyes than they have when they lie in the normal position of one in each homologue. The phenomenon was a very peculiar one; in a homozygous Bar stock, normal individuals appeared at the same time as individuals with the extremely reduced eyes which were attributed to the fact that both Bar genes had entered the same chromosome. Sturtevant showed that the phenomenon involved an

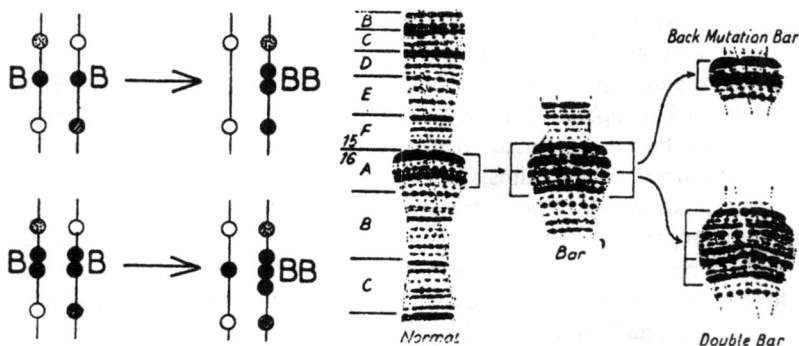

Fig. 151. **Mutations of the Bar Locus.**—At the left are diagrams of the mutation from homozygous Bar to heterozygous double-Bar (the latter reduces the size of the eye still more than the former). Above is the old interpretation, when Bar was thought of as a definite gene; the mutation involves a crossing-over and was supposed to be due to this crossing-over being unequal. Below is the modern scheme, according to which Bar is itself a duplication; the unequal crossing-over causes three of the duplicated segments to come into the double-Bar chromosome. At the right are the salivary gland chromosomes.

(From Bridges.)

unequal crossing-over, in which one chromosome broke to the right of Bar while the other broke to the left, so that one of the cross-over chromatids would apparently have no Bar gene while the other had two. There was considerable discussion as to the possible bearings of this process on the origin of new genes during evolution, but recent work has somewhat reduced the importance of the case in this connection. It turns out that the Bar "gene" is actually a small duplication, the duplicated fragment lying immediately beside its homologue.[1] The unequal crossing-over is therefore not difficult to understand, and the chromatid with "two Bar genes" really has three of the duplicated segment, that with "no Bar gene" is perfectly normal with the segment represented only once.

[1] Bridges 1936, Muller, Prokofieva, and Kossikov 1936.

This discovery does not alter the fact that a fly homozygous for Bar, with four of the segments arranged two in each chromosome, has a larger eye than "a heterozygous double Bar" fly with four segments arranged three in one chromosome and one in the other. It seems that the order of the chromosome sections may have a developmental effect, and Muller[1] suggested that this may be the reason for the commonly observed fact that phenotypic effects are associated with chromosome rearrangements in which no gross losses or gains of chromatin can be shown to be involved. This explanation turned out to be satisfactory, and the phenomenon, known as the position effect, is now generally accepted. A classical case is that of *cubitus interruptus*, a IVth chromosome recessive studied by Dubinin and Sideroff.[2] The position effects consist in alterations of the degree of dominance of the normal allelomorph in translocations of the IVth to the X, Y and IInd chromosomes. Muller[3] has also described some very striking examples in which different breakages near the locus of the gene scute altered the pattern of bristles which were affected by the gene. Moreover, the same breakage turned up on two or more different occasions, and was always found to have the same effect.

We must conclude that the expression of a gene may be affected by the nature of the genes in its neighbourhood; it is not known exactly how far along the chromosome the influence extends, but its range is certainly short. The most usual interpretation of this phenomenon is that the connections between genes are not purely physical, but involve chemical changes in the united genes. In a sense, then, the whole chromosome should be considered as a single chemical unit. We cannot, however, exclude the alternative and perhaps more commonsense view which attributes the position effect not to a direct interaction between the genes, but to reactions between the primary products of gene action. If, for instance, two genes both act on the same substrate during development, their rates of reaction may well not be the same when they are lying side by side in the same chromosome as when they are widely separated in different regions of the chromatin.

Muller[4] has drawn attention to the fact that inversions and translocations may occur which are too small to be detected by cytological or crossing-over investigations. Their position effects would then be falsely interpreted as gene effects. Several cases are already known

---

[1] Muller 1932b. [2] Dubinin and Sideroff 1934.
[3] Muller and Prokofieva 1935, cf. Grüneberg 1936.
[4] Cf. Muller and Prokofieva 1935, Muller, Prokofieva, and Raffel 1935.

which come near to falling into this category; one may mention Bar, which was only recently discovered to be a duplication by very careful examination of salivary gland chromosomes. Very minute inversions are likely to be even more difficult to identify. If the affected segment were below the limit of resolving power of the microscope, there is at

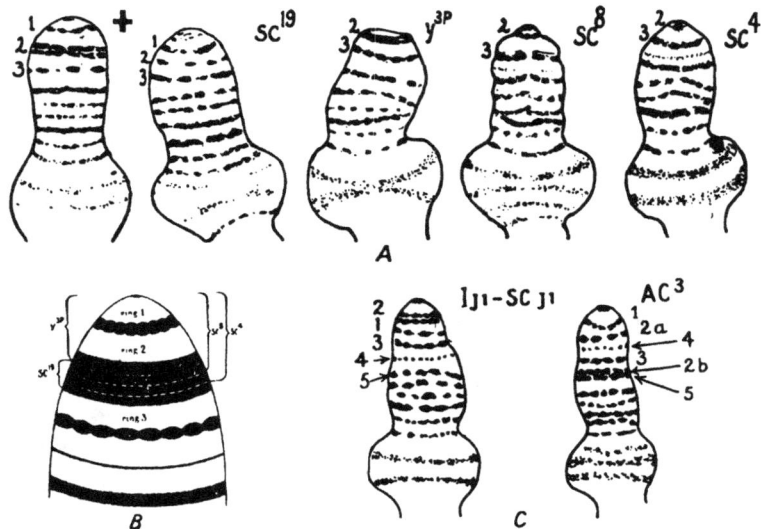

Fig. 152. **Small Rearrangements** in the left end of the X chromosome, as seen in salivary gland chromosomes. The "factors" scute 19, yellow 3P, scute 8 and scute 4 are cases in which minute sections of rings 2 and 4 have been deleted or translocated to other regions. The evidence indicates that ring 2 must have a compound structure, as in the diagram, and it is actually possible, by ultra-violet photography, to demonstrate this. Lethal J1-scute J1 is a minute inversion of rings 1 and 2, while achaete 3 is an inversion stretching from within ring 2 to just above ring 5.

(From Muller and Prokofieva, and Muller, Prokofieva, and Raffel.)

present no known way in which a position effect could be distinguished from a gene mutation. Possibly many of the so-called gene mutations will turn out to be minute rearrangements.

Some authors have concluded from this that all mutations are really rearrangements of the chromatin, and that we must give up the idea of the gene and substitute for it the concept of the chromosome as a whole.[1] This pessimism appears extremely premature. The position effect is a comparatively rare one, known only in Drosophila, where it only

[1] Cf. Goldschmidt 1938.

modifies the general picture of independent and particulate hereditary units. Moreover, it is only conceivable if different regions of the chromosome have different properties; it is meaningless to talk about the rearrangement of a linear series of identical units. These differences at different points in the chromosome must have arisen in the past, and it is therefore unreasonable to deny that they may arise to-day; but the process by which they arise is exactly what we mean by gene mutation.

4. The Size of the Gene

Various attempts have been made to estimate the size of the gene. It is, of course, by no means clear that a gene has a definite and constant size; if the gene is a compound structure, it might grow gradually between two divisions. But the estimates of gene size are all rather rough at present, and it is very unlikely that the size of the gene is sufficiently variable to affect the order of magnitude.

All estimates of the dimensions of the gene start from considerations of the observed size of chromosomes or parts of chromosomes. Chromosomes certainly change in apparent dimensions during the processes of growth and division. Many of these variations can be explained as results of the changing degree of spiralization of the chromonema; and the relation between the gene and the chromosome, and thus the estimates of the size of the former, always depends on the structure which we find or assume for the chromosome thread. The most securely founded of all statements about the condition of the chromonema is probably that it is completely straight and uncoiled in salivary gland chromosomes. An estimate of the length of the chromonema associated with a gene can be made on this assumption. Muller and Prokofieva[1] studied seven chromosome breaks occurring in a short region at the left end of the X chromosome in D. melanogaster. They found that the breaks apparently occurred in only four different places; and they concluded that these four places were the connections between five successive genes in the linear sequence. Further, Muller[2] obtained a case of a minute deficiency which was viable in a homozygous condition and showed a phenotype indicating the absence of only two genes, yellow and achaete. By cytological examination of the salivary gland chromosomes in these cases they deduced the interval between successive genes to be about 125 mμ; the yellow-achaete deficiency was just visible as the loss of a part of a dark band. But it should be noted that the chromosomes are extensible, in the usual cytological preparations, by at least

[1] Muller and Prokofieva 1935.　　　　　　　[2] Muller 1935a.

twice, so that the measurements are subject to at least this error. More-over, as the chromosome is stretched, apparently single bands some-times become resolved into a group of distinct bands.

The yellow-achaete deficiency shows that several genes may be present in each dark band of a salivary chromosome. It is not known whether genes are also present in the non-staining regions. There may also be several genes in each of the chromomeres of pachytene chromo-somes, though Belling[1] claimed that there was only a single one, visible as a minute particle embedded in the larger chromomere. However, the multiple nature of the pachytene chromomere is strongly suggested by the fact that chromosomes at pachytene seem to be eight or ten times shorter than in salivary glands, and in view of this the relation between the size of the gene and of the pachytene chromomere is very obscure. The actual diameter for the latter is from 200 to 600 mμ.

Estimates of the widths of the genes cannot be obtained directly from measurements of the widths of the chromosomes. In salivary glands, the individual chromonemata cannot be distinguished, being pre-sumably below the resolving power of the microscope. By ultra-violet light, at least 64 parallel threads can be seen in Drosophila, and up to 400 in the thicker chromosomes of Chironomus;[2] there may be even more. In other chromosomes, the apparent thickness is dependent partly on the spiralization and perhaps also on the amount of nucleic acid condensed on the thread. Muller[3] has made an estimate of the maximum thickness of the chromonema and therefore presumably of the gene by assuming that when the chromosomes are most contracted at meiotic metaphase they are entirely filled with the coiled-up thread. The active part of the X in D. *melanogaster* has a volume of about $1/12$ cu. μ. The length of the thread, when uncoiled, is found to be about 200 μ in the salivary glands. If the thread has a square cross section with a side x, we can determine x, on the above assumptions, from the relation

$$200 . x^2 = \frac{1}{12}, \text{ whence } x = 20 \text{ m}\mu.$$

This again is a maximum estimate, since there is no proof that the chromonema does fill the whole of the metaphase chromosome, part of which may be occupied by accessory material. Cases are known in which the dimensions of metaphase chromosomes are under genetic control;[4] for instance, in Matthiola[5] there is a race with abnormally long chromosomes. In this case there is no very great variation in volume,

---

[1] Belling 1926, 1928, 1931a.   [2] Bauer 1935.   [3] Muller 1935c.
[4] *Rev*, Darlington 1932b, 1938.   [5] Lesley and Frost 1927.

and the difference is probably one of the degree or kind of spiralization, but in some species hybrids, chromosomes may appear up to ten times larger than they normally do, and this increase is almost certainly due to the accretion of extra chromatic material on to their surface. Moreover, it is probable that the inert regions of the chromosome are not tighly coiled at metaphase, yet they appear, with most stains, to be of the same thickness as the active parts; this thickness may be due to the condensation of material on to the surface. There are then considerable grounds for supposing that a metaphase chromosome contains other material in addition to the chromonema, and that the estimate of the thickness of the chromonema given above is considerably too large.

Various other estimates have been made of the sizes of genes, but they rest on even less secure assumptions.[1] One attempt was based on an estimate of number of genes in an organism, which is an interesting question for its own sake. The total length of the cross-over maps in *D. melanogaster* is about 285 units. The minimum cross-over distance measured in breeding experiments is about 0·2 units, whence one can deduce the total number of genes as about 1,425.[2] Dividing this into the total length of the active chromonemata, one can arrive at an estimate of the length associated with a single gene. But the estimate of gene-number is almost certainly too small. There are about 5,000 to 6,000 separate bands in the salivary chromosome complement of *D. melanogaster*,[3] and we have seen that it is possible that each band contains more than one gene. In other organisms the number of chromomeres cannot be found so accurately. In liliaceous plants there may be up to 2,000 in the haploid complement, but each may contain rather more genes than a salivary band. Perhaps both Drosophila and lilies contain about 10,000 genes.

The two best estimates we have obtained above are maximum estimates for the dimensions of parts of the chromonema; in the first place, a length of about 125 mμ seems to be associated with each gene, and secondly, the maximum thickness of the thread is about 20 mμ. If we can accept these two estimates, arrived at in different ways, as of about equal accuracy, it is perhaps significant that even the units of the chromonema are elongated, fibre-like bodies (cf. p. 393). But we must note that, quite apart from the possibility that these figures may over-estimate the size of the chromonema-unit-particle, there is no proof that the whole of this particle is occupied by the gene. It is possible to assume that the gene is a very much more minute body embedded

[1] Cf. Gowen and Gay 1933.   [2] Muller 1926.   [3] Cf. Bridges 1935, 1938.

within and always associated with the larger particle whose size we have been estimating. We shall discuss this possibility further on page 401. (For a discussion of the "sensitive volume" of the gene, see p. 391).

D. GENE MUTATION[1]

The word mutation is used to cover any sudden change in hereditary constitution, and is therefore applied to the origin of chromosome rearrangements or of polyploidy as well as to gene mutation. We shall here consider mainly gene mutation, which seems likely to provide one of the most fruitful lines of approach to the problem of the nature of the gene.

1. *Spontaneous Mutation*

Gene mutations may occur spontaneously, that is to say, with no apparent cause. The total frequency of spontaneous mutation in a species cannot be stated accurately, since there may be many small mutations which easily escape detection. The total rate of easily detectable spontaneous mutations (including lethals) in the X chromosome of *D. melanogaster* is about 0·1 per cent per life cycle (i.e. about one gamete per thousand bears a mutation). Allowing for the other chromosomes, and doubling the figure to allow for undetected mutations, one may guess the total mutation rate at about 2 or 3 per cent.[2] In other organisms, the facts are even less known. There seems no reason why the mutation rate should not be of the same order of magnitude in terms of life cycles, but it certainly cannot be so in terms of absolute time units, since if it were practically every gamete of a long living form such as an elephant or Sequoia would carry one or more new mutations, which is certainly not the case. In what follows, mutation rates will always be quoted in terms of life cycles unless the contrary is explicitly stated.

The spontaneous rates of individual steps of mutation in Drosophila are usually of the order of 0·0005 per cent. In man, Haldane[3] has calculated the mutation rate from the normal allelomorph to the hæmophilia gene as about 1 in 50,000. In all well-investigated organisms, there are, however, great differences in the mutabilities of different loci and even of different allelomorphs. At one extreme are the so-called mutable

---

[1] *General reference:* Timofeeff-Ressovsky 1937.
[2] Timofeeff-Ressovsky 1937.   [3] Haldane 1935.

genes, which can mutate hundreds of times in a single individual, while at the other end of the scale there are certainly genes of which no mutation has been noticed in all the millions of Drosophilas which have been examined.

Spontaneous mutations may occur in somatic tissues as well as in germinal tissues. It can, of course, only be discovered in the former if the mutant gene is dominant or is sex-linked in the heterogametic sex and therefore has no normal allelomorph to mask it. There is considerable evidence that certain genes mutate at different rates in different tissues. This is certainly so for some of the highly mutable genes studied by Demerec[1] in *Drosophila virilis*. Moreover, Stadler[2] has recorded the mutation rates of some genes affecting the grains in maize.

Fig. 153. Frequency of Spontaneous Mutation of some Genes in Maize.—The genes affect the grains so that very large numbers can be easily tested.

(Data of Stadler.)

Gene	Gametes Tested	Mutations	Mutations per Million
R	554,786	273	492
I	265,391	28	106
Pr	657,102	7	11
Su	1,678,736	4	2·4
Y	1,745,280	4	2·2
Sh	2,469,285	3	1·2
Wx	1,503,744	0	0

These genes mutate in germinal tissues at rates which would give several mutations in every plant, if they occurred in somatic tissue, and since such somatic patches are not found, one must suppose that the somatic mutation rates are considerably lower than the germinal.

The spontaneous mutation rate of a given class of genes (e.g. sex-linked lethals) in a given tissue appears to be constant; each gene is just as likely to mutate at one time as at any other, and the number of mutations is simply proportional to the lapse of time.[3] The rate is also dependent on temperature, being higher at higher temperature. This increase in rate occurs even though the length of the life cycle is shorter at higher temperatures. The mutation rate, in fact, has a higher temperature coefficient than the developmental processes; its value, corrected for the shortening of the life-span, is about 5.[4] The mutable genes are an exception to this, showing no greater mutation rate (in

---

[1] Demerec 1933, 1935.　　　　　　　[2] Quoted Demerec 1933.
[3] For genetic factors which affect mutation-rate, see Demerec 1937.
[4] Muller 1928.

terms of life cycles) at higher temperatures. The significance of this is discussed later (p. 389).

2. The Induction of Mutation

The first successful attempt to induce mutations in controlled experiments was made by Muller in 1927, using X-rays on Drosophila. The effect of such radiations in causing an increased mutation rate has since been amply confirmed on other organisms; Timofeeff-Ressovsky lists forty-three species, ranging from Protozoa to the higher plants and animals, in which mutations have been induced. Among the susceptible species, man (and in fact all species) must presumably be included. The sociological results of the exposure of human gonads to high frequency radiation, which is now a not uncommon occurrence in the course of medical treatment or in industrial processes, has hardly yet been considered seriously enough, except in Germany.[1] The majority of mutations induced are likely to be both deleterious and recessive; the latter character will prevent them showing in the immediate off-spring of rayed individuals, and may obscure the responsibility of radiation for their occurrence, but their injurious effects will not be so easily overlooked. (From figures given later in this section it will be seen that the magnitude of the effect is that the mutation rate is about doubled by a dose of 30 r units.)

The fullest data on induced mutations relate to Drosophila, mainly to D. melcnogaster. There are two standard genetic techniques used in this work, the ClB and attached-X methods. In the former, males are rayed and crossed to females containing one X chromosome which carries a cross-over suppressor C, a lethal recessive l and the dominant eye-gene Bar. Her Bar daughters must contain the ClB chromosome, and also a rayed X chromosome from their father. They are crossed to any male; half the F2 sons will die (because of the lethal in ClB) while the other half will show any recessive sex-linked gene which has been produced in the rayed X, or will die if a lethal has been produced. This method thus enables one to detect both recessive sex-linked lethals or visibles; the detection of the former is particularly easy, since they give F2 families containing no males. The attached X method only reveals sex-linked genes with visible effects. Rayed males are crossed with females in which both X chromosomes are attached to one another, and which therefore give 100 per cent non-disjunction; the sons of the cross have received their single X from their father, and will show any

[1] Stubbe 1934, Pickham 1936.

genes produced in it. Similar methods have been elaborated for detecting mutations in the other chromosomes, but the C*lB* and attached-*X* methods are the most commonly used.

Since Muller's success in inducing mutations by X-rays, many attempts have been made to achieve the same result by other methods. Other types of high frequency radiation (gamma rays, α and β rays, etc.) have been shown to have the same effect, but short wireless waves (3 to 6 metres) and visible light are ineffective. The upper limit of effective radiation is the ultra-violet region of the spectrum.

In his first paper on the subject, Muller raised the possibility that spontaneous mutations may be produced by the natural radiation on the earth's surface. Very slight increases of mutation rate have in fact been found in regions with a high natural radiation intensity. But calculations, based on the measurements of the effects of radiation discussed in the next section, have shown that the natural radiation is much too small (perhaps five hundred times)[1] to account for the observed spontaneous mutation rates. It is possible that some peculiar result is produced by cosmic ray "showers," but none has yet been demonstrated.

We have seen that the mutation rate is increased by raising the temperature. This is similar to the effect of temperature on a normal chemical process. It has also been suggested that just sub-lethal temperatures might have an effect of a different order of magnitude. Goldschmidt[2] and Jollos claim to have demonstrated this by submitting larvae to temperatures of about 37° C. for periods of twelve to twenty-four hours. Jollos states that if the procedure is repeated in successive generations, a given locus may mutate successively to a more and more extreme allelomorph. The experiments were unfortunately not performed in a way which allows of quantitative results. Other workers, using the standard C*lB* method, have found only a rather slight increase in mutation rate at sublethal temperatures (which may, however, be rather greater than can be accounted for by the temperature coefficient) and no sign of progressive mutations. If the results of Goldschmidt and Jollos can be confirmed by future research, they may have great theoretical importance, but at present opinion about them must be held in suspense.

[1] Muller and Mott-Smith 1930.
[2] Goldschmidt 1929, Jollos 1934, 1935, cf. Timofeeff-Ressovsky 1937.

3. Mutations and X-rays[1]

The effect of X-rays is a general increase in mutation rate; there is apparently no specificity in the response, and it is impossible to stimulate the production of any particular allelomorph. It is possible that such a specificity may be obtainable by the use of monochromatic ultra-violet radiation, since on theoretical grounds one might suppose that particular genes may be sensitive to particular wave lengths, but as yet the technical difficulty of getting the radiation into the gonads without injuring the more superficial tissues has prevented much work being done on these lines.[2]

The types of mutation produced by X-rays are of the same general nature as those occurring spontaneously; in fact, many, though not all,

Fig. 154. The Proportion of Lethal and Visible Sex Linked Mutations in
Drosophila melanogaster

	Lethal	Visible	Total	% Visible
Control (spontaneous mutation)..	52	7	59	$11 \cdot 8 \pm 4 \cdot 2$
Soft X-rays (6–10 kV)	143	12	155	$7 \cdot 7 \pm 2 \cdot 1$
Hard X-rays (50–160 kV)	647	51	698	$7 \cdot 3 \pm 1 \cdot 0$
Gamma rays	234	19	253	$8 \cdot 1 \pm \cdot 6$

(From Timofeeff-Ressovsky.)

the X-ray induced mutations were known previously. It is more difficult to decide whether the different types of mutation occur with the same relative frequencies under X-rays as they do spontaneously. At first sight, one is struck by the high percentage of lethals in X-rayed flies (over 90 per cent of all mutations produced) but when the spontaneous mutation is carefully investigated by the same methods, it is found that there too lethals make up as high, or nearly as high, a proportion of the total. Timofeeff-Ressovsky[3] has recently shown that there is an even more frequent class of sublethals whose only effect is a slight lowering of viability. These are normally missed by ordinary methods of investigation (they are only detectable by their effect on the sex ratio in F2's from ClB cultures), but there is no reason to doubt that they occur in an equally large proportion in untreated flies. They have only recently been detected, and are not included in the figures for mutation rates given in other parts of this section. Chromosome aberrations are also plentifully produced by X-rays.

---

[1] *General references:* Muller 1934, Timofeeff-Ressovsky 1934b, 1937.
[2] Altenburg 1934, Reuss 1935. Extensive work on maize by Stadler is not yet published. Cf. Stadler and Sprague 1936. [3] Timofeeff-Ressovsky 1935.

The effect of X-rays on the genes appears to be a more or less direct one. Certainly it is not a secondary consequence of an enduring alteration of the cytoplasm, since unrayed chromosomes introduced into rayed cytoplasm show no increase in mutability. Further, the effect seems to be on the gene itself and not on the process by which the gene reproduces; if the effect is direct on to the gene, a rayed gamete may contain either one mutated gene, or one mutated and one normal, according as the raying takes place before or after the gene-reproduction; while if the effect is on the process of reproduction, there must always be one normal gene accompanying the mutated one. Actually, when mature sperm are rayed, most of those which bear mutated genes do not also bear the normal allelomorphs, though in about one-seventh of the cases, the mutation-bearing sperm is heterozygous and gives rise to a mosaic individual; in these cases one must suppose that the irradiation took place after the reduplication of the gene.

It is still rather doubtful how far the mutation rate under X-rays can be influenced by the biological condition of the cells or by the type of cells in which the gene is located. In Drosophila, somatic and germinal mutations of the white locus occur with the same order of frequency for a given dose of X-rays, and the difference, such as it is, can probably be explained by the greater certainty with which somatic mutations can be detected. The X-ray induced mutation rate is usually higher in mature than in young sperm, but the difference is probably due to the elimination in the latter case of some lethal genes which have a deleterious effect on the development of the sperm. On the other hand, Stadler[1] claims that in barley the general mutability under X-rays is higher in germinating than in resting seeds.

Even with X-rays, mutation remains a rather rare occurrence and quantitative data are laborious to collect. But in fairly numerous collections of genes the rate is high enough to be accurately measured; for the whole of the X chromosome, it may be raised as high as 10 per cent. The most important results of the X-ray investigations are derived from careful quantitative studies. We will first summarize the three main conclusions which have been arrived at, and then discuss the deductions to which they lead.[2]

(1) The rate of mutation produced is linearly proportional to the amount of ionization caused in the tissue. This linear proportionality, between mutation rate and the dosage measured in ionization or

[1] Stadler 1928.
[2] Discussion follows Timofeeff-Ressovsky, Zimmer, and Delbrück 1935.

"*r*-units," has been found by many authors, both for germinal and somatic mutations. It applies to individual mutations, or to small groups of mutations as well as to the total rate.

(2) The mutation rate produced by a given ionization dose is independent of the wave-length of the rays used. This result holds for all wave-lengths from gamma rays to very soft X-rays (10 KV) and also

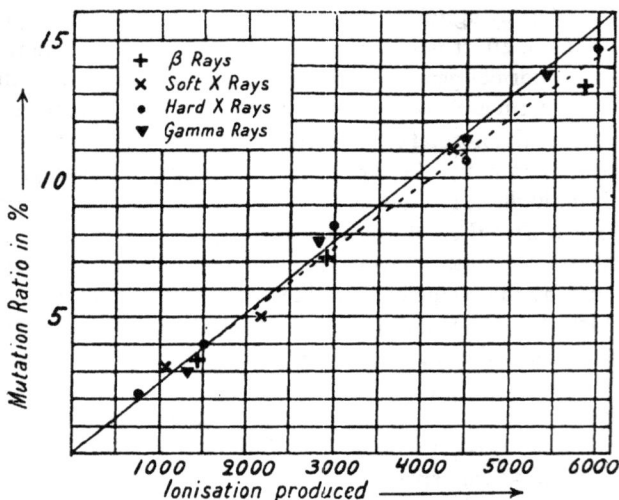

Fig. 155. The Relation between Ionization and Mutation Rate.—The mutation rate is directly proportional to the ionization produced, and is not affected by the nature of the ionization-producing rays (whether β particles, soft X-rays (with long wave length), hard X-rays, or gamma-rays).
(From Zimmer, Griffith, and Timofeeff-Ressovsky.)

for β rays. It probably does not hold for ultra-violet, and it may not hold for rays of α particles or neutrons, but data for the latter, whose importance we shall discuss later (p. 388), are not yet available.

(3) The mutation rate produced by a given dose is independent of the time over which the dose is given. A short exposure to intense radiation is no more effective than a longer exposure to weaker radiation; the only important variable is the total dosage received.

These three results clearly amount to the single datum that the mutation rate is linearly proportional to the ionization produced and is independent of any other characteristic of the radiation. From this two conclusions can be drawn: (*a*) The fact that the relation is linear shows that a mutation is produced by a single event, and not by the concurrence of two or more events. The argument is as follows. Suppose we have a large number of particles and produce among them a number of processes which can cause the particles to change. We can plot curves relating the frequency of the processes and the frequency of changes. If only one process is necessary to cause a change in a particle, the curve will at first be linear, falling off later when the

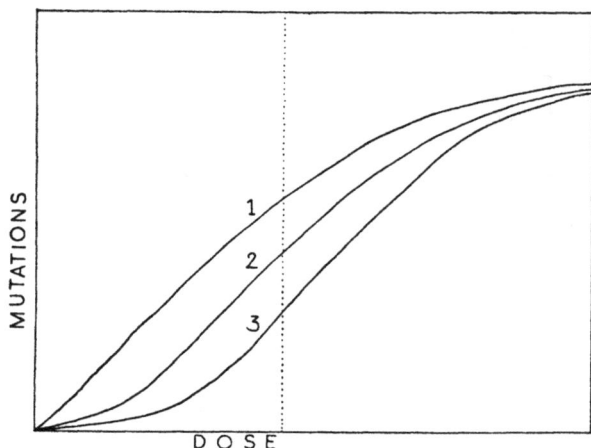

Fig. 156. **The Unitary Character of the Mutation Process.**—The curves show (diagrammatically) the relation to be expected between dosage and mutation rate on the suppositions that 1, 2, or 3 ionizations are needed to cause a gene to mutate. The dotted line indicates the maximum dosage which can be employed in biological experiments.

processes become so frequent that there is a considerable chance that the same particle may be affected twice. On the other hand, if two or more processes are required to cause a change the curve will at first rise very slowly, since at low frequencies of the processes it is unlikely that any one particle will be affected by the required number; at rather higher frequencies the curve will rise more steeply, falling off finally when many particles are affected by more than the necessary number of processes. In X-ray mutation work, only the first part of the curve is available, since at higher doses the flies are rendered sterile or killed. Thus the statement that the curve relating mutation rate to dosage is a straight line means that it corresponds to the early part of the curve which is appropriate when a change is produced by a single process. A mutation is therefore produced when a single thing happens to a gene.

(b) The fact that the mutation rate is linearly proportional to the ionization, and is independent of other variables such as wave-lengths, shows that the single event which causes mutation is an ionization. As examples of the possibilities which can be excluded by this fact we may mention the absorption of a whole quantum of energy from the in-

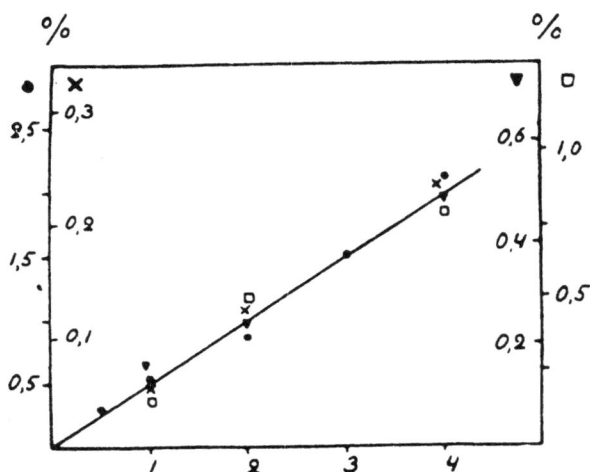

Fig. 157. **Ionization and Rates of Visible Mutations.**

• Somatic $w^e \to w$ mutations.
× y, w^e, w, v, m, g or f mutations.
▼ Visible mutations (attached X method).
□ Visible mutations (ClB method).

(From Timofeeff-Ressovsky.)

coming radiation, or the passage of a single electron or ion through a certain volume.

In drawing these two conclusions, we have assumed that the events (ionizations) which cause the mutations are scattered at random through the tissues. For the types of X-rays usually worked with, this assumption is justified. But quite different effects on mutation should be produced by rays of a particles or neutrons, when the ionizations instead of being thinly spread over a long track, occur very much nearer together in a small volume. In such case the ionizations may be so close to one another that it is likely that two or more would occur in the neighbourhood of a single gene, and we should expect that fewer mutations would be produced for a given number of ions, owing to the wastage of hitting a single gene twice. Work on these lines is proceeding.[1]

We derive, from the work outlined above, a picture of the mutation process as one involving a single event of ionization. Now in a complicated molecule, an ionization is not to be regarded simply as the loss of an electron but rather as a transference of an electron to a new position. Such transferences can also occur spontaneously, and the mutation process deduced from X-ray work can therefore also be used to account for spontaneous mutation. The agent which brings about the spontaneous transference of an electron is heat. The atoms and electrons in a molecule can be represented as vibrating about their equilibrium positions, the energy of the vibration being the heat energy. The magnitude of the vibration is only statistically constant, and is subject to random fluctuations which in extreme cases may carry an electron over its limits of stability, when it will, as it were, click over into a new position in the molecule and begin vibrating around a new equilibrium. The frequency with which this occurs naturally increases as the temperature, and therefore the average amplitude of the vibrations, is raised. It is a general rule that the rarer a process is, the greater will be its dependence on temperature, other things being equal; and it is thus understandable that the natural mutation rate has a temperature coefficient higher than that of the developmental processes, and increases with the rise of temperature even when measured in terms of life cycles. On the other hand, if the mutation process is more frequent for any reason, the effect of temperature should be less. And in fact we find no effect of temperature on the mutation rate (measured in life cycles) of mutable genes or on X-ray induced mutation; that is to say, these processes have temperature coefficients of the same order as those of the processes of development.

The scheme of mutation therefore applies both to the spontaneous and the induced process. A mutation occurs by a single alteration in the equilibrium position of an atom or electron in a molecule; and this change may occur spontaneously as a result of heat vibrations or under radiation as an ionization.

It has been mentioned that chromosome aberrations are also produced by X-rays. The relation between the dosage of radiation and the number of aberrations is still not entirely clear.[2] On *a priori* grounds, two mechanisms have been suggested for the formation of translocations and inversions; on the one hand, they may arise when

[1] Cf. Nagai and Locker 1938.   [2] *Rev.* Muller 1938.

the chromosomes are lying in contact with one another (in the way indicated in Fig. 43) or on the other hand two chromosomes lying some distance apart may be broken and subsequently unite in a new configuration. If the first alternative is true, it is possible to imagine that a single ionization may be sufficient to produce a rearrangement, but on the second alternative, where the chromosomes lie some distance apart, at least two ionizations would be necessary. As far as it is known, however, the curve relating the frequency of breaks to dosage does not follow the course predicted either for single or double ionizations, but lies somewhere between the two. It is likely, therefore, that both mechanisms of rearrangement may occur.

One class of rearrangement has a somewhat special theoretical importance. These are the very small rearrangements, which, if they are not detected cytologically, may be mistaken for gene mutations, It is highly probable, for instance, that many of the so-called lethals produced by X-rays (and spontaneously) are actually small deficiencies. It was at one time hoped that the dosage-frequency relation would make possible a distinction between these and true gene mutations, but recent evidence[1] shows that the small aberrations are apparently produced by single ionizations and therefore cannot be distinguished in this way. The mechanism by which a single ionization produces an effect of the size of a minute deficiency (i.e. some hundreds of mμ at least) is unknown.

4. Rates of Single Mutations

It is not possible to assess accurately the total mutation rate of a single locus, since we can never be certain of identifying all the allelomorphs; mutations with small effects may be missed altogether, and lethals can only be identified with great labour. It is, however, fairly certain that some loci are more unstable than others.

More exact results can be obtained if we investigate the frequency of particular mutation-steps, that is to say, the frequency with which a certain gene A mutates to a particular allelomorph A'. The first important point which has been discovered is that although the mutation rate from A to A' is usually different from that from A' to A, they may in some cases be of the same order of magnitude.[2] (Mutation from the wild type to a mutant form is often spoken of as forward mutation, that from the mutant to the wild type as back mutation.) The mere fact that a hypomorph mutant can mutate back to the wild type shows that the

[1] Belgovsky and Muller 1938, Muller 1938.   [2] Timofeeff-Ressovsky 1932.

mutant is not a mere absence (p. 368). The fact that the mutation rates may be of the same order of magnitude demonstrates the further fact that these genes at least cannot be aggregates of identical parts only one of which has to change in order to produce the mutation; it is easy to see that if the gene was made up in this way, it would be much easier to produce an ionization in any one of the parts, to give a forward mutation, than to hit exactly the one part which had been altered, as would be necessary to change the mutant back into normal.

The relation between dosage and induced mutation rate is accurately known for a few particular mutation steps. Since the relation is linear, we can give for each such step a constant (the mutation constant)

Fig. 158. **Mutation Constants and Sensitive Volume for some Mutation Steps** in *D. melanogaster.*—The mutation constant is given in mutations per *r* unit. (From Timofeeff-Ressovsky; the sensitive volume of $m \rightarrow +$ is wrongly given in the original as 1·0.)

Mutation Step	Mutation Constant	Sensitive Volume in cu. mμ.
$+ \rightarrow w^e$	$2·6 \times 10^{-8}$	6·5
$w^e \rightarrow +$	$0·8 \times 10^{-8}$	2·0
$w \rightarrow w^e$	$0·3 \times 10^{-8}$	0·75
$+ \rightarrow m$	$2·4 \times 10^{-8}$	6·0
$m \rightarrow +$	$1·0 \times 10^{-8}$	2·5
$+ \rightarrow f$	$6·6 \times 10^{-8}$	16·5
$f \rightarrow +$	$2·4 \times 10^{-8}$	6·0

which expresses the chance that a mutation will be produced by a unit dose of ionization. This can also be expressed in another way: we can calculate a volume such that ions are produced in it at exactly the same rate as the mutation is produced in the gene. This volume is known as the "sensitive volume."

The relation between the sensitive volume and the size of the gene cannot be assumed to be straightforward. If every ion produced in the sensitive volume caused the particular step of mutation we are concerned with, the sensitive volume could perhaps be taken as a measure of that part of the gene which is altered by the mutation. But actually we cannot be certain that every ionization is effective, and even if this could be decided, it is possible that the disturbance due to the ionization might be conducted to some smaller region within the sensitive volume before it produced its definitive effect; and finally, it may be only a small part of the gene which is altered in any one mutation-step. The sensitive volume, in fact, cannot be taken to give any information about the size of the gene; it is simply another method of expressing

N*

the mutation constant. The values actually obtained for it are very small in experiments on mutation in germinal tissue; Timofeeff-Ressovsky gives figures varying from sensitive volumes which contain from 1,650 to as few as 75 atoms. Larger values have been obtained by Haskins[1], who measured the mutation rate from eosin to white in somatic cells, where there is a greater chance of detecting all mutations which occur. Marschak[2] measured not particular mutations rates but the frequency of the less restricted class of visible chromosome aberrations produced by X-rays in Gasteria, and thus calculated the "sensitive width" (i.e. width which must be traversed by an ion) of the chromatid as 5 mμ. If this is taken as the diameter of the sensitive volume of individual genes, this comes out as about a hundred times as large as the volumes for particular mutation-steps found by Timofeeff-Ressovsky. (Fig. 160, p. 399).

E. THE CHROMOSOME AND THE GENE

1. The Chemical Nature of Chromosomes

Chromosomes as such have never been chemically analysed; they are too small for present methods. The nearest material which can be collected in quantities large enough for ordinary chemical investigation is sperm, particularly fish sperm. The head part of the sperm consists almost entirely of nuclear material, and analyses[3] of this show that the two main constituents are thymonucleic acid (about 60 per cent) and simple proteins of the kind known as protamines (35 per cent).[4] Nucleic acid combines very easily with proteins to form complex nucleo-proteins, and it probably occurs in this combined form in the nucleus.

The distribution of nucleic acid in the nucleus can be investigated by means of ultra-violet spectroscopy,[5] since it has a characteristic strong absorption at wave-lengths near 2600 A. During division stages, when the chromosomes are contracted and can be seen as separate bodies, almost all the nucleic acid is attached to the chromosomes; it may also be in the chromosomes in resting stages, but the fully extended chromonemata are not separately distinguishable from the nuclear sap, and the evidence is therefore not clear. The metaphase chromosomes can also be shown to contain protein, since they are attacked by proteolytic enzymes. In the salivary gland nuclei, the chromosomes contain

---

[1] Cf. Haskins and Enzman 1936.
[2] Marschak 1935. [3] Quoted Caspersson 1936.
[4] Caspersson and Hammarsten 1935. [5] Caspersson 1936.

the same two constituents, which perhaps makes it likely that the resting stage chromosomes are built up in the same way, and that the chromosome constitution is constant throughout the division cycle.

The chemical make-up of protein is not yet fully understood.[1] It is known that some of the chemically rather inert proteins such as hair and silk are formed from fibrous elements consisting of chains of "polypeptide links," each link having the constitution $-CO-CHR-NH-$, where R is a group (the side chain) which may be a simple hydrocarbon, an alcohol, or a base such as arginine. In the fibrous proteins we have mentioned, the links are arranged in linear chains, the chains being connected together by means of the side chains. There are, however, other types of proteins in which the molecules seem to be spherical rather than elongated; these are known as the globular proteins, and there are others intermediate between the fibrous and globular types. We do not know to which of these types the chromosome proteins belong, since the protein isolated from sperm (clupein) has not been examined from this point of view. The thread-like appearance of the extended chromosome, and the two-ended nature of the chromomeres (p. 367), suggests that the chromosome consists of protein fibres arranged more or less parallel to its length. But this does not by any means necessitate the assumption that the chromosome protein is itself fibrous, since the orders of magnitude are quite different. The polypeptide links in a fibrous protein chain are about $0 \cdot 35$ mμ long by $0 \cdot 45$ mμ thick by 1 mμ wide in the direction of the side chains. The visible chromonema is a few hundred mμ thick, while Muller's estimate, based on its length in salivary gland chromosomes, gives it a width of about 20 mμ. Fibres as large as this can just as well be formed from globular as from strictly fibrous proteins, since the units (molecules or repeat cells in crystals) of the former are about 6 mμ in diameter, and cases are known, for instance in some of the virus proteins, in which these units unite to form fibres a few tens of mμ thick.

Studies on the extensibility, and particularly the reversible extensibility, of the chromosomes give some, and could probably give much more, information about globular or fibrous nature of the chromosome proteins. In completely fibrous proteins the polypeptide chains lie fairly parallel and are more or less unfolded; they can only be stretched by actual straining of the chemical bonds. It is probable, however,

[1] Cf. Astbury 1933, 1936, 1938, Bergmann and Niemann 1937, Crowfoot et al. 1938, Wrinch 1938, Cold Spring Harbor Symp. 1938.

that in globular proteins, the same or very similar polypeptide chains exist in a folded configuration, so that extension of a fibre constructed of globular protein involves only the unfolding of the chains and can proceed much farther before the fibre is ruptured. Duryee[1] has shown that the lampbrush chromosomes of amphibian oocytes (p. 101) can be reversibly extended to about $3\frac{1}{2}$ times their normal length, at least under favourable conditions (in absence of calcium or other heavy metallic ions). Salivary gland chromosomes easily stretch to at least twice their length, and probably can be stretched farther when special efforts are made to do so by microdissection methods. Thus even in chromosomes in which the chromosome thread or chromonema is apparently uncoiled, the thread itself has considerable elasticity, and may perhaps be constructed of globular proteins in which the polypeptide chains are folded on a molecular scale. Much further study is required, however, before this can be taken as more than a suggestion.

When we turn to consider the other main constituent of chromosomes, the nucleic acid, a series of facts emerge which are extremely suggestive of an essential connection between nucleic acid and proteins, but whose exact significance cannot yet be stated. Nucleic acid itself easily forms fibres, and X-ray studies[2] have shown that these consist of a chain of phosphoric acid residues to the side of which are attached a series of flat, plate-shaped groups each of which contains a purine base attached to a sugar. The first remarkable fact is that the repeat distance along the chain, i.e. the distance between neighbouring phosphoric acid residues, is almost exactly the same as the repeat distance in a polypeptide chain; $0 \cdot 336$ for the nucleic acid, $0 \cdot 334$ mμ for the polypeptide. The difference, which may not be significant, is at least so small that it is easy to imagine that the polypeptide and nucleic acid chains might unite parallel to one another to give protein-nucleate chains. This can in fact actually be observed; Astbury[3] has prepared the nucleate of clupein, the protein isolated from fish sperm, and shown that it is a fibrous material. Further confirmation comes from a study of the double refraction. Protein fibres have a somewhat weak double refraction which is positive in the direction of the fibre, while nucleic acid, in which there are large flat plate-like groups sticking out at right angles to the length of the fibres, has a much stronger double refraction which is negative in the direction of the fibre axis. The clupein-nucleate shows a double refraction negative in the fibre direc-

---

[1] Duryee 1938.
[2] Astbury and Bell 1938. [3] Cf. C. S. H. Symp. 1938.

tion due to the nucleic acid. So do fully uncoiled chromosomes,[1] such as those of salivary glands and zygotene stages; when the chromonema is presumably coiled in a single coil (e.g. mitotic metaphase), so that it runs perpendicularly to the length of the chromosome, the sign of the double refraction changes and becomes positive in the direction of the *chromosome* axis, while in meiotic metaphase, where the minor spiral is coiled again in a major spiral, the double refraction reverses again and becomes once more negative in the direction of the chromosome axis. All these data fit in very well with the idea that the protein and nucleic acid have combined to form composite fibres in which the two constituent fibres lie parallel to one another.

The cytological evidence makes it quite clear that the chromosomes are not homogeneous structures. In the first place, there is a differentiation in salivary gland chromosomes between the darkly staining bands and the non-staining inter-band regions. The property of stainability depends on the content of nucleic acid, and the concentration of this substance in the bands can be demonstrated directly by studies of ultra-violet absorption. One must suppose that the proteins in the band regions have a particular affinity for nucleic acid, and Wrinch[2] has suggested that this may be due to a higher concentration of basic groups, particularly arginine, which is known to be present in remarkably high amounts in clupein. The difference between the bands and inter-bands appears, however, rather larger and more sharply defined than would be expected if it were due to a merely quantitative difference; but this appearance may turn out to be illusory when actual measurements of nucleic acid content become available.

The extensibility of the bands seems to differ sharply from that of the interbands, the former being much the more rigid. There are two factors to be taken into consideration here. Firstly, the nucleic acid fibre itself appears to be inelastic, and the rigidity of the bands may be due simply to their nucleic acid content. Secondly, while it is easy to see how nucleic acid may combine with fully extended polypeptides, it is not so clear how it can fit on to a globular protein; it is possible then that the proteins of the bands, when combined with nucleic acid, are in the extended form, and thus have themselves lost much of their extensibility; but whether this should be regarded as a result or as a contributory cause of their affinity for nucleic acid is as yet quite unknown.

[1] Kuwada and Nakamura 1934*b*, Nakamura 1937, Schmidt 1936, 1937.
[2] Wrinch 1936.

The differential staining behaviour of the heterochromatic regions presumably depends on a chemical composition or physical state different to that of the euchromatin; but in general very little is known about this. In salivary glands of Drosophila, the inert[1] regions show a structure of longitudinal striations and transverse bands which is somewhat similar to that of the euchromatic regions, except that the bands are more feebly staining and the whole structure less clear-cut. Although the inert region of the X chromosome, for instance, is fairly short in the salivary gland chromosomes and has only a small number of bands, it occupies a large proportion of the whole chromosome at metaphase of mitosis. This may perhaps be partly due to a lesser degree of

Fig. 159. **The Structure of the Chromocentre (Inert Region) in Salivary Gland Nuclei.**—The right arm of the third chromosome is missing. The figure is from a male and the Y chromosome can be seen as a small lump pairing with the inert region of the X.

(From Prokofieva-Belgovskaya.)

spiralization in mitosis, although, since the region at that time is definitely shorter than it is in salivary chromosomes, some spiralization must occur. It is probable that the large relative volume of the inert region in mitosis is at least partly due to an abnormally large concentration of nucleic acid on to it at this stage. Muller[2] claims that the greater bulk of the mitotic region is produced under the influence of only two loci in the region, and it is conceivable that the region contains loci specially concerned with the synthesis of nuclei acid.

The physical and chemical basis of this structure is unknown, but it is remarkable to find that the conditions underlying it appear to be transmissible; when inert regions are brought, by translocation, into contact with euchromatic parts of the chromosomes, there is a tendency for the latter to be modified in their appearance in salivary glands, so as to assume more nearly the inert structure.[3] This suggests that the

[1] Bridges 1935, 1938, Prokofieva 1935.
[2] Muller 1938, Muller and Gershenson 1935.
[3] Prokofieva 1935, Schultz and Caspersson 1938.

inert regions differ from the euchromatic regions only in some general condition which overlies the same basic differentiation into band and interband.

The centromeres are probably quite differently constituted from the rest of the chromosome. They seem to be unable to transmit torsional stresses, since the directions of coiling at metaphase are apparently independent of one another in the two arms of a chromosome with a central centromere; and similarly interference in crossing over does not extend across a centromere. It has also been shown that when centromeres divide at metaphase they do not always split along a plane parallel to the length of the chromosome, but may occasionally be divided transversely or at any angle.[1] All these facts tend to suggest that the centromeres, unlike the rest of the chromosome, are not fibrous structures.

2. *The Nature of the Gene*

Before we can discuss the chemical nature of the gene, we must reconsider the definition of the word.[2] The Mendelian factor was originally defined simply as an entity which obeyed Mendel's laws and had an action in determining the characters of the adult organism. This definition could apply to whole chromosomes, or large sections of chromosomes. With the discovery of linkage and crossing-over, the definition of the factor, or, as it was now called, the gene, became narrowed down by adding the property that genes act as units in crossing-over, which occurs between them but not through them. At the present time, this definition has become unsatisfactory because there are cases to which it is inapplicable. We know that genes may be guarded from crossing-over, for instance in the parts of the Y chromosome which have no homologue or in the complexes in a ring-forming heterozygote such as Oenothera. Moreover, we know of inert sections of the chromosomes, which appear to consist of particles much like those in the other regions of the chromosome, sharing with them the properties of multiplication and attraction, but lacking any effect on development; we may wish to stretch our definition to cover inert genes. Finally, the position effect shows that genes which may cross over independently of one another may not be independent in their developmental actions; in this case the two parts of the definition do not tally.

It is clear that the old picture of the chromosome, as a linear array

[1] Upcott 1938.   [2] Cf. Darlington 1938.

of individual indivisible particles, each of which is a gene, is too simple. In attempting to work out a more adequate picture, one can start from the fundamental fact that the chromosome is an elongated structure which, whenever we can analyse it, has differences arranged in a linear order along it; these differences can be detected by linkage studies, chromosome structures, etc. The units, between which differences are noted, may be of different sizes according to the different methods of investigation; there are, in roughly descending order, inert or precociously condensing regions, large chromomeres, ultimate chromomeres or salivary gland chromomeres, and the units of cross-over and X-ray breakage. One might symbolically represent the chromosome thus: abcd′e′f′g′hijkl<u>MNOPQRSTU</u>′V′W′ where there are differences on three scales, between the capitals and lower-case letters; normal, underlined and dashed letters; and finally the letters themselves. The smallest units of this scheme, symbolized by the individual letters, are the units of crossing-over and X-ray breakage, and probably measure, as we have seen, about 100 mμ in length.

If we view the chromosome as it were through the other end of the telescope, attempting to build it up from chemical units, we arrive at a somewhat similar scheme of a linear order of units of different orders of magnitude. The ultimate units now are the links in a polypeptide chain, with a length of only 0·334 mμ. Exactly what the larger units are is more doubtful, but we have a range of possibilities; there are the periodicities along the chains,[1] the repeat units out of which protein crystals are built, the protein molecules such as they exist in solution, and finally virus particles, all of which may be considered as providing suggestions as to the kinds of units which may be involved. These units range in size nearly up to the 100 mμ which we took as an estimate of the smallest units to be considered when we approached the chromosome structure from the other end. It is, then, possible to conceive of the chromosome as a linear array of units, the units themselves forming a hierarchy all the way from heterochromatic and euchromatic regions, some tens of thousands of mμ long, to polypeptide links only a few tenths of a mμ long.

This apparent homogeneity in the type of formal order exemplified by the chromosome on different scales should not tempt one to suppose that other properties may be just as easily conceived of in any of these scales. For instance, it is sometimes suggested that because the nature of one link in a polypeptide chain may chemically affect the properties

[1] Cf. Bergmann and Niemann 1938.

of a neighbouring link, the same type mechanism may explain the phenomenon of position effect. But in the latter case, the influence is between neighbouring genes (i.e. breakage units) and extends over distances about a hundred times as great as in the former case. No direct analogy between mechanisms of two phenomena is possible; and in fact no example of a direct chemical influence extending throughout such a distance appears to be known in protein chemistry.

Certain of the properties of the genes give some hints as to the

Fig. 160. **Table of sizes.**—The sizes are given in $m\mu$ ($= 10A° = 10^{-6}$ mm.). Where only one dimension is given, it is the diameter of a spherical unit.

(Partly after Stanley.)

Vaccinia virus	175
Rous sarcoma virus	100
Tobacco mosaic virus	$430 \times 12\cdot3 \times 12\cdot3$
Bushy stunt virus	28
Haemocyanin molecule	$59 \times 13\cdot2 \times 13\cdot2$
S13 Bacteriophage	10 (? shape)
Repeat unit of virus crystal	$15 \times 15 \times 7$
Haemoglobin molecule	$2\cdot8 \times 0\cdot6 \times 0\cdot6$
Protein fibre (repeat unit)	$0\cdot334 \times 0\cdot45 \times 1\cdot0$
Nucleic acid (repeat along fibre)	$0\cdot336$

Gene (estimated maximum dimensions)	$100 \times 20 \times 20$

Sensitive volumes:	
gene mutations (Timofeeff-Ressovsky)	c. 1
,, (somatic, Haskins and Enzmann)	15
cytological effects (Marschak)	5

possible kind of units which may fill the gap between the $0\cdot334\, m\mu$ polypeptide links and the $100\, m\mu$ genes. The most important is the property of identical reproduction. Between two cell divisions, each gene causes the formation of another gene exactly like it; if the gene mutates into an abnormal form, it is the mutated gene which is reduplicated. The gene, then, must in some way act as a model on which the new gene is formed. This can only occur if chemical forces originating in the radicals in the gene can extend far enough to influence the nature of radicals formed in the equivalent places in the new gene. The thickness which we can postulate for the gene is therefore limited by the distance through which we can imagine such chemical forces extending. Probably the maximum estimate which is chemically reasonable is about $10\, m\mu$, which is the order of magnitude of the thickness of the repeat units out of which protein crystals are built. This is of the same order of magnitude as the estimate given above for the maximum thickness of the chromosome thread (p. 378). It is therefore impossible to reject, from consideration of gene reduplication, the idea that the gene is a single unit. On the other hand, a further difficulty arises in this connection, namely the necessity to find some mechanism which accounts for the fact that only two genes, the old one and the new, are present at the end of each intermitotic period. The reduplication occurs only once. No plausible hypothesis to account for this has been put forward.

Alternatively, we may assume that the gene is compound, consisting of a number of identical subunits. Such a supposition probably simplifies the task of accounting for gene reproduction. The chemical forces on which the identity of the new and old gene depend would not have to extend so far from the radicals to which they were due, since the thickness of the subunits would be less than that estimated for the whole chromosome thread. Similarly the reproduction might continue gradually, and the gene grow until it eventually split into two by reason of some instability which increased with increasing size, such as that which causes a drop to break up when it passes a certain size limit. The difficulty of this hypothesis, as was pointed out before, is the fact that some genes (though only a few) show more or less equal rates of back and forward mutation.

It appears not unlikely that nucleic acid plays some important role in the process of gene reduplication. For instance, the most rapid synthesis of nucleic acid occurs just before the prophase of mitosis, at the time when the chromosome appears to split or reduplicate. Again, it is remarkable that the virus proteins, which share with the genes the property of identical reproduction in living systems,[1] and of mutation also,[2] contain large quantities of nucleic acid. Conceivably there is some connection here with the remarkable fact recently revealed by Schultz and Caspersson,[3] that nucleic acid is in some way connected with the stability of the gene; when parts of the inert region in Drosophila are translocated into the euchromatic regions, they frequently cause the neighbouring loci to become unstable and undergo somatic mutations which give rise to phenotypic spotting such as that found with other mutable genes; and this instability appears to be correlated with an increase in the nucleic acid content of the corresponding bands in the salivary gland chromosomes.

All the above considerations apply to genes considered as units of

[1] Cf. Stanley 1938.
[2] Cf. McKinnery 1937.   [3] Schultz and Caspersson 1938.

crossing-over and X-ray breakage. It is quite posssible that only a small part of the gene defined in this way is actually active in the control of development. We cannot rule out the possibility that this activity is due to some particular group within the large protein-nucleic acid complex we have been discussing. In fact, the small size of the "sensitive volumes" found for particular steps of mutation might suggest that only quite restricted regions are concerned in producing the phenotypic differences between two allelomorphs; but we have pointed out the many uncertainties in the interpretation of the sensitive volume measurements.

On the other hand, it is quite possible that all primary gene products are enzymes and therefore probably proteins, which may be similar in composition to the genes themselves. It would then be in order to suggest a connection between gene activity, in which enzymes were produced and liberated into the cytoplasm, and gene reproduction, in which similar bodies were formed but retained in the neighbourhood to form a new chromosome.

It will be apparent from the above discussion that the exact knowledge at our disposal is so meagre that very many alternative hypotheses are still possible as to the nature of the chromosome, and the gene in its different senses. However, the enormously important effects of the genes on development, their capacity for identical reproduction, and the fact that they, rather than the cells of an earlier time, seem to be the most ultimate units into which we can analyse living organisms, make the problem of their constitution one of the most fundamental questions of biochemistry, well worthy of discussion even long before it can be fully answered.

APPENDIX

LABORATORY METHODS FOR CLASS WORK ON DROSOPHILA

The fruit fly, *Drosophila melanogaster*, is very easy to breed in the laboratory. By its use it should be possible to arrange practical work for genetics classes within the time allowed in a single term. With no other organism is it possible for students to obtain F2s in the time available. The notes which follow will, it is hoped, give adequate instructions as to the technique to use.[1]

THE ANIMAL

Drosophila melanogaster is a cosmopolitan fly, easiest collected from the neighbourhood of rotting or fermenting fruit; it is common round fruit stalls, warehouses, breweries, etc. The fly is about 3 mm. in length, the body colour being yellow with black bars, and the eyes red. The sexes can be easily distinguished by the abdomen; in the female it is pointed, while in the male it is rounded and the posterior bars are fused so that the tip is solid black (Fig. 102). The development from egg to adult is very rapid. At 25° C. it takes about a fortnight from the deposition of eggs to the mating and egg-laying of the females which hatch from them; at room temperature the period is nearer three weeks. A single female lays eggs throughout a period of about three weeks or more. In order to avoid overcrowding of the cultures, it is best to make matings, for any particular cross, between about three females and six males, and to transfer the flies to a new bottle or vial every third day.

CULTURE METHODS

a. *Containers*

The stocks of flies are best kept in half-pint milk bottles. Crosses and single families may be kept in flat-bottomed glass tubes or vials, about 4 in. by 1 in. in diameter. The containers are closed by wads of cotton-wool, which may, if desired, be enclosed in cheesecloth. When

[1] For further details see Bridges 1932c, and D.I.S. Brochures.

closing the containers it is important to see that there are no folds against the glass, which may form channels through which the flies can creep out and escape. Cardboard stoppers may also be used, but must be punched with small holes to provide ventilation.

b. *Food*

Standard food mixtures are as follows:

1. Banana-agar food (for stocks).

Water	..	..	..	1,000 c.c.
Chopped agar	..	..		25 grm.
Mashed bananas	..		..	1,000 grm.

Boil the agar and water until dissolved, and add the thoroughly mashed bananas. The latter may be made from overripe fruit, which can often be purchased very cheaply.

2. Treacle-semolina food (for stocks).

Water	..	..	..	750 c.c.
Black treacle	..	..		135 c.c.
Semolina	..	..	..	100 grm.
Agar	..	..	..	15 grm.

Dissolve agar before adding the other constituents, and boil all together a short time.

3. High productivity food (for getting maximum offspring from crosses).

Water	..	..	..	800 c.c.
Agar	..	..	..	15 grm.
Dry yeast	..	..	..	15 grm.
Mashed raisins	..	..		40 grm.
Syrup or treacle	..	..		50 grm.
Semolina	..	..	..	50 grm.

Dissolve agar in boiling water and add yeast, boil for a further ten minutes before adding raisins, syrup, and semolina.

c. *Preparation of Culture Vessels*

To keep down pests (mites and molds) it is advisable to sterilize the vessels before use, but this is not essential. It can easily be done by heating to 130° C. for half an hour in an oven, but temperature changes should be gradual or there is danger of the bottles breaking. The cotton wool stoppers may be sterilized at the same time.

The food is poured into the bottles while still hot and fluid, preferably through a fairly wide funnel of thin metal which gets hot and does not cool the food and cause it to solidify. The layer of food should be about $\frac{1}{2}$ in. thick. Stopper the bottles and allow to cool, or place a (sterile) duster over them. When food has solidified, put on the surface of the food one or two drops of a thin suspension of live brewer's yeast. At this stage two further refinements can be added (1). A strip of absorbent paper (e.g. toilet paper) is pressed into the surface of the food so as to form a cone with the open end upwards; when putting etherized flies into a bottle they are placed on the paper, which prevents them getting stuck to the surface of the food, and the paper is also used by the larvae for pupation (2). A hole may be punctured in the food to assist the escape of fermentation gases.

d. *Control of Pests*

The only pests are molds and mites. It is best to prevent these by addition of preservative to the food. The best are Nipagin-M or Moldex-A, about 0·15 per cent, which should be boiled with the food for about three minutes. If cultures get mouldy, the flies may be cleaned by transferring every day to fresh food.

e. *Temperature*

The optimum is about 25° C., but the flies do quite well at ordinary room temperature. At 25° C. the life cycle is about a fortnight.

Stocks kept at room temperature should be transferred to new food about every six weeks, or rather oftener. It is best to keep the old bottle in case the new one does not take. Stocks can be transferred by simply shaking the flies from one bottle into the next. It is often convenient to remember that the flies are strongly positively phototropic, and do not escape quickly from an open bottle if the bottom is held towards the light.

TECHNIQUE OF MAKING CROSSES
AND SCORING PROGENY

a. *Etherizing*

For sorting flies or selecting parents for crosses, it is necessary to anaesthetize with ether. This is done most simply in an etherizer bottle with mouth wide enough to fit against the mouth of a milk bottle. The flies can be shaken straight into the etherizer, which is closed with a cork, to which is attached a pad of cotton wool on which the ether is placed.. The same bottle can be used for flies from vials, though a better, but more elaborate arrangement is to have a funnel fitted into the cork, so that both milk bottles and vials fit when placed against it. The flies should, of course, be quite still after etherization, but they should not be allowed to get to the point when their wings are folded backwards above the body, as they then rarely recover.

When opening a bottle of flies, for transferring or etherizing, it is necessary first of all to shake them off the cotton wool plug by bumping the bottle sharply downwards on to the palm of the hand. Then remove the plug and immediately place the mouths of the two bottles in contact. Hold them together and vertical with the flies in the upper bottle, then by jarring the side of the upper bottle, shake the flies out of it into the lower bottle.

b. *Sorting*

The flies are tipped from the etherizing bottle on to a plate for sorting. Note that *all* flies are removed from the bottle; if too much ether has been used the sides of the bottle may get damp and some flies stuck, remaining in the bottle to come out later and upset some other result. The counting plate may be white cardboard, white tile, or unused photographic plates. The flies are handled with a "pusher"; either a small paint brush or a pointed piece of stiff card or sheet metal, cut so as to be convenient to hold. It is easiest to arrange the flies in a line along the centre of the plate, and, working from one end of the line, sort the flies into two or more parallel rows according to how many sorts are present. Some hereditary characters can be sorted by the naked eye, but many require a hand lens.

c. *Discarded Flies*

Discarded flies should be killed; it is dangerous to have mutant flies loose in a laboratory where they may contaminate experimental cultures. They can be poured, still etherized, into a "fly-morgue," i.e. a wide-mouthed jar containing 70 per cent alcohol, or, better, car engine oil.

d. *Making Crosses*

In crosses for experiments on heredity, the female must be virgin, since sperm from one mating are stored for a considerable time. If all flies are removed from a culture, the females which are found four to six hours later, having emerged from the pupae in the interval, will still be virgin. Virgins can also be obtained by taking newly hatched flies from ordinary cultures; they are recognized by their crumpled and unstretched wings, and elongated bodies with pale soft chitin.

The male (which, of course, need not be virgin) and female for a cross should be placed gently into a vial, care being taken that they do not get stuck on the food. It is best to put them on a piece of paper, and push this into the vial which is laid on its side until the flies have recovered from the anaesthetic, when it can be stood on end. Always label all crosses immediately.

SPECIAL TECHNIQUES: CYTOLOGICAL

The ordinary chromosomes of Drosophila are so small that they cannot be well used even for demonstrations. On the other hand, the nuclei of the salivary glands contain giant chromosomes which can be seen with a low-power microscope (very well with $\frac{1}{6}$th objective) and are so easy to prepare that they could be stained by fairly advanced classes.

In making the preparation it is first necessary to obtain the glands. They lie just behind the head of the larva. Take old, fat larvae, which have crawled up on the side of the bottle to pupate. Hold the body of the larva with a needle or fine forceps, in the left hand, and pull off the head into a drop of water or salt solution. The two glands will be seen as rather long, very transparent sacs attached along their sides to highly refringent fat bodies. It will probably be necessary to use a lens to isolate them. As much as possible of the debris should be removed, and the salt solution sucked off with a pipette or filter paper. The glands are then simultaneously fixed and stained in aceto-carmine (boil excess carmine with 40 per cent acetic acid, filter). Leave the glands in acetocarmine for up to two hours, not allowing it to dry up. Then blot off excess carmine, put on a cover glass, lay blotting paper on the over glass and squeeze firmly to flatten the glands and expel fluid. The cells are then quite disrupted, and the tangles of chromosomes, which are stained red, are spread out. The usual fault is not to squeeze the preparation hard enough. The time of staining, which also hardens the tissue and affects the flattening and spreading, varies with different batches of stain and should be tried out first. For best results, the larvae should be overfed with yeast and grown at a low temperature.

Preparations made as above can be kept for some days if the cover-glass is sealed with vaseline. Permanent preparations can be made by the following method. The smears are made on slides which have been thinly wiped with albumen solution, which is then allowed to dry completely. After staining and pressing out the glands, place the slide in a closed stain jar with enough 95 per cent alcohol just to cover one or two millimetres of the lower part of the cover-glass. In about half an hour, the alcohol vapour will have permeated the liquid film between cover-glass and slide. Transfer to a full jar of 95 per cent alcohol, and after some time remove the cover-glass carefully, when the stained glands should remain attached to the slide. Mount in Euparal.

Similar methods can be used for the demonstration of mitotic and meoitic chromosomes. For the latter, pollen mother cells of liliaceous plants or spermatocytes of Orthoptera are recommended.

b. *Collecting Eggs*

Valuable experiments on competition and selective advantages can be made by setting up overcrowded cultures of mixtures of two pheno-types, and determining the ratio of the different types which eventually emerge. This can be done in a somewhat imprecise way by varying the number of parents or of the time given for egg laying, but more satisfactorily by planting out definite numbers of eggs. Eggs can be most easily collected in either of the following ways: Put the flies in an empty bottle or vial with a microscope slide which is thinly (1 mm.) smeared with heavily yeasted food, or close the bottle with a cardboard cap carrying a smear of food and stand the bottle upside down, so that the food is below the flies. The eggs can be more easily seen if the food is darkened by admixture of carbon. Eggs are only about 0·2 mm. long and must be handled under a binocular microscope. The stage of development at the time of laying is variable, depending on how long they have been retained in the oviduct and uterus of the female.

BIBLIOGRAPHY AND AUTHOR-INDEX

The figure in heavy type, which will be found immediately following the name of the journal, gives the volume number of the journal. The figures in ordinary type refer to the pages in this book on which the paper is mentioned.

For general bibliographies of plant genetics, see:

MATSUURA, H.   1933.   A bibliographical monograph of plant genetics. Hokkaido University, Sapporo

WARNER, M. F., SHERMAN, M. A., and COLVIN, E. M.   1934.   A bibliography of plant genetics. *U.S. Dept. Agric. Misc. Publ.* 164

AGAR, W. E.   1920.   The Genetics of a Daphnia hybrid during parthenogenesis. *J. Gen.* 10.   61

ALLEN, C. E.   1926.   The direct results of Mendelian segregation. *P.N.A.S.* 12. 102
1932.   Sex inheritance and sex determination. *Am. Nat.* 66.   207, 214
1935.   The genetics of bryophytes. *Bot. Rev.* 1.   54, 102

ALTENBURG, E.   1934.   Theory of hermaphroditism. *Am. Nat.* 67.   232
1934a.   The artificial production of mutations by ultra-violet light. *Am. Nat.* 68.   384

ANDERSON, E. G.   1925.   Crossing over in the case of attached x chromosomes in *Drosophila melanogaster. Gen.* 10.   105
and RHOADES, M. M.   1931.   The distribution of interference in the x chromosome of Drosophila. *Pap. Mich. Acad. Sci.* 13.   91

ANDERSON, E.   1934.   Origin of Angiosperms. *Nat.* 133.   254

ANDERSSON-KOTTÖ, I.   1931.   The genetics of ferns. *Bibliog. Gen.* 8.   54

ANKEL, W. E.   1927, 1929.   Neuere Arbeiten zur Cytologie der natürlichen Parthenogenese der Tiere. Sammelreferat. *Z.I.A.V.* 45, 52.   59

ANON.   1935.   Vernalization and phasic development of plants. *Imp. Bureau Plant Gen.*   323

ASHBY, E.   1937.   The physiology of heterosis. *Am. Nat.* 71.   317

ASTBURY, W. T.   1933.   Fundamentals of fibre structure. Oxford.   393
1936.   X-ray studies of protein structure. *Nat.* 137.   393
1938.   Protein structure from the viewpoint of X-ray analysis. *C.R. Lab. Carlsberg. Sørensen Jubilee Vol.*   393
and BELL, F. O.   1938.   X-ray studies of thymonucleic acid. *Nat.* 141.   394

BABCOCK, E. B. and CLAUSEN, R. E.   1927.   Genetics in relation to agriculture. New York.   310, 317, 320
and NAVASHIN, M.   1930.   The genus Crepis. *Bibliog. Gen.* 6.   270

BALKASCHINA, E. L.   1929.   Ein Fall der Erbhomoösis (die Genovariation *Aristopedia*) bei *Drosophila melanogaster. Arch. Entw. Mech.* 115.   206

BALTZER, F.   1933.   Über die Entwicklung von Tritonbastarden ohne Eikern. *Verh. deutsch Zool. Ges.*   151
1937.   Analyse des Goldschmidtschen Zeitgesetzes der Intersexualität auf Grund eines Vergleiches der Entwicklung der *Bonellia* und *Lymantria* Intersexe. *Arch. f. Entw. mech.* 136.   218, 224

BATHER, F. A.   1927.   Biological classification in past and future. *Q.J. Geol. Soc.* 83.   241

BAUCH, R. 1930. Über multipolare Sexualität bei Ustilago longissima. *Arch. f. Prot. Kunde*, 70. 215

1931. Geographische Verteilung und funktionelle Differenzierung der Faktoren der multipolaren Sexualität von Ustilago longissima. *Arch. f. Prot. Kunde*, 75. 215

BAUER, H. 1935. Die Speicheldrüsenchromosomen der Chironomiden. *Naturwiss*. 23. 378

BAUR, E. 1932. Artumgrenzung und Artbildung in der Gattung Antirrhinum, Sektion Antirrhinastrum. *Z.I.A.V.* 63. 271

FISCHER, E., and LENZ, F. 1931. Human heredity. London. 325

BATESON, W. 1894. Materials for the study of variation. London. 207

1930. Mendel's principles of heredity, 4th ed. Cambridge. 24, 33, 160

BEADLE, G. W. 1932. The relation of crossing over to chromosome association in Zea-Euchleana hybrids. *Gen*. 17. 123

1933. Further studies of asynaptic maize. *Cytologia* 4. 92

and EMERSON, S. 1935. Further studies of crossing-over in attached x chromosomes of Drosophila melanogaster. *Gen*. 20. 129

and EPHRUSSI, B. 1937. Development of eye colours in Drosophila: diffusible substances and their interrelations. *Gen*. 22. 177

BECHER, E. 1938. Die Gen-Wirkstoff-Systeme der Augenausfärbung bei Insekten. *Naturwiss*. 26. 177

BELAR, K. 1926. Zur Cytologie von Aggregata eberthi. *Arch. Prot*. 53. 42

1927. Beiträge zur Kenntnis des Mechanismus der indirekten Kernteilung. *Naturwiss*. 36. 35, 364

1929. Beiträge. II. *Arch. Entw. mech*. 118. 364

1928. Die Cytologischen Grundlagen der Vererbung. Berlin. 34, 113

BELGOVSKY, M. L., and MULLER, H. J. 1938. Further evidence of the prevalence of minute rearrangements, etc. *Gen*. 23. 390

BELLING, J. 1931. Chiasmas in flowering plants. *U.C.P. Bot*. 16. 121

1927. The attachment of chromosomes at the reduction division in flowering plants. *Gen*. 18. 109

1926. The structure of chromosomes. *J. Exp. Biol*. 3. 378

1928. The ultimate chromomeres of Lilium and Aloe with regard to the numbers of genes. *U.C.P. Bot*. 14. 378

1931a. Chromomeres of Lilliaceus plants. *U.C.P. Bot*. 16. 378

1933. Crossing over and gene rearrangement in flowering plants. *Gen*. 18. 121, 133

BERGMANN, M., and NIEMANN, C. 1937. Newer biological aspects of protein chemistry. *Sci*. 86. 393, 398

BERNSTEIN, F. 1931. Die geographische Verteilung der Blutgruppen und ihre anthropologische Bedeutung. *Int. Cong. Stud. Pop. Prob*. Rome. 343

BLACKER, C. P. 1934. The chances of morbid inheritance. London. 325

BLAKESLEE, A. F. 1929. Cryptic types in Datura due to the chromosome interchange and their geographical distribution. *J. of Heredity*, 20. 268

1930. Extra chromosomes: a source of variations in the Jimson weed. *Smithsonian Report*. 84, 108

1932. The species problem in Datura. *Proc. 6th Int. Cong. Gen*. 1. 269

1934. New Jimson weeds from old chromosomes. *J. Hered*. 25. 84, 108, 170

and AVERY, A. 1937. Methods of inducing doubling of chromosomes in plants. *J. Hered*. 28. 255

BLAKESLEE, A. F., BELLING, J., and FARNHAM, M. E. 1923. Inheritance in tetraploid Datura. *Bot. Gaz.* 76. 105

BERGNER, A. D., and AVERY, A. 1937. Geographical distribution of chromosomal prime types in *Datura stramonium. Cytologia, Fujii Jub. Vol.* 268

and FOX, A. L. 1932. Our different taste worlds. *J. Hered.* 23. 325

MORRISON, G., and AVERY, A. G. 1927. Mutations in a haploid Datura. *J. Hered.* 18. 271

BOAS, FRANZ. 1928. Anthropology and modern life. New York. 342

BODENSTEIN, D. 1936. Das Determinationsgeschehen bei Insekten mit Ausschluss der frühembryonalen Determination. *Erg. d. Biol.* 13. 143

BONNEVIE, K. 1934 Embryological analysis of gene manifestation in Little and Bagg's abnormal mouse tribe. *J. exp. Zool.* 67. 202

BOVERI, T. 1907. Zellenstudien VI. Die Entwicklung dispermer Seeigeleier. *Jena Zeitsch.* 37. 38

1909. Die Blastomerenkerne von Ascaris megalocephala und die Theorie der Chromosomenindividualität. *Arch. f. Zellforsch.* 3. 42

BOYCOTT, A. E., and DIVER, C. 1923. On the inheritance of sinistrality in *Limnea peregra. Proc. Roy. Soc. B.* 195. 143

DIVER, C., GARSTANG, S. L., and TURNER, F. M. 1930. The inheritance of sinistrality in *Limnea peregra. Phil. Trans. Roy. Soc.* 209. 143

BRACHET, J. 1937. La différentiation sans clivage dans l'oeuf de Chétoptère envisagée aux points de vue cytologique et métabolique. *Arch. de Biol.* 48. 138

BRAGG, W. H 1933. Liquid crystals. Proc. Royl. Inst. *G.B.* 28. 363

BREWER, H. 1935. Eutelegenesis. *Eugenics Rev.* 27. 358

BRIDGES, C. B. 1914. Direct proof through non-disjunction that the sex-linked genes of Drosophila are borne by the x chromosome. *Science* 40. 81

1916. Non-disjunction as a proof of the chromosome theory of heredity. *Genetics* 1. 81, 105

1917. Deficiency. *Gen.* 2. 96

1919a Duplications. *Anat. Rec.* 1919. 96

1919. Specific modifiers of eosin eye colour in *Drosophila melanogaster. J.E.Z.* 28. 161

1921 Genetic and cytological proof of non-disjunction of the fourth chromosome of *D. melanogaster. Proc. N.A.S.* 7. 82

1925. Sex in relation to chromosomes and genes. *Am. Nat.* 59. 107

1925a. Elimination of chromosomes due to a mutant (Minute-N) in *D. melanogaster. Proc. Nat. Acad. Sci.* 11. 86

1927. The relation of the age of the female to crossing over in the third chromosome of *D. melanogaster. J. Gen. Phys.* 8. 93

1929. Variation in crossing over in relation to age of female in *D. melanogaster. Carn. Inst. Wash.* 399. 93

1932. Genetics of sex in Drosophila. Allen's *Sex and Internal Secretions*, Baltimore. 207, 219, 222, 228

1932a. The suppressors of purple. *Zeit. ind. Abst. u. Ver.* 60. 160

1932b. Specific suppressors in Drosophila. *Quart. Rev. Biol.* 160

1932c. Apparatus and methods for Drosophila culture. *Am. Nat.* 66. 403

1935. Salivary chromosome maps. *J. Hered.* 26. 99, 267, 379, 396

1936. The Bar gene a duplication. *Sci.* 83. 374

1938. A revised map of the salivary gland x chromosome in Drosophila. *J. Hered.* 29. 99, 379, 396

BRIDGES, C. B., 1938a. Chapter on Drosophila in new edition of Allen's *Sex and Internal Secretions*. 221

and ANDERSON, E. G. 1925. Crossing over in the x chromosome of triploid females of *Drosophila melanogaster*. *Gen.* 10. 105

and DOBZHANSKY, TH. 1933. The mutant "proboscipedia" in *D. melanogaster*—a case of hereditary homoösis. *Arch. f. Ent. mech.* 127. 206

BRIEGER, F. 1930. Selbstsserilität und Kreuzungssterilität. *Monog. Gesamt. Phys. Pfl. u. Tiere.* 21. 56

BRIGHAM, C. C. 1922. A study of American intelligence. Princeton. 344

1930. Intelligence tests of immigrant groups. *Psychol. Rev.* 37. 345

BRINK, R. A. 1929. Studies on the physiology of a gene. *Q. Rev. Biol.* 4. 55, 179

BRYDEN, W. 1935. Some observations on the mitotic and meiotic divisions in the Wistar rat. *Cyt.* 6. 123

BULMAN, O. M. B. 1933. Programme evolution in the graptolites. *Biol. Rev.* 8. 247

BURNS, R. K. 1938. Hormonal control of sex differentiation. *Am. Nat.* 22. 225

CAROTHERS, E. E. 1926. The maturation divisions in relation to the segregation of homologous chromosomes. *Qt. Rev. Biol.* 1. 46, 47

CASPERSSON, T. 1936. Über den chemischen Aufbau der Strukturen des Zellkerns. *Skand. Arch. f. Physiol.* 73. *Supp.* 8. 34, 392

and HAMMARSTEN, E. H. 1935. Interaction of proteins and nucleic acid. *Trans. Farad. Soc.* 31. 392

and SCHULTZ, J. 1938. Nucleic acid metabolism of the chromosomes in relation to gene reproduction. *Nat.* 142. 396, 400

CASTLE, W. E. 1916. Further studies on piebald rats and selection, etc. *Carn. Inst. Wash.* 241. 313

1926. Biological and social consequences of race crossing. *Am. J. Phys. Anthrop.* 9. 347

1930. Genetics and eugenics. 4th ed. *Harvard.* 340

and GREGORY, P. W. 1929. The embryological basis of size inheritance in the rabbit. *J. Morph. Physiol.* 48. 147, 318

CATCHESIDE, D. G. 1937. Secondary pairing in *Brassica oleracea*. *Cytologia, Fujii Jub. Vol.* 74

CHAMBERS, R. 1925. The physical structure of protoplasm as determined by microdissection and injection. Cowdry's *General Cytology*. 363

CHARLES, E. 1934. The Twilight of Parenthood. London. 287, 351, 353

CHESLEY, P. 1935. Development of the short-tailed mutant in the house mouse. *J. exp. Zool.* 70. 201

CHILD, G. 1935. Phenogenetic studies on scute-1 of *D. melanogaster*. *Gen.* 20. 371

CLAUSEN, J. 1927. Chromosome number and the relationship of some North American species of Viola. *Ann. Bot.* 43. 263

1931a. *Viola canina* L. a cytologically irregular species. *Hereditas* 15. 263

1931b. Cytogenetic and taxonomic investigations in Melanium violets. *Hereditas* 15. 263

and GOODSPEED, T. H. 1925. Interspecific hybridization in Nicotiana. II. A tetraploid glutinosa-tabacum hybrid, and experimental verification of Winge's hypothesis. *Gen.* 10. 256

1928. Interspecific hybridization and the origin of species in Nicotiana. *5th Int. Cong. Vererb.* 1. 258

COLD SPRING HARPER SYMPOSIUM. 1938. The structure of proteins. 393, 394

CLELAND, R. E. 1931. Cytological evidence of genetical relationships in Oenothera. *Am. Jour. Bot.* **18.** 111, 112

1922. The reduction divisions in the pollen mother cells of *Oenothera franciscana*. *Am. J. Bot.* **9.** 111

COLLINS, J. L., HOLLINGSHEAD, L., and AVERY, P. 1929. Interspecific hybrids in Crepis III. *Gen.* **14.** 262

CONKLIN, E. G. 1905. Mosaic development in ascidian eggs. *J. exp. Zool.* **2.** 141

1924. Cellular differentiation in Cowdry's *General Cytology*. Chicago. 141

COOK, R. 1932. A chronology of genetics. *U.S. Dept. Agric. Yrbk.* 1937. 29

COOPER, K. W. 1938. Concerning the origin of polytene chromosomes in Diptera. *P.N.A.S.* **24.** 98

CORRENS, C. 1907. Bestimmung und Vererbung des Geschlechtes. Leipzig. 75, 207

1921. Versuche bei Pflanzen das Geschlechtsverhältniss zu verschieben. *Hereditas* **2.** 55

1924. Abhandlungen. Berlin. 207

1928a. Bestimmung, Vererbung und Verteilung des Geschlechtes bei den höheren Pflanzen. *Handb. der Vererb.* **2.** 207, 215

1937. Nichtmendelnde Vererbung. *Handb. d. Vererb.* **22.** 277

CRANE, M. B., and LAWRENCE, W. J. C. 1936. The genetics of garden plants. London. 320

CREIGHTON, H. B., and McCLINTOCK, B. 1931. A correlation of cytological and genetical crossing over in Zea mays. *P.N.A.S.* **17.** 50

CREW, F. A. E. 1927. Genetics of sexuality in animals. 207

1933. A case of non-disjunction in the fowl. *Proc. Roy. Soc. Edinb.* **53.** 82

1936. A repetition of McDougall's Lamarckian experiment. *J. Gen.* **33.** 305

and ROBERTS, A. F. 1933. Heredity as a factor in animal disease. *Proc. Roy. Soc. Med.* **26.** 324

CROWFOOT, D. et al. 1938. Crystal structures of the proteins. *Nat.* **141.** 365, 393

DALCQ, A. 1932. Étude des localizations germinales dans l'oeuf vierge d'Ascidie par des expériences de mérogonie. *Arch. Anat. micr.* **28.** 141

1935. L'organization de l'œuf chez les Chordés. Paris. 137

1935a. La regulation dans le germ et son interpretation. *C.R. Soc. Biol.* **119.** 141

1938. Form and causality in early development. Cambridge. 137

et PASTEELS, J. 1937. Une conception nouvelle des lois physiologiques de la morphogénèse. *Arch. de Biol.* **48.** 148

and SIMON, S. 1932. Contribution à l'analyse des fonctions nucléaires dans l'ontogénèse de la grenouille II. *Protoplasma* **14.** 150

DANFORTH, C. H. 1929. The effect of foreign skin on feather pattern in the common fowl. *Arch. Ent. mech.* **116.** 198

1932. Artificial and hereditary suppression of sacral vertebrae in the fowl. *Proc. Soc. Exp. Biol. Med.* **30.** 202

DANNEEL, R. 1938. Die Wirkungsweise der Grundfaktoren für Haarfärbung beim Kaninchen. *Naturwiss.* **26.** 200

DARLINGTON, C. D. 1930a. Cytological demonstration of genetic crossing over. *Proc. Roy. Soc. B.* **107.** 122

1930b. Chromosome studies in Fritillaria. III. *Cyt.* **2.** 121

1929a. Meiosis in polyploids. II. *J. Gen.* **21.**

DARLINGTON, C. D., 1929b.   Ring formation in Oenothera and other genera. *J. Gen.* 20.   111

1931c.   Cytological theory of inheritance in Oenothera. *J. Gen.* 26.   111

1932a.   Recent advances in cytology. London, 1st ed.   72, 126, 229, 230, 253

1932b.   The control of the chromosomes by the genotype and its bearing on some evolutionary problems. *Am. Nat.* 66.   378

1934a.   Anomalous chromosome pairing in the male *Drosophila pseudo-obscura. Gen.* 19.   123

1934b.   The origin and behaviour of chiasmata VII. Zea Mays. *Z.I.A.V.* 67.   122

1935a.   The internal mechanics of the chromosome. *Proc. Roy. Soc. B.* 118.   44, 131

1935b.   The time, place, and action of crossing over. *J. Gen.* 31.   131

1936.   Crossing over and its mechanical relationships in Chorthippus and Stauroderus. *J. Gen.* 33.   266

1936a.   The external mechanics of the chromosomes. *Proc. Roy. Soc. B.* 118.   364

1937.   Recent advances in cytology. 2nd ed. London.   34, 52, 58, 64, 75, 95, 109, 113, 121, 122, 129, 264, 361, 364, 367

1938.   The evolution of genetic systems. Cambridge.   52, 230, 378, 397

and MOFFETT, A. A.   1930.   Primary and secondary chromosome balance in Pyrus. *J. Gen.* 22.   261

and MATHER, K.   1932.   The origin and behaviour of chiasmata III. *Cyt.* 4.   125

DAVENPORT, C. R.   1917.   The effect of race intermingling. *Proc. Am. Philosoph. Soc.* 56.   346

and STEGGERDA, M.   1929.   Race crossing in Jamaica. *Publ. Carnegie Inst.* 395.   344

DAVIES, A. M.   1937.   Evolution and its modern critics. London.   241

DAVIS, A.   1938.   The distribution of the blood groups and its bearing on the concept of race. *Political Arithmetic,* ed. L. Hogben. London.   342

DAVIS, R. A.   1928.   The influence of heredity on the mentality of orphan children. *Brit. J. Psych.* 19.   337

DEARBORN, W. F.   1928.   Intelligence tests. Boston.   337

DEMEREC, M.   1933.   What is a gene? *J. Hered.* 24.   372, 381

1934a.   Biological action of small deficiencies of x chromosome in *D. melanogaster. P.N.A.S.* 20.   163

1934b.   The gene and its role in ontogeny. *C.S.H. Symposium on Quant. Biol.* 2.   163

1935.   Unstable genes. *Bot. Rev.* 1.   372, 381

1937.   A mutability stimulating factor in the Florida stock of *D. melanogaster. Gen.* 22.   381

DE VRIES, H.   1907.   Die Mutationstheorie. Leipzig.   110

DE WINTON, D.   1931.   Linkage in the tetraploid *Primula sinensis. J. Gen.* 24.   107

and HALDANE, J. B. S.   1932.   Genetics of *Primula sinensis. J. Gen.* 25.   159

DETLEFSON, J. A.   1914.   Genetic studies on a cavy species cross. *Carnegie Inst. Publ.* 205.   291

1925.   The inheritance of acquired characters. *Physiol. Rev.* 5.   302

DIGBY, L.   1912.   The cytology of *Primula Kewensis* and of other related Primula hybrids. *Ann. Bot.* 26.   256

DOBZHANSKY, T. 1927. Studies on the manifold effects of certain genes in D. melanogaster. Z.I.A.V. 43. 162

1929. Genetical and cytological proof of translocations involving the third and fourth chromosomes of D. melanogaster. Biol. Zbl. 49. 97

1930. Time of development of the different sexual forms in D. melanogaster. Biol. Bull. 59. 221

1930a. The manifold effects of the genes Stubble and Stubbloid in D. melanogaster. Z.I.A.V. 54. 162

1930b. Cytological map of the second chromosome in D. melanogaster. Gen. 16. 97

1931. The decrease in crossing over observed in translocations and its probable explanation. Am. Nat. 65. 130

1936a. Position effect of genes. Biol. Rev. 11. 373

1936b. Studies on hybrid sterility. II. Localization of sterility factors in D. pseudoobscura hybrids. Gen. 21. 283

1937a. Genetic nature of species differences. Am. Nat. 71. 253

1937b. Genetics and the origin of species. Columbia University Press. 253, 266

and KOLLER, P. 1938. Sexual isolation between two species of Drosophila. Gen. 22. 283

1938a. In press. Gen. 22. 295

and SCHULTZ, J. 1934. The distribution of sex factors in the x chromosome of D. melanogaster. J. Gen. 28. 223

and STURTEVANT, A. H. 1935. Further data on maternal effects in Drosophila pseudoobscura hybrids. P.N.A.S. 21. 281

and TAN, C. C. 1936. Studies on hybrid sterility III. A comparison of the gene arrangement in two species, D. pseudoobscura and D. miranda. Z.I.A.V. 72. 267

and QUEAL, M. L. 1938. Genetics of natural populations. Gen. 23. 286

DODGE, B. O. Reproduction and inheritance in Ascomycetes. Sci. 83. 54

DONCASTER, L. 1908. Sex inheritance in the moth Abraxas grossulariata and its variety lacticolor. Rep. Evol. Cttee. 4. 79

DROSOPHILA INFORMATION SERVICE. Privately circulated.

DUBININ, N. P. 1929. Allelenmorphentreppen bei D. melanogaster. Biol. Zbl. 49. 369

1932a. Step allelomorphism in D. melanogaster. J. Gen. 25. 369

1932b. Step allelomorphism and the theory of centres of the genes achaete and scute. J. Gen. 26. 369

1934, 1936. Experimental alteration of the number of chromosome pairs in D. melanogaster. Biol. Zhurn. 3 and 5. 268

and fourteen collaborators. 1934. Experimental study of the ecogenotypes of D. melanogaster. Biol. Zhurn. 3. 286

and SIDEROV, B. N. 1934. Relation between the effect of a gene and its position in the system. Am. Nat. 68. 168, 375

and SOKOLOV, N. N., and TINIAKOV, G. G. 1936. Occurrence and distribution of chromosome aberrations in nature. Nat. 138. 286

DUERDEN, J. E. 1920. The inheritance of the callosities in the ostrich. Am. Nat. 54. 303

DUNCAN, H. G. 1929. Race and population problems. London. 351

DUNN, L. C. 1925. The inheritance of rumplessness in the domestic fowl. J. Hered. 16. 202

1934. Analysis of a case of mosaicism in the house mouse. J. Gen. 29. 86

DURKEN, B. 1923. Über die Wirkung farbigen Lichtes auf die Puppen des Kohlweisslinges (*Pieris brassicae*) und das Verhalten der Nachkommen. *Arch. Entw. mech.* 99. 304

DURYEE, W. R. 1938. A microdissection study of amphibian chromosomes. *Collecting Net.* 13. 101, 394

EAST, E. M. 1910. A mendelian interpretation of variation that is apparently continuous. *Am. Nat.* 44. 159

1929. Self-sterility. *Bibliog. Gen.* 56

1934. The Nucleus-plasma problem. *Am. Nat.* 68. 275

1936. Heterosis. *Gen.* 21. 317

and JONES, D. F. 1919. Inbreeding and outbreeding. Philadelphia. 314

1920. Genetic studies on the protein content of maize. *Gen.* 5. 317

ELDERTON, E. M. 1922. A summary of the present position with regard to the inheritance of intelligence. *Biometrica* 14. 337

ELLES, G. L. 1922. The graptolite faunas of the British Isles. *Proc. Geol. Assn.* 33. 247

ELOFF, G. 1932. A theoretical and experimental study on the changes in crossing over value, etc. *Genetica* 14. 78

EMERSON, R. A. 1932. The present status of maize genetics. *Proc. 6th Int. Cong. Gen.* 223

BEADLE, G. W., and FRASER, A. C. 1935. A summary of linkage studies in maize. *Cornell Mem.* 180. 91

ENGELSMEIER, W. 1935. Nachweis der alternativen Modifikabilität der Haarfärbung beim Russenkaninchen. *Z.I.A.V.* 68. 200

EPHRUSSI, B. 1938. The physiology of gene action. *Am. Nat.* 72. 177

and BEADLE, G. W. 1937. Development des couleurs des yeux chez la drosophile, revue des expériences de transplantation. *Bull. Biol. Fr. et Belg.* 71. 177

KHOUVINE, Y., and CHEVAIS, S. 1938. Genetic control of a morphogenetic substance in *D. melanogaster*. *Nat.* 142. 184

EYSTER, W. H. 1925. Mosaic pericarp in maize. *Gen.* 10. 372

1928. The mechanism of variagation. *Verh. 5. Kong. Vererb. Z.I.A.V. Suppl.* 372

1934. Genetics of Zea mays. *Bibliog. Gen.* 11. 159

FELL, H. B., and LANDAUER, W. 1935. Experiments on skeletal growth and development in vitro in relation to the problem of avian phokomelia. *Proc. Roy. Soc. B.* 118. 203

FISHER, R. A. 1918. Correlation between relatives on the supposition of mendelian inheritance. *Trans. Roy. Soc. Edin.* 52. 334

1928. The possible modification of the response of the wild type to recurrent mutations. *Am. Nat.* 62. 185, 297

1928a. Statistical methods for research workers. Edinburgh. 333

1930. The genetical theory of natural selection. Oxford. 287, 291, 340, 352

1931. The evolution of dominance. *Biol. Rev.* 6. 185, 297

1932. Family allowances in the contemporary economic situation. 357

1935. Dominance in poultry. *Phil. Trans. Roy. Soc. B.* 225. 298

FORD, E. B. 1931. Mendelism and evolution. London. 297

1937. Problems of heredity in the Lepidoptera. *Biol. Rev.* 12. 162

and HUXLEY, J. S. 1927. Mendelian genes and rates of development in *Gammarus chevreuxi*. *J. exp. Biol.* 5. 173

1929. Genetic rate-factors in Gammarus. *Arch. f. Entw. Mech.* 117. 173

FREEMAN, F. S. 1934. Individual differences. London. 335, 340, 344, 345, 347

HOLTZINGER, K. J., and MITCHELL, B. C. 1928. The influence of environment on the intelligence, school achievement, and conduct of foster children. 27th year book, Nat. Soc. Study Educ. 337

FREUNDLICH, H. 1932. Kapillarchemie II. 362

FRIESEN, H. 1936. Roentgenmorphosen bei Drosophila. Arch. f. Entw. Mech. 134. 191

—— 1936a. Crossing over in male Drosophila. Nature 137. 93

—— 1936b. Auslösung von crossing over bei Drosophila Männchen durch Roentgenisierung der Imago. Genetica 18. 93

FRYER, J. C. F. 1913. Preliminary notes on some experiments with the polymorphic Phasmida. J. Gen. 3. 61

GABRITSCHEWSKY, E., and BRIDGES, C. B. 1928. The giant mutation in D. melanogaster. Z.I.A.V. 46. 190

GAIRDNER, A. E. 1929. Male sterility in flax II. J. Gen. 21. 275

—— and DARLINGTON, C. D. 1931. Ring formation in diploid and polyploid Campanula persicifolia. Genetica 13. 268

GATES, R. R. 1908. A study of reduction in Oenothera rubrinervis. Bot. Gaz. 46. 110

—— 1929. Heredity in man. London. 325, 341, 346

GAUSE, G. F. 1934. The Struggle for Existence. Baltimore. 299

GEIGY, R. 1931a. Action de l'ultraviolet sur le pôle germinale dans l'œuf de D. melanogaster. Rev. Suisse Zool. 38. 145

—— 1931b. Erzeugung rein imaginaler Defekte durch ultraviolette Eibestrahlung bei D. melanogaster. Arch. f. Entw. Mech. 125. 145

GEITLER, L. 1934. Grundriss der Cytologie. Berlin. 34, 113, 364

GELEI, J. 1913, 1921, 1922. Ueber die Ovogense von Dendrocoelum lacteum. Arch. f. Zellf. 11, 16, 16. 45

GEROULD, J. H. 1911. The inheritance of polymorphism and sex in Colias philodice. Am. Nat. 45. 162

GOLDSCHMIDT, R. 1923. The mechanism and physiology of sex determination. London. 207, 215

—— 1927. Physiologische Theorie der Vererbung. Berlin. 173, 193, 201

—— 1929. Experimentelle Mutationen und das Problem der sogenannten Parallel-Induktionen. Biol. Zbl. 49. 383

—— 1931. Die entwicklungsphysiologische Erklärung des Falls der sogenannten Treppenallelomorphie des Gens Scute bei Drosophila. Biol. Zbl. 51. 371

—— 1931a. Analysis of intersexuality in the gypsy moth. Q. Rev. Biol. 6. 216

—— 1931b. Die sexuellen Zwischenstufen. Berlin. 207, 215

—— 1932. Genetics and development. Biol. Bull. 63. 173, 186

—— 1934a. Lymantria. Bbl. Gen. 11. 216, 271

—— 1934b. Influence of the cytoplasm upon gene-controlled heredity. Am. Nat. 68. 275

—— 1934c. Die Genetik der geographischen Variation. Proc. 6th Cong. Gen. 271

—— 1935. Gen und Ausseneigenschaft. Z.I.A.V. 69. 191

—— 1935a. Multiple sex genes in Drosophila. J. Gen. 31. 222

—— 1935b. Geographische Variation und Artbildung. Naturwiss. 23. 271

—— 1937. Gene and character. Univ. Calif. Publ. Zool. 41. 167, 197

—— 1938. Physiological Genetics. New York. 173, 191, 197, 376

—— 1938a. The time law of intersexuality. Genetica 20. 218

GOLDSCHMIDT, R., and PARISER, K. 1923. Triploide Intersexe bei Schmetterlingen. *Biol. Zbl.* 43. 219

GORDON, C. 1936. Frequency of heterozygosis in free living populations of *D. subobscura. J. Gen.* 33. 286

GORDON, H. 1923. Mental and scholastic tests among retarded children. *Board of Education Educ. Pamph.* 44. 350

GOWEN, J. W. 1929. The cell division at which crossing over takes place. *P.N.A.S.* 15. 101

— 1933. Anomalous human sex-linked inheritance of colour-blindness in relation to attached sex chromosomes. *Human. Biol.* 5. 82

— 1933a. Meiosis as a genetic character in *D. melanogaster. J. Exp. Zool.* 65. 92

— 1937. Contributions of genetics to the understanding of animal diseases. *J. Hered.* 28. 334

— and GAY, E. H. 1933. Gene number, kind and size in Drosophila. *Gen.* 18. 379

GOWEN, M. S., and GOWEN, J. W. 1922. Complete linkage in *D. melanogaster. Am. Nat.* 56. 92

GRAY, J. 1931. Experimental cytology. Cambridge. 35, 39

GRAY, J. L. 1935. The nation's intelligence. London. 347

GROSS, F. 1932. Untersuchungen uber die Polyploidie und die Variabilität bei *Artemia salina. Naturwiss.* 20. 254

GRUENEBERG, H. 1936. The position effect proved by a spontaneous reversion of the X-Chromosome in *D. melanogaster. J. Gen.* 32. 375

HADORN, E. 1935. Chimärische Tritonlarven mit bastardmerogonischen und normalkernigen Teilstücken. *Rev. Suisse Zool.* 42. 151

— 1936. Uebertragung von Artmerkmalen durch das entkernte Eiplasma beim merogonischen Tritonbastard, *palmatus-Plasma* × *cristatus-Kern. Verh. d. Deutsch. Zool. Gens.* 151

— 1937. Die entwicklungsphysiologische Auswirkung einer disharmonischen Kernkombination beim Bastardmerogen *Triton palmatus* ♀ × *Triton cristatus* ♂. *Arch. f. Entw. mech.* 136. 151

HAEMMERLING, J. 1932. Entwicklung und Formbildungsvermögen von *Acetabularia mediterranea. Biol. Zbl.* 52. 153

— 1934. Ueber formbildende Substanzen bei *Acetabularia mediterranea. Arch. f. Entw. Mech.* 131. 153

— 1937. Fortpflanzung und Sexualität. *Fortschr. Zool. N.F.* 1. 207, 216

HALDANE, J. B. S. 1922. Sex ratio and unisexual sterility in hybrid animals. *J. Gen.* 12. 78

— 1927. Comparative genetics of colour in rodents and carnivora. *Biol. Rev.* 2. 270

— 1929. The species problem in the light of genetics. *Nat.* 124. 253

— 1930. A note on Fisher's theory of the origin of dominance. *Am. Nat.* 64. 186, 298

— 1930a. Theoretical genetics of autopolyploids. *J. Gen.* 22. 71

— 1931. The cytological basis of genetical interference. *Cyt.* 3. 119

— 1932. The inheritance of acquired characters. *Nat.* 129, 130. 302

— 1932a. The time of action of genes. *Am. Nat.* 46. 234

— 1932b. The causes of evolution. London. 186, 241, 253, 287, 288

— 1932c. Genetic evidence for a cytological abnormality in man. *J. Gen.* 26. 82

— 1934. Anthropology and human biology. *C.R. Cong. Int. Sci. Anthrop. et Ethnol.* 340

HALDANE, J. B. S. 1935. The rate of spontaneous mutation of a human gene. *J. Gen.* 31. 355, 380

1935*a*. Contributions de la génétique à la solution de quelques problems physiologiques. *C.R. Soc. Biol.* 175

1936. The amount of heterozygosity to be expected in an approximately pure line. *J. Gen.* 32. 315

1936*a*. A search for incomplete sex linkage in man. *Ann. Eug.* 7. 333

1936*b*. Some principles of causal analysis in genetics. *Erkenntnis* 6. 189

1938. Heredity and politics. London. 347, 353, 355, 356

HAMAKER, H. C. 1937. The London–van der Waals attraction between spherical particles. *Physica IV.* 362

1938. London–van der Waals forces in colloidal systems. *Rec. Trav. Chim. Pays. Bas.* 57. 362

HAMBURGER, V. 1936. The larval development of reciprocal species hybrids of *Triton taeniatus* (and *T. palmatus*) × *T. cristatus. J. Exp. Zool.* 73. 150

HARDER, R. 1934. Ueber die Musterbildung an Petunienblüten. *Nach. Ges. Wiss. Gött. N F.* 1. 206

HARDY, G. H. 1908. Mendelian proportions in a mixed population. *Sci.* 28. 189

HARLAND, S. C. 1936. The genetical conception of the species. *Biol. Rev.* 11. 274, 298

HARRISON, J. W. H. 1920. Genetical studies in the moths of the Geometrid genus Oporabia (Operina). *J. Gen.* 9. 299

1927. Experiments on the egg laying of the sawfly *Pontamia salicis. Proc. Roy. Soc. B.* 101. 304

and GARRETT, F. 1926. The induction of melanism in the Lepidoptera. *Proc. Roy. Soc. B.* 99. 304

HARTMANN, M. 1929*a*. Fortpflanzung und Befruchtung als Grundlage der Vererbung. *Handb. Vererbungswiss.* 1. 52

1929*b*. Verteilung, Bestimmung und Vererbung des Geschlectes bei Protisten und Thallophyten. *Handb. Vererbungswiss.* 2. 207

1931. Relative Sexualität und ihre Bedeutung für eine allgemeine Sexualitäts und Befruchtungstheorie. *Naturwiss.* 19. 209, 230, 232

1932. Neue Ergebnisse zum Befructungs- und Sexualitäts-problem. *Naturwiss.* 20. 209

1934. Beiträge zur Sexualitätstheorie. *Sitzber. Preuss. Akad. Wiss.* 20. 209

HASKINS, C. P., and ENZMANN, E. V. 1936. A determination of the magnitude of the cell "sensitive volume" associated with the white-eyed mutation in X-rayed *D. melanogaster.* II. *P.N.A.S.* 22. 392

HEARNE, E. M., and HUSKINS, C. L. 1935. Chromosome pairing in *Melanoplus femur-rubrum. Cyt.* 6. 37, 119, 129

HEILBORN, O. 1924. Chromosome number and dimensions, species formation and phylogeny in the genus Carex. *Hereditas.* 16. 364

HEILBRUNN, L. V. 1928. The colloid chemistry of protoplasm. *Protopl. Monog.* 1. 39

HEITZ, E. 1931. Nukleolen und Chromosomen in der Gattung Vicia. *Planta.* 15. 39

1935. Chromosomenstruktur und Gene. *Z.I.A.V.* 70. 40

and BAUER, H. 1933. Beweis für die Chromosomennatur der Kernschleifen in den Knauelkernen von *Bibio hortulanus L. Zeits. Zellf. u. mikr. Anat.* 17. 98

HENKE, K. 1933. Zur Morphologie und Entwicklungsphysiologie der Tier-zeichnungen. *Naturwiss.* 21. 192

Entwicklung und Bau tierischer Zeichnungsmuster. *Verh. d. Deutsch. Zool. Ges.* 192

HERBST, C. 1935. Untersuchungen zur Bestimmung des Geschlectes IV. *Arch. f. Entw. Mech.* 132. 224

HERRMANN, L., and HOGBEN, L. 1933. The intellectual resemblance of twins. *Proc. Roy. Soc. Edin.* 53. 337, 338

HERTWIG, P. 1936. Artbastarde bei Tiere. *Handb. Vererbungswiss* 21. 271

HOGBEN, L. 1931. Genetic principles in medicine and social science. London. 325, 327, 335, 355

1933. Nature and nurture. London. 189, 325, 327, 355

1933a. A matrix notation for mendelian populations. *Proc. Roy. Soc. Edin.* 53. 334

1938. Political Arithmetic. London. 340, 350

HOLLINGSHEAD, L., and BABCOCK, E. B. 1930. Chromosomes and phylogeny in Crepis. *Univ. Calif. Publ. Agric. Sci.* 6. 270

HOLMES, S. J. 1934. Human genetics and its social import. New York. 325, 335, 340, 347, 354, 355

HOLTFRETER, J. 1935. Ueber das Verhalten von Anurenektoderm in Urodelen Keimen. *Arch. f. Entw. Mech.* 133. 149

HOLTZINGER, K. J. 1929. The relative effect of nature and nurture influences on twin differences. *J. Educ. Psych.* 20. 338

HORSTADIUS, S. 1928. Ueber die Determination des Keimes bei Echinodermen. *Acta Zool.* 9. 145

1935. Ueber die Determination im Verlaufe der Eiachse bei Seeigeln. *Publ. Staz. Zool. Nap.* 14. 145

1936. Determination in the early development of the sea-urchin. *Collecting Net* 11. 145

HOWLAND, R. B., and CHILD, G. P. 1935. Experimental studies on development in *D. melanogaster. J. Exp. Zool.* 70. 145

and SONNENBLICK, B. P. 1936. Experimental studies on development in *D. melanogaster* II. *J. Exp. Zool.* 73. 145

HUDSON, P. S. 1937. Genetics in its application to plant breeding. *Biol. Rev.* 12. 309, 311, 321, 323

HUGHES, A. MCK. 1932. Induced melanism in Lepidoptera. *Proc. Roy. Soc. B.* 110. 305

HUNTER, H., and LEAKE, H. M. 1933. Recent advances in agricultural plant breeding. London. 316, 322, 309

HUSKINS, C. L. 1930. The origin of *Spartina Townsendii. Genetica* 12. 261

1937. The internal structure of chromosomes. *Cyt. Fujii Jub.* 2. 117

and SMITH, S. G. 1934. Chromosome division, and pairing in *Frittilaria Meleagris. J. Gen.* 28. 115

1935. Meiotic Chromosome structure in *Trillium erectum L. Ann. Bot.* 49. Pl. 5

HUXLEY, J. S. 1929. Sexual difference of linkage in *Gammarus chevreuxi. J. Gen.* 20. 78

1932. Problems of relative growth. London. 243

1936. Eugenics and Society. *Eugen. Rev.* 28. 358

and HADDON, A. C. 1936. We Europeans. London. 340

and DE BEER, G. 1934. Elements of experimental embryology. Oxford. 137

IMAI, Y. 1937. The behaviour of the plastid as a hereditary unit. *Cyt. Fujii Jub. Vol.* 277

IRWIN, M. R., and COLE, L. J. 1936. Immunogenetic studies of species and species hybrids. *J. Exp. Zool.* 73. 271

ISELY, F. B. 1938. Survival value of Acridian protective coloration. *Ecology.* 19. 300

ITARD. 1932. The wild Boy of Aveyron. New York. 348

JANSSENS, F. A. 1909 Spermatogénèse dans les Batrachiens V. La théorie de la chiasmatypie. *Cellule* 25. 48, 121
    1924. La chiasmatypie dans les insectes. *Cellule* 34. 121

JENNINGS, H. S. 1930. Genetics of the Protozoa. *Bibliog. Gen.* 59

JOHANNSEN, W. 1911. The genotype conception of heredity. *Am. Nat.* 45. 155
    1903. Ueber Erblichkeit in Populationen und in reinen Linien. *Jena.* 315

JOLLOS, V. 1921. Experimentelle Protistenstudien I. *Arch. Protistenk.* 43. 278
    1934. Inherited changes produced by heat treatment in *D. melanogaster. Genetica* 16. 279, 383
    1935. Studien zum Evolutionsproblem. *Biol. Zbl.* 55. 278, 383
    1938. In Handbuch der Vererbungswiss. 278

JONES, D. F. 1928. Selective fertilization. Chicago. 55
    1934. Unisexual maize plants and their bearing on sex differentiation in other plants and animals. *Gen.* 19. 223

JONES, W. N. 1935. Plant chimaeras and graft hybrids. London. 87

JORDAN, D. S., and JORDAN, H. E. 1914. War's aftermath. Boston. 354

JØRGENSEN, C. A. 1927. Cytological and experimental studies in the genus Lamium. *Hereditas* 9. 79
    1928. The experimental formation of heteroploid plants in the genus Solanum. *J. Gen.* 19. 255
    and CRANE, M. B. 1927. Formation and morphology of *Solanum* chimaeras. *J. Gen.* 18. 87

JUHN, M., and FRAPS, R. M. 1936. Developmental analysis in plumage. *Physiol. Zool.* 8. 198

JUST, E. E. 1937. The significance of experimental parthenogenesis for the cell biology of to-day. *Cyt. Fujii Jub.* 1. 62

KARPECHENKO, G. D. 1924. Hybrids of *Raphanus sativus* (♀) × *Brassica oleracea* (♂). *J. Gen.* 14. 170, 256
    1928. Polyploid hybrids of *Raphanus sativus* and *Brassica oleracea.* Z.I.A.V. 48. 170, 256

KAUFMAN, B. P. 1936. Chromosome structure in relation to the chromosome cycle. *Bot. Rev.* 2. 116
    1938. Nucleolus organizing regions in the salivary gland chromosomes of *D. melanogaster. Zeits. Zellf.* 28. 40

KENNAWAY, E. L., and KENNAWAY, N. M. 1937. Some factors affecting carcinogenesis. *Acta int. Union against Cancer* 2. 343

KERKIS, J. 1934. On the mechanism of development of triploid intersexuality in *D. melanogaster. C.R. Acad. Sci. U.R.S.S.* 221
    1933. Development of gonads in hybrids between *D. melanogaster* and *D. simulans. J. Exp. Zool.* 66. 283

KIHARA, H. 1930. Karyologische Studien an Fragaria. *Cyt.* 1. 79

KIHARA, H., and LILIENFELD, F. 1932. Untersuchungen an Aegilops × Triticum and Aegilops × Aegilops Bastarden. *Cyt.* 3. 261

KING, H. D. 1918, 1919. Studies on inbreeding, I—IV. *J. Exp. Zool.* 26. 27, 29 319

KLINEBERG, O. 1931. A study of psychological differences between "racial" and national groups in Europe. *Arch. Psychol.* 132. 345

KNAPP, E. 1936. Heteroploidie bei Sphaerocarpus. *Ber. Deutsch. Bot. Ges.* 54. 214

KNIEP, H. 1928. Die Sexualität der niederen Pflanzen. Jena. 207, 214

1929. Vererbungserscheinungen bei Pilzen. *Bibliog. Gen.* 5. 54, 214

KÖHLER, W., and FELDOTTO, W. 1935. Experimentelle Untersuchungen über die Modifikabilität der Flugelzeichnung u.s.w. *Arch. Jul. Klaus Stift.* 10. 196

KOLLER, P. C. 1936. Structural hybridity in *D. pseudo-obscura. J. Gen.* 32. 266

KOLTZOFF, N. K. 1938. The structure of the chromosomes and their participation in cell metabolism. *Biol. Zhurn.* 7. 101

KOMAI, T. 1934. Pedigrees of hereditary diseases and abnormalities found in the Japanese race. *Kyoto.* 343

KOSSIKOV, K. V., and MULLER, H. J. 1935. Invalidation of the genetic evidence for branched chromosomes. *J. Hered.* 26. 367

KOSSWIG, C. 1935. Idiotypus und Geschlecht. *Z.I.A.V.* 70. 224

1929. Ueber die veränderter Wirkung von Farbgenen des Platypoecilus in der Gattungskreuzung mit Xiphophorus. *Z.I.A.V.* 50. 298

KRALLINGER, H. 1928. Gibt es einen Spermatozoendimorphismus beim Hausrind? *Munich Techn. Hochsch.* 58

KÜHN, A. 1932. Zur Genetik und Entwicklungsphysiologie des Zeichnungsmuster der Schmetterlinge. *Nach. Ges. Wiss. Gött.* 6. 194

1936. Versuche über die Wirkungsweise der Erbanlagen. *Naturwiss.* 24. 177, 194, 280

KÜHN, E. 1937. Befructungsphysiologische Untersuchungen zur Problem der Vererbung der Blütenfullung bei Matthiola. *Z.I.A.V.* 72. 55

KULESHOV, N. N. 1933. World's diversity of phenotypes of maize. *J. Am. Soc. Agric.* 25. 252

KUWADA, Y. 1927. On the spiral structure of chromosomes. *Bot. Mag. Tokyo* 41. 364

and NAKAMURA, T., 1934a. Behaviour of chromonemata in mitosis II. Artificial unravelling of coiled chromonemata. *Cyt.* 5. 43

1934b. Do. IV. Double refraction of chromosomes in *Tradescantia reflexa. Cyt.* 6. 395

LAMMERTS, W. E. 1934. On the nature of the association in *N. tabacum* haploids. *Cyt.* 6. 72

LANDAUER, W. 1928. The morphology of intermediate rumplessness in the fowl. *J. Hered.* 19. 202

1932. Studies on the creeper fowl III. *J. Gen.* 25. 202

1933. Untersuchungen über das Kruperhühn. *Zeits. mikr. Anat.* 32. 202

1933a. Russian methods of artificial insemination. *J. Hered.* 24. 312

LANG, W. D. 1923. Evolution; a resultant. *Proc. Geol. Ass.* 34. 241

LANGE, J. 1930. Crime as Destiny. New York. 338

LEHMANN, F. E. 1936. Selektive Beeinflussbarkeit frühembryonaler Entwicklungsvorgänge. *Naturwiss* 26. 202

LESLEY, J. W. 1928. The cytological and genetical study of progenies in triploid tomatoes. *Gen.* 13. 84

1932. Trisomic types of the tomato and their relation to the genes. *Gen.* 17. 84

LESLEY, M. M., and FROST, H. B. 1927. Mendelian inheritance of chromosome shape in Matthiola. *Gen.* **12.** 378

LEVIT, S. G. 1936. The problem of dominance in man. *J. Gen.* **31.** 329, 332

L'HÉRITIER, P., and TEISSIER, G. 1937. Elimination des formes mutantes dans les populations des Drosophiles. *C.R. Soc. Biol.* **124.** 302

— and NEEFS, Y. 1937. Aptérisme des insectes et sélection naturelle. *C.R. Acad. Sci.* **204.** 302

LILLIE, F. R. 1902. Differentiation without clearage in the egg of the Annellid, *Chaetopterus pergamentaceus. Arch. f. Entw. Mech.* **14.** 138

— and JUHN, M. 1932. The physiology of development of feathers I. *Physiol. Zool.* **5.** 198

LINDAHL, P. E. 1936. Zur Kenntnis der physiologischen Grundlagen der Determination in Seeigelkeim. *Acta. Zool.* **17.** 146

LINDEGREN, C. C. 1933. The Genetics of Neurospora III. *Bull. Torrey Bot. Club* **60.** 102

— 1936a. A six-point map of the sex chromosome of *Neurospora crassa. J. Gen.* **32.** 102

— 1936b. The structure of the sex chromosome of *N. crassa. J. Hered.* **27.** 102

— and LINDEGREN, G. 1937. Non-random crossing over in Neurospora. *J. Hered.* **28.** 129, 133

LITTLE, C. C., and BAGG, H. J. 1924. The occurrence of four inheritable morphological variations in mice, etc. *J. Exp. Zool.* **41.** 202

LJUNGDAHL, H. 1924. Ueber der Herkunft der in der Meiosis konjugierenden Chromosomen bei Papaver Hybriden. *Sv. Bot. Tidschr.* **18.** 73

LORIMER and OSBORN. 1934. Dynamics of population. London. 347, 351

LOTSY, J. P. 1911. Hybrides entre espèces d'Antirrhinum. *C.R. IV. Cong. Gen.* 271

— 1928. Hybridization among human races in S. Africa. *Genetica* **10.** 346

LUCAS, F. F., and STARK, M. B. 1931. A study of living cells of certain grass-hoppers by means of the ultra-violet microscope. *J. Morphol.* **52.** 34

LUTHER, W. 1935. Entwicklungsphysiologische Untersuchungen an Forellen-keim. *Biol. Zbl.* **55.** 152

MCCLINTOCK, B. 1932. A correlation of ring-shaped chromosome with variegation in Zea mays. *P.N.A.S.* **18.** 87

— 1933. The association of non-homologous parts of chromosomes in the midprophase of meiosis in Zea mays. *Z. Zellforsch. mikr. Anat.* **19.** 367

— 1934. The relation of a particular chromosomal element to the development of nucleoli in Zea mays. *Z. Zellforsch. mikr. Anat.* **21.** 39

MCCLUNG, C. E. 1902. The accessory chromosome—sex determinant? *Biol. Bull.* **9.** 75

— 1927. The chiasmatype theory of Janssens. *Q. Rev. Biol.* **2.** 126

MCDOUGALL, W. 1927. An experiment for testing the hypothesis of Lamarck. *Brit. J. Psychol.* **17.** 305

— 1930. A second report on a Lamarckian experiment. *Brit. J. Psychol.* **20.** 305

MCFADDEN, E. S. 1930. A successful transfer of emmer characters to vulgare wheat. *J. Am. Soc. Agron.* **22.** 323

MCKINNERY, H. H. 1937. Virus mutation and the gene concept. *J. Hered.* **28.** 400

MACKLIN, M. T. 1933, 1934. Inherited abnomalies of metabolism. *J. Hered.* **24, 25.** 326

O*

MAINX, F. 1933. Die Sexualität als Problem der Genetik. Jena. 206, 210, 214, 215, 230

1937. Analyse der Genwirkung durch Faktorenkombination (Die Augenfarbe von D. melanogaster). Z. ind. Abst. Vererb. 73. 176.

MANGELDORF, P. C., and FRAPS, G. S. 1931. A direct quantitative relationship between vitamin A in corn and the number of genes for yellow pigmentation. Sci. 73. 116

MARCHAL, E. L., and MARCHAL, E. M. 1911. Aposporie et sexualité chez les Mousses III. Bull. Acad. Roy. Belg. 9–10. 213, 255

MARGOLIS, O. S. 1935. Studies on the Bar series of Drosophila. Gen. 20. 183

and ROBERTSON, C. N. 1937. Studies on the Bar series of Drosophila. Gen. 22. 183

MARSCHAK, A. G. 1935. The sensitive volume of the meiotic chromosomes of Gasteria as determined by irradiation with X-rays. P.N.A.S. 21. 392

MATHER, K. 1933. The relation between chiasmata and crossing over in diploid and triploid D. melanogaster. J. Gen. 27. 71, 129

1933a. Interlocking as a demonstration of the occurrence of genetical crossing over during chiasma formation. Am. Nat. 67. 127

1936a. Competition between bivalents during chiasma formation. Proc. Roy. Soc. B. 120. 126

1936b. Segregation and linkage in autotetraploids. J. Gen. 32. 71, 107

1936c. The determination of position in crossing over I. J. Gen. 33. 125

1937. Do. II. Cyt. Fujii Jub. 125

1937a. The experimental determination of the time of chromosome doubling. Proc. Roy. Soc. B. 124. 116

1938. Crossing over. Biol. Rev. 13. 89

and LAMM, R. 1935. The negative correlation between chiasma frequencies. Hereditas 20. 126

and STONE, L. H. A. 1933. The effect of X-rays on somatic chromosomes. J. Gen. 28. 116

MAVER, J. W. 1923. An effect of X-rays on the linkage of mendelian characters in the 1st chromosome of Drosophila. Gen. 8. 93

MEISTER, G. K. 1930. The present purposes of the study of interspecific hybrids. Proc. U.S.S.R. Cong. Gen. and Plant. Br. 2. 261, 322

MENDEL, G. Papers translated in Bateson 1930. 24, 30

METZ, C. W. 1935. Structure of the salivary gland chromosomes in Sciara. J. Hered. 26. 98

and MOSES, M. S. 1923. Chromosomes of Drosophila. J. Hered. 14. 264

MICHAELIS, P. 1937. Untersuchungen zur Problem der Plasmavererbung. Protopl. 27. 275

MICHURIN. Cf. Plant Breeding Abstracts, passim. 321

MOEWUS, F. 1935. Ueber die Vererbung des Geschlectes bei Polytoma Pascheri und beim Polytoma uvella. Z.I.A.V. 69. 210

1936. Faktorenaustausch, insbesondere der Realisatoren bei Chlamydomonas Kreuzungen. Ber. Deutsch. Bot. Ges. 54. 210

MOFFETT, A. A. 1931. The chromosome constitution of the Pomoideae. Proc. Roy. Soc. B. 108. 261

MOHR, O. 1934. Heredity and disease. New York. 324

MONTGOMERY, T. H. 1901. A study of the chromosomes of the germ cells of Metazoa. Trans. Am. Phil. Soc. 20. 33

MOORE, A. R. 1933. Is cleavage rate a function of the cytoplasm or the nucleus? J. Exp. Biol. 10. 147

MORANT, G. M. 1928. A preliminary classification of European races based on cranial measurements. *Biometrika*. 20b. 341

MORGAN, L. V. 1922. Non-criss-cross inheritance in *D. melanogaster*. *Biol. Bull.* 42. 82

MORGAN, T. H. 1911. Random segregation versus coupling in mendelian inheritance. *Sci.* 34. 49

— 1915. The predetermination of sex in phylloxerans and aphids. *J. Exp. Zool.* 19. 229

— 1926. The theory of the gene. New Haven. 91, 121, 229

— 1927. Experimental embryology. New York. 137

— 1934. Embryology and Genetics. New York. 147

— 1938. The genetic and the physiological problems of self-sterility in Ciona. I and II. *J. Exp. Zool.* 78. 58

— and BRIDGES, C. B. 1919. The origin of gynandromorphs. *Carn. Inst. Publ.* 278. 86

BRIDGES, C. B., and STURTEVANT, A. H. 1925. The genetics of Drosophila. *Bibliog. Gen.* 2. 76, 264

MORTIMER, C. H. 1935. Untersuchungen über den Generationswechsel bei Cladoceren. *Naturwiss.* 23. 60, 231

MULLER, H. J. 1916. The mechanism of crossing over. *Am. Nat.* 50. 89

— 1918. Genetic variability, twin hybrids, and constant hybrids in a case of balanced lethal factors. *Gen.* 3. 93, 110

— 1925. Why polyploidy is rarer in animals than in plants. *Am. Nat.* 59. 79, 254

— 1925a. The regionally differential effects of X-rays on crossing over in the chromosomes of Drosophila. *Gen.* 10. 93

— 1926. The gene as the basis of life. *Proc. Int. Cong. Plant Sci. Ithaca.* 379

— 1927. Artificial transmutation of the gene. *Sci.* 66. 382

— 1928. The measurement of the gene mutation rate in Drosophila. *Gen.* 13. 381

— 1930. Types of visible variations induced by X-rays in Drosophila. *J. Gen.* 22. 49

— 1932a. Some genetic aspects of sex. *Am. Nat.* 66. 230, 231

— 1932b. Further studies on the nature and causes of gene mutations. *Proc. 6th Int. Cong. Gen.* 1. 163, 166, 186, 375

— 1933. On the incomplete dominance of the normal allelomorphs of white in Drosophila. *J. Gen.* 33. 164, 168

— 1934. The effects of Roentgen rays on hereditary material. *Sci. Radiol.* 384

— 1935. Out of the night. London. 358

— 1935a. A viable two-gene deficiency. *J. Hered.* 26. 377

— 1935b. Human genetics in Russia. *J. Hered.* 26. 340

— 1935c. On the dimensions of chromosomes and genes in the Dipteran salivary glands. *Am. Nat.* 69. 378

— 1936. On the variability of mixed races. *Am. Nat.* 70. 347

— 1938. The remaking of chromosomes. *Collecting Net.* 13. 389, 390, 396

— and DIPPEL, A. L. 1926. Chromosome breakage by X-rays and the production of eggs from genetically male tissue in Drosophila. *J. Exp. Biol.* 3. 209

— and GERSHENSON, S. M. 1935. Inert regions of chromosomes as temporary products of individual genes. *P.N.A.S.* 21. 396

— and PAINTER, T. S. 1932. The differentiation of the sex chromosomes of Drosophila into genetically active and inert regions. *Z.I.A.V.* 62. 92

MULLER, H. J., and MOTT-SMITH, L. M. 1930. Evidence that natural radio-activity is inadequate to explain the frequency of natural mutations. *P.N.A.S.* 16. 383

and PROKOFIEVA, A. A. 1935. The individual gene in relation to the chromomere and the chromosome. *P.N.A.S.* 21. 375, 377

and SETTLES, F. 1927. The non-functioning of the genes in spermatoza. *Z.I.A.V.* 43. 58

PROKOFIEVA, A. A., and KOSSIKOV, K. V. 1936. Unequal crossing over in the Bar mutant as a result of a duplication of a minute chromosome section. *C.R. Acad. Sci. U.S.S.R.* 1. 374

PROKOFIEVA, A. A., and RAFFEL, D. 1935. Minute intergenic rearrangement as a cause of apparent gene mutation. *Nat.* 135. 375

MUNTZING, A. 1932. Cytogenetic investigations on the synthetic *Galeopsis tetrahit*. *Hereditas* 16. 256

1936. The evolutionary significance of autopolyploidy. *Hereditas* 21. 64

NABOURS, R. K. 1919. Parthenogenesis and crossing over in the grouse locust Apotettix. *Am. Nat.* 53. 61

1929. The genetics of the Tettigidae (grouse locusts). *Bibliog. Gen.* 5. 91

NAGAI, M. A., and LOCKER, G. L. 1938. The production of mutations in Drosophila with neutron radiation. *Gen.* 23. 389

NAKAMURA, T. 1937. Double refraction of chromosomes in paraffin sections. *Cyt. Fujii Jub.* 1. 395

NAVASHIN, M. 1932. The dislocation hypothesis of the evolution of chromosome numbers. *Z.I.A.V.* 63. 263

NEBEL, B. R. 1936. Chromosome structure X. An X-ray experiment. *Gen.* 21. 117

and RUTTLE, M. L. 1938. The cytological and genetical significance of colchicine. *J. Hered.* 29. 255

NEEDHAM, J. 1934. A History of Embryology. Cambridge. 29

1936. New Advances in the chemistry and biology of organized growth. *Proc. Roy. Soc. Med.* 29. 149

NEWMAN, H. H., FREMAN, F. N., and HOLTZINGER, K. J. 1937. Twins. Chicago. 339

NEWTON, W. C. F., and PELLEW, C. 1929. *Primula kewensis* and its derivatives. *J. Gen.* 20. 256

NILSSON-EHRLE, H. 1908. Einige Ergebnisse von Kreuzungen bei Hafer und Weizen. *Bot. Notiser.* 160

NOUJDIN, N. I. 1936. Genetic analysis of certain problems of the physiology of development of *D. melanogaster*. *Biol. Zhurn.* 5. 86

NOWINSKI, W. W. 1934. Die vermännlichende Wirkung fraktionieter Darmextrakte des Weibchens auf die Larven der *Bonellia viridis*. *Publ. Staz. Zool. Nap.* 14. 224

OFFERMANN, C. A. 1935. The position effect and its bearing on genetics. *Bull. Acad. Sci. U.S.S.R.* 373

1936. Branched chromosomes as symmetrical duplications. *J. Gen.* 32. 95

STONE, W. S., and MULLER, H. J. 1931. Causes of inter-regional differences in cross-over frequency. *Anat. Rec.* 51. 92, 132

ONO, T. 1935. Chromosomen und Sexualität von *Rumex acetosa*. *Sci. Rep. Tokoka Imp. Univ.* 10. 219

and SHIMOTOMAI, N. 1928. Triploid and tetraploid intersexes of *Rumex acetosa*. *Bot. Mag. Tokyo.* 42. 219

OPPENHEIMER, J. M. 1936. Transplantation experiments in developing teleosts. *J. Exp. Zool.* 72. 152

PAINTER, T. S. 1934. The morphology of the X chromosomes in the salivary glands of *D. melanogaster* and a new type of chromosome map for this element. *Gen.* 19. 98, 99

—— 1935. Do. IIIrd chromosome. *Gen.* 20. 98, 99

—— 1939. The structure of salivary gland chromosomes. *Am. Nat.* 73. 98.

—— and MULLER, H. J. 1929. Parallel cytology and genetics of induced translocations and deletions in Drosophila. *J. Hered.* 20. 97

—— and STONE, W. S. 1935. Chromosome fusion and speciation in Drosophila. *Gen.* 20. 95

PARKS, H. B. 1936. The relationship between the first cleavage spindle and mosaic formation. *J. Hered.* 27. 85

PARNELL, F. R. 1921. A note on the detection of segregation by examination of the pollen of rice. *J. Gen.* 11. 55

PÄTAU, K. 1935. Chromosomenmorphologie bei *D. melanogaster* und *D. simulans* und ihre genetische Bedeutung. *Naturwiss.* 23. 265

PATTERSON, J. T., STONE, W., and BEDICHEK, S. 1935. The genetics of X-hyperploid females. *Gen.* 20. 223

PEACOCK, A. D. 1925. Animal parthenogenesis in relation to chromosomes and species. *Am. Nat.* 59. 58, 61

PEARL, R. 1930. Introduction to medical biometry and statistics. Philadelphia. 333

PEARSON, K., and LEE, K. 1903. The law of ancestral heredity. *Biometrika* 2. 334

PELLEW, C. 1929. The genetics of unlike reciprocal hybrids. *Biol. Rev.* 4. 275

—— and SANSOME, E. R. 1931. Genetical and cytological studies on the relation between European and Asiatic varieties of *Pisum sativum*. *J. Gen.* 25. 268

PENROSE, L. S. 1936. Mental defect. London. 335

PHILIP, U. 1935. Crossing over between X and Y chromosomes in *D. melanogaster*. *J. Gen.* 31. 78

PHILP, J., and HUSKINS, C. L. 1931. The cytology of *Matthiola incana* R. Br., especially in relation to the inheritance of double flowers. *J. Gen.* 24. 55

PICKHAN, A. 1936. Welche Strahlendose dürfen bei der Röntgendiagnostik als unschädlich betrachtet werden? *Fortschr. Rontgenstr.* 53. 382

PINCUS, G., and WADDINGTON, C. H. 1939. The effect of colchicine on rabbit eggs. *J. Hered.* 255

—— and WHITE, P. 1934. On the inheritance of *diabetes mellitus III*. *Am. J. Med. Sci.* 6. 326

PLOUGH, H. H. 1917. The effect of temperature on crossing over in Drosophila. *J. Exp. Zool.* 24. 93, 101

POPENHOE, P., and JOHNSON, R. S. 1933. Applied Eugenics, 2nd ed. London. 355

POULSON, D. F. 1937. The embryonic development of *D. melanogaster*. *Actualités Sci. et Ind.* Paris. 145

PRATT. 1932. In Allen's *Sex and Internal Secretions*. London. 332

PROKOFIEVA (BELGOVSKAIA), A. A. 1935. The structure of the chromocenter. *Cyt.* 6. 396

PUNNETT, R. C. 1923. Heredity in poultry. London. 205

PUNNETT, R. C. 1927. Linkage groups and chromosome number in Lathyrus *Proc. Roy. Soc. B.* **102.** 91

1931. Ovists and Animalculists. *Am. Nat.* **62.** 29

RAFFEL, D. 1932. The occurrence of mutations in *Paramecium aurelia. J. Exp. Zool.* **63.** 59, 278

REDFIELD, H. 1930. Crossing over in the IIIrd chromosome of triploids of *D. melanogaster. Gen.* **15.** 105, 125

1932. A comparison of triploid and diploid crossing over for chromosome II of *D. melanogaster. Gen.* **17.** 105, 125

REED, S. C. 1938. Determination of hair pigments. *J. Exp. Zool.* **79.** 179

RENNER, O. 1925. Untersuchungen über die faktorielle Konstitution einiger komplexheterozygotischer Oenotheren. *Bibliog. Gen.* **9.** 110

1929. Artbastarde bei Pflanzen. *Handb. Vererbungswiss.* **2.** 110

1934. Die pflanzlichen Plastiden als selbstständige Elemente der genetischen Konstitution. *Acad. Wiss. Leipzig.* **86.** 277

RENSCH, B. 1929. Das Prinzip geographischer Rassenkreise und das Problem der Artbildung. Berlin. 303

1936. Studien über klimatische Parallelität der Merkmalausprägung bei Vogeln und Saugern. *Arch. Naturgesch. N.F.* **5.** 303

REUSS, A. 1935. Über die Auslosung von Mutationen durch Bestrahlung erwachsener Drosophila Männchen mit ultravioletten Licht. *Z.I.A.V.* **70.** 384

RHOADES, M. M. 1933. An experimental and theoretical study of chromatid crossing over. *Gen.* **27.** 275

1935. The cytoplasmic inheritance of male sterility in Zea mays. *J. Gen.* **23.** 275

and MCCLINTOCK, B. 1935. The cytogenetics of maize. *Bot. Rev.* **1.** 100

RICE, V. A. 1934. The breeding and improvement of farm animals. New York. 309, 311, 312, 324

RICHARDS, A. G., and MILLER, A. 1937. Insect development analysed by experimental methods. *J. New York Entom. Soc.* **45.** 143

RILEY, H. P. 1936. The effect of X-rays on the chromosomes of *Tradescantia gigantea. Cyt.* **7.** 116

ROBB, R. C. 1935. A study of mutations in evolution, I and II. *J. Gen.* **31.** 244

ROBERTSON, C. W. 1936. The metamorphosis of *Drosophila melanogaster. J. Morph.* **59.** 145

ROBERTSON, M. 1929. Life-cycles in Protozoa. *Biol. Rev.* **4.** 231

ROBSON, G. C., and RICHARDS, O. W. 1936. The variation of animals in nature. London. 299, 302

ROSENBERG, O. 1909. Cytologischmorphologische Studien an *Drosera longifolia* × *rotundifolia. K. Sv. Vet. Handl.* **43.** 258

1930. Apogamie und Parthenogenesis bei Pflanzen. *Handb. Vererbungswiss.* **2.** 58

ROTMANN, E. 1935a. Der Anteil von Induktor und reagierendem Gewebe an der Entwicklung des Haftfadens. *Arch. f. Entw. Mech.* **133.** 149

1935b. Do. Kiemes. *Arch. f. Entw. Mech.* **133.** 149

ROWE, A. W. 1899. An analysis of the genus Micraster. *Q.J. Geol. Soc.* **55.** 244

SACHAROFF, W. 1935. Jod als chemischer Mutationsauslösungfaktor bei Drosophila. *Biol. Zhurn.* **4.** 304

SALAMAN, R. N. 1934. The raising of blight-resistant varieties and virus-free stocks. *Rothampsted Conf.* **16.** 323

SANSOME, E. R. 1932. Segmental interchange in Pisum. *Cyt.* **3.** 268

SANSOME, F. W., and PHILP, J. 1932. Recent Advances in plant genetics. London. 64, 92, 258

SATO, H. 1932. Die postembryonale Differenzierung der Gonaden von *Lymantria dispar. Z. Zellf. mikr. Anat.* 16. 228

SAUNDERS, E. R. 1928. Matthiola. *Bibliog. Gen.* 55

SAX, K. 1930. Chromosome structure and the mechanism of crossing over. *J. Arnold Arboretum.* 2. 121, 126

1931. The mechanism of crossing over. *Science* 74. 121, 126

1936. Chromosome coiling in relation to miosis and crossing over. *Gen.* 21. 129

1937. Effects of variations in temperature on nuclear and cell division in Iradescantia. *Am. J. Bot.* 24. 255

SCHLEIP, W. 1929. Die Determination der Primitiventwicklung. Leipzig. 137, 139, 147, 278

SCHMIDT, W. J. 1936. Doppelbrechung von Chromosomen und Kernspindel in der lebenden Zelle. *Naturwiss.* 24. 395, 363

1937. Die Doppelbrechung von Karyoplasma, Zytoplasma und Metaplasma. Berlin. 363, 395

SCHRADER, F. 1928. The sex chromosomes. Berlin. 75

and HUGHES-SCHRADER, S. 1931. Haploidy in Metazoa. *Q. Rev. Biol.* 6. 229

SCHULTZ, J. 1929. The minute reaction in the development of *D. melanogaster. Gen.* 14. 186

1934. See Morgan, Bridges, and Schultz. *Yearb. Carn. Inst.* 33. 164

1935. Aspects of the relation between genes and devlopment in Drosophila. *Am. Nat.* 69. 175

1936. Variegation in Drosophila and the inert chromosome regions. *P.N.A.S.* 22. 373

and BRIDGES, C. B. 1932. Methods for distinguishing between duplications and specific suppressors. *Am. Nat.* 66. 95, 160

SCHULTZ and REDFIELD, see Morgan, Bridges, and Sturtevant. 1932, 1933. The constitution of germinal material in relation to heredity. *Yearb. Carn. Inst.* 31, 32. 126

SCHWEITZER, M. D. 1935. Analytical study of crossing over in *D. melanogaster. Gen.* 20. 92

SCOTT, J. P. 1937. The embryology of the guinea-pig III. The development of the polydactylous monster. *J. Exp. Zool.* 77. 203

SCOTT-MONCRIEFF, R. 1936. A biochemical survey of some mendelian factors for flower colour. *J. Gen.* 32. 175

1937. The biochemistry of flower colour variation. In Perspectives in Biochemistry. Cambridge. 175

SEIDEL, F. 1936. Entwicklungsphysiologie des Insekten-Keimes. *Verh. d. Deutsch. Zool. Ges.* 143

SEXTON, E. W., and PANTIN, C. F. A. 1927. Inheritance in *Gammarus chevreuxi. Nat.* 119. 281

SHULL, A. F. 1929. Determination of types of individuals in aphids, rotifers, and cladocera. *Biol. Rev.* 4. 60

SINNOTT, E. W., and DUNN, L. C. 1935. The effect of genes on the development of size and form. *Biol. Rev.* 10. 207, 318

SIRKS, M. J. 1932. Beiträge zu einer genotypischen Analyse der Ackerbohne. *Genetica* 13. 278

SMITH, H. F. 1936. Influence of temperature on crossing over in Drosophila. *Nat.* 138. 94

430 AN INTRODUCTION TO MODERN GENETICS

SONNEBORN, T. M., and LYNCH, R. S. 1934. Hybridization and segregation in *Paramecium aurelia*. *J. Exp. Zool*. 67. 59

SPATH, L. F. 1933. The evolution of the Cephalopoda. *Biol. Rev*. 8. 245

SPEMANN, H. 1938. Organizers and Induction in embryonic development. Yale. 137

— and MANGOLD, H. 1924. Ueber Induktion von Embryonalanlagen durch Implantation artfremder Organisatoren. *Arch. Mehr. Anat. Entw*. 100. 148

— and SCHOTTE, O. 1932. Ueber xenoplastische Transplantationen als Mittel zur Analyse der embryonalen Induktion. *Naturwiss*. 20. 149

SQUIRES, P. C. 1927. Wolf children of India. *Am. J. Psychol*. 38. 348

STADLER, L. J. 1928. The rate of induced mutation in relation to dormancy, temperature, and dosage. *Anat. Rec*. 41. 385

— and SPRAGUE, G. F. 1936. Genetic effects of ultra-violet radiation in maize. *P.N.A.S*. 22. 384

STANDFUSS, M. 1908. Experimentelle zoologishe Studien. *Deutsch. Schweiz Ges. Naturwiss*. 219

STANLEY, W. M. 1938. The reproduction of virus proteins. *Am. Nat*. 72. 400

STERN, C. 1929. Ueber die additive Wirkung multipler Allele. *Biol. Zbl*. 49. 164, 166

— 1930. Multiple Allelie. *Handb. Vererbungswiss*. 14. 168

— 1928. Fortschritte der Chromosomentheorie der Vererbung. *Erg. Biol*. 4. 52, 58

— 1931. Zytologisch-genetische Untersuchungen als Beweis für die Morgansche Theorie des Faktorenaustausches. *Biol. Zbl*. 51. 50

— 1936. Somatic crossing over and segregation in *D. melanogaster*. *Gen*. 21. 373

— 1938. During which stage in the nuclear cycle do the genes produce their effects? *Am. Nat*. 72. 179

— and SEKIGUTI, K. 1931. Analyse eines Mosaikindividuums bei *D. melanogaster*. *Biol. Zbl*. 51. 87

STEVENS, W. L. 1936. The analysis of interference. *J. Gen*. 32. 91

STOCKHARD, C. R. 1921. Developmental rate and structural expression, etc. *Am. J. Anat*. 28. 202

STUBBE, H. 1934. Erbschädigung durch Röntgenbestrahlung und Chemikalien. *Med. Welt*. 2. 382

STURTEVANT, A. H. 1913. The linear arrangement of six sex-linked factors as shown by their mode of association. *J. Exp. Zool*. 14. 49

— 1921. Genetic studies on *D. simulans* II. *Gen*. 6. 223

— 1923. Inheritance of direction of coiling in Limnea. *Sci*. 58. 143

— 1925. The effects of unequal crossing over in the Bar locus in Drosophila. *Gen*. 10. 373

— 1926a. A cross-over reducer in *D. melanogaster* due to an inversion of a section of the IIIrd chromosome. *Biol. Zbl*. 46. 93, 131

— 1926b. Renner's studies on the genetics of Oenothera. *Q. Rev. Biol*. 1. 110

— 1929a. The genetics of *D. simulans*. *Carn. Inst. Publ*. 399. 265

— 1929b. The claret type of mutant in *D. simulans*; a study of chromosome elimination and of cell-lineage. *Z. wiss Zool*. 135. 86

— 1937. Autosomal lethals in wild populations of *D. pseudo-obscura*. *Biol. Bull*. 73. 286

— 1938. Essays in evolution. III. *Q. Rev. Biol*. 13. 284

— and BEADLE, G. W. 1936. The relation of inversions in the X chromosome of *D. melanogaster* to crossing over and disjunction. *Gen*. 21. 131

STURTEVANT, A. H., and DOBZHANSKY, T. 1936. Inversions in the IIIrd chromosome of wild races of D. pseudo-obscura. P.N.A.S. 22. 266

and SCHULTZ, J. 1931. Inadequacy of the subgene hypothesis of the nature of the scute allelomorphs of Drosophila. P.N.A.S. 17. 161, 371

and TAN, C. C. 1937. The comparative genetics of D. pseudo-obscura and D. melanogaster. J. Gen. 34. 265

SUKATSCHEV, W. 1928. Einige experimentelle Untersuchungen über den Kampf ums Dasein zwischen Biotypen derselben Art. Z.I.A.V. 47. 300

SUMNER, F. B. 1930. Genetic and distributional studies of three sub-species of Peromyscus. J. Gen. 23. 273

1932. Genetic, distributional and evolutionary studies of the sub-species of deer mice (Peromyscus). Bibliog. Gen. 9. 273

1935. Evidence of the protective value of changeable coloration in fishes. Am. Nat. 49. 300

SUTTON, W. S. 1902. On the morphology of the chromosome group in Brachystola magna. Biol. Bull. 4. 33

SWINNERTON, H. H. 1923. Outlines of palaeontology. London. 241

1932. Unit characters in fossils. Biol. Rev. 7. 241

THOMPSON, W. S. 1935. Population problems. 2nd ed. New York. 351

TIMOFEEFF-RESSOVSKY, N. W. 1931. Gerichtetes Varieren in der phanotypischen Manifestierung einiger Genovariationen von D. funebris. Naturwiss. 19. 188, 200

1932. Mutations of the gene in different directions. Proc. 6th Cong. Gen. 165, 390

1932a. Die heterogene Variationsgruppe Abnormes Abdomen bei D. funebris. Z.I.A.V. 62. 187

1933. Ueber die relative Vitalität von D. melanogaster und D. funebris u.s.w. Arch. Naturgesch. N.F. 2. 299

1933a. Rückgenmutationen und die Genmutatbilität in verschiedenen Richtungen V. Z.I.A.V. 66. 168

1934. Ueber den Einfluss des genotypischen Milieus und der Aussenbedingungen auf die Realization des Genotyps. Nach. Ges. wiss. Gött. N.F. 1. 163

1934a. Ueber die Vitalität einiger Genmutationen bein D. funebris u.s.w. Z.I.A.V. 66. 307

1934b. The experimental production of mutations. Biol. Rev. 9. 384

1935. Auslösung von Vitalitätsmutationen durch Röntgenbestrahlung bei D. melanogaster. Nach. Ges. wiss. Gött. N.F. 1. 384

1937. Mutationsforschung in der Vererbungslehre. Dresden. 279, 305, 380, 383, 384

and TIMOFEEFF-RESSOVSKY, H. A. 1927. Genetische Analyse einer freilebenden Population D. melanogaster. Arch. Entw. Mech. 109. 286

1934. Polare Schwankungen in der phänotypischen Manifestierung einiger Genmutationen bei Drosophila. Z.I.A.V. 67. 188

ZIMMER, K. G., and DELBRUCK, M. 1935. Ueber die Natur der Genmutation und der Genstruktur. Nach. Ges. wiss. Gött. N.F. 1. 385

TOYAMA, K. 1909. A sport of the silkworm and its genetical behaviour. Z.I.A.V. 1. 281

TRUEMAN, A. E. 1922. The use of Gryphaea in correlation in the lower Lias. Geol. Mag. 59. 242

TRUEMAN, A. E. 1930. Results of recent statistical investigations of invertebrate fossils. *Biol. Rev.* 5. 242

TURESSON, G. 1925. The plant species in relation to habitat and climate. *Hereditas* 5. 274, 300

1930. The selective effect of climate on plant species. *Hereditas* 14. 274, 300

1931. The geographical distribution of the alpine ecotypes of some Eurasiatic plants. *Hereditas* 15. 252, 300

TWITTY, V. C. 1936. Correlated genetic and embryological experiments on Triturus. *J. Exp. Zool.* 74. 149

V. UBISCH, L. 1937. Untersuchungen über Formbildung. *Auch. Entw. mech.* 134. 147

UPCOTT, M. B. 1937. The genetic structure of Tulipa II. Structural hybridity. *J. Gen.* 34. 131, 266

1938. Behaviour of the centromeres at meisois. *Proc. Roy. Soc. B.* 125. 397

VANDEL, A. 1931. La Parthenogenese. Paris. 59

1927. Gigantisme et Triploidie chez l'isopode Trichoniscus. *C.R. Soc. Biol.* 97. 59, 254

VAVILOV, N. I. 1922. The law of homologous series in variation. *J. Gen.* 12. 270

1928. Geographische Genzentren unserer Kulturpflanzen. *Verh. 5ten Kong. Vererb. Z.I.A.V. Suppl.* 250

1932. The process of evolution in cultivated plants. *Proc. 6th Cong. Gen.* 250

VERSHUER, O. V. 1931. Ergebnisse der Zwillingsforschung. *Verh. d. Gesf. Phys. Antrop.* 6. 340

1932. Die biologische Grundlage der menschlichen Mehrlingsforschung. *Z.I.A.V.* 61. 338

WADA, B. 1935. Mikrurgische Untersuchungen lebender Zellen in der Teilung. II. *Cyt.* 6. 363

WADDINGTON, C. H. 1929. Pollen germination in stocks, etc. *J. Gen.* 21. 55

1932. Experiments on chick and duck embryos cultivated in vitro. *Phil. Trans. Roy. Soc. B.* 221. 152

1933. Induction by the endoderm in birds. *Arch. f. Entw. Mech.* 128. 152

1937. Experiments on determination in the rabbit embryo. *Arch. Biol.* 48. 152

NEEDHAM, J., and BRACHET, J. 1936. Studies on the nature of the Amphibian organization centre III. *Proc. Roy. Soc. B.* 120. 149

WALTON, A. 1933. The technique of artificial insemination. *Imp. Bur. Am. Gen.* 312

WATKINS, A. E. 1930. The Wheat Species; a critique. *J. Gen.* 23. 258

WEISS, F. E. 1930. The problem of graft hybrids and chimaeras. *Biol. Rev.* 5. 87

WEISS, P. 1930. Entwicklungsphysiologie der Tiere. Berlin. 137

WETTSTEIN, F. V. 1920. Kunstliche haploide Parthenogenese bei Vaucheria u.s.w. *Ber. Deutsch. Bot. Ges.* 38. 62

1924. Morphologie und Physiologie des Formwechsels der Moose auf genetische Grundlage 1. *Z.I.A.V.* 33. 102

1927. Die Erscheinung der Heteroploidie. *Erg. Biol.* 2. 170, 255, 276

1926. Ueber plasmatische Vererbung, sowie Plasma und Genwirkung. *Nach. Ges. wiss. Gött.* 68, 276

WETTSTEIN, F. V. 1928a. Ueber plasmatische Vererbung u.s.w. *Ber. Deutsch. Bot. Ges.* 46. 276

1928b. Morphologie und Physiologie des Formwechsels der Moose. II. *Bibliog. Gen.* 10. 102

1932. Bastardpolyploidie als Artbildungsvorgang bei Pflanzen. *Naturwiss.* 257

1936. Gesichertes und Problematisches zur Geschlectsbestimmung. *Ber. Deutsch. Bot. Ges.* 54. 207

1937. Experimentelle Untersuchungen zum Artbildungsproblem. *Z.I.A.V.* 74. 68

1937a. Die genetische und entwicklungsphysiologische Bedeutung des Cytoplasma. *Z.I.A.V.* 73. 275

WHEELER, W. M. 1937. Mosaics and other anomalies among ants. Cambridge, Mass. Cf. Review Whiting. *J. Hered.* 29. 190

WHITE, M. J. D. 1934. The influence of temperature on chiasma frequency. *J. Gen.* 29. 93, 123

1936. Chiasma localization in *Mecostethus grossus* L. and *Merioptera brachyptera* L. (Orthoptera). *Zeits. f. Zellforsch.* 24. 120

1937. The Chromosomes, London. 34, 113

WHITING, P. W. 1932. Mutants in Habrobracon. *Gen.* 17. 62

1932a. Reproductive reactions of sex mosaics of a parasitic wasp, *Habrobracon juglandis. J. Exp. Psychol.* 14. 86

1935. Sex determination in bees and wasps. *J. Hered.* 26. 229

1935a. Selective fertilization. *J. Hered.* 26. 230

WHITTINGHILL, M. 1937. Induced crossing over in Drosophila males and its probable nature. *Gen.* 21. 94

WIESNER, B. P. 1935. The post-natal development of the genital organs in the albino rat. *J. Obst. Gyn. Brit. Emp.* 41. 228

WILLIER, B. H., GALLAGHER, T. F., and KOCH, F. C. 1937. The modification of sex development in the chick embryo by male and female sex hormones. *Physiol. Zool.* 10. 228

RAWLES, M. E., and HADORN, E. 1937. Skin transplants between embryos of different breeds of fowls. *P.N.A.S.* 23. 199

WILLIS, J. C. 1922. Age and area. Cambridge. 248

WILSON, E. B. 1928. The cell in development and heredity. 3rd ed. New York. 33, 52, 75

WINGE, O. 1917. The chromosomes, their number and general importance. *C.R. Trav. Carlsb.* 13. 254

1923. Crossing over between X and Y chromosomes in Lebistes. *J. Gen.* 13. 80, 223

1927. The location of 18 genes in *Lebistes reticulatus. J. Gen.* 18. 80

1931. X and Y linked inheritance in Melandrium. *Hereditas* 15. 80

1932. The nature of sex chromosomes. *Proc. 6th Cong. Gen.* 223

WINKLER, H. 1916. Ueber die experimentelle Erzeugung von Pflanzen mit abweichenden Chromosomenzahlen. *Zeits. f. Bot.* 8. 255

WINTER, F. L. 1929. The mean and variability as affected by continuous selection for composition in corn. *J. Agric. Res.* 39. 310

WITSCHI, E. 1929. Bestimmung und Vererbung des Geschlectes bei Tieren. *Handb. Vererbungswiss.* 207

1934. Genes and inductors of sex differentiation in Amphibians. *Biol. Rev.* 9. 208, 225, 230

1937. Stimulative and inhibitive induction in the development of the primary and secondary sex characters. *P.N.A.S.* 23. 225, 228

434 AN INTRODUCTION TO MODERN GENETICS

WODSEDALEK, J. E. 1913. Spermatogenesis of the pig, with special reference to the accessory chromosomes. *Biol. Bull.* **25.** 58

WOLTERECK, R. 1932. Grundzüge einer allgemeinen Biologie. Stuttgart.

1934. Artdifferenzierung (insbesandere Gestaltänderung) bei Cladoceren. *Z.I.A.V.* **67.** 279

WOODGER, J. H. 1937. The axiomatic method in biology. Cambridge. 29

WOODRUFF, L. L. 1935. Physiological significance of conjugation in *Blepharisma undulans. J. Exp. Zool.* **70.** 231

WRIGHT, S. 1917, 1918. Colour inheritance in mammals. *J. Hered.* **8, 9.** 179, 198

1922. The coefficient of inbreeding. *Am. Nat.* **56.** 312

1927. The effects in combination of the major colour-factors of the guinea-pig. *Gen.* **12.** 179

1929. Fisher's theory of dominance. *Am. Nat.* **63.** 186, 298

1931. Evolution in Mendelian populations. *Gen.* **16.** 231, 293

1932. The role of mutation, inbreeding, crossbreeding, and selection in evolution. *Proc. 6th Cong. Gen.* 293

1934. Physiological and evolutionary theories of dominance. *Am. Nat.* **68.** 185

1935. Evolution in populations in approximate equilibrium. *J. Gen.* **30.** 293

1935a. A mutation in the guinea-pig, tending to restore the pentadactyl foot when heterzygous, etc. *Gen.* **20.** 203

WRINCH, D. M. 1936. On the molecular structure of chromosomes. *Protopl.* **25.** 395

and LANGMUIR, I. 1938. The structure of insulin. *Science.* 393

YULE, G. U. 1929. An introduction to the theory of statistics. London. 333

ZELENY, C., and FAUST, E. C. 1915. Dimorphism in size of spermatozoa and its relation to the chromosomes. *P.N.A.S.* **1.** 58

ZIMMER, K. G., GRIFFITH, H. D., and TIMOFEEFF-RESSOVSKY, N. W. 1937. Mutationsaulösung durch Betastrahlung des Radiums bei *D. melanogaster. Strahlenther.* **59.** 386

SUBJECT INDEX

In the case of some common technical terms (e.g. locus) only one reference is given to a page on which a definition will be found.

For Product Safety Concerns and Information please contact our EU
representative GPSR@taylorandfrancis.com
Taylor & Francis Verlag GmbH, Kaufingerstraße 24, 80331 München, Germany

www.ingramcontent.com/pod-product-compliance
Lightning Source LLC
Chambersburg PA
CBHW060744220326
41598CB00022B/2325

9 781138 956971